Prior Art

Prior Art
Patents and the Nature of Invention in Architecture

Peter H. Christensen

THE MIT PRESS
Cambridge, Massachusetts . London, England

Introduction 1

1 *The Home* 25
2 *The Studio* 67
3 *The Corporate Lab* 125
4 *The Repository* 167
5 *The Patent Office* 217
6 *The Courtroom* 255
7 *The Commons* 287
   Coda 319

   Acknowledgments 321
   Notes 323
   Bibliography 345
   Image Credits 373
   Index 375

Introduction

Invention is a keyword of the modern epoch. Modern men and women hailed the pursuit of invention, fueled as it was by the Industrial Revolution, as the fount of progress and the primary means of personal emancipation consonant with Enlightenment values. Invention, in this context, was a form of liberation, and patents were the sedimentary building blocks of invention's promise not only to create a new world but also a better one. Guilds, kingly privileges, and other premodern regimes of exclusivity certainly laid the groundwork for the world's myriad patent systems today. Yet it was not until the second third of the nineteenth century that inventors embraced the practice of patenting in significant numbers. Patents could ensure the economic and cultural benefits afforded by exclusive property rights. Architecture and engineering—two of the most obvious and ubiquitous domains of humankind's inventive spirit—would surely feature in the cascade of industrial innovation that patents churned forth.

Or would they? As it turns out, patent culture was never directly translatable to the field of architecture, which always needed to negotiate the issues of technological innovation associated with patent culture in the context of the more abstract issues of artistic influence and formal expression. Consequently, since the Industrial Revolution, architects and engineers participating in the consistent but never commonplace practice of patenting aspects of design and building construction have had a complex and often fraught relationship with the kind of "invention" that a patent marks. This volume is the first full-scale monographic treatment of that complex relationship, one that grapples with the full range of technology related to architecture and buildings.

The geographic scope of this investigation is global to the extent that patents, as a legal construct, have existed globally. At the time of writing, several nations do not have national patent systems, and many more have only fledgling systems. What this effectively means is that the history of patents in architecture is the history of industrialization, a history that originates in England, spreads to Germany and France, and by the twentieth century is well under way in Europe, North America, and Japan, as well as in numerous enclaves in the so-called Global South. This familiar, axiomatic history of modernity is rightfully the source of frustration and pushback in numerous historical studies today in which scholars challenge the teleological assumptions behind the dominance of the Global North in the modern period. To be clear, my usage of this geography of industrialization is not acritical. By examining patents, one is automatically limiting one's geography to the places where patents exist, be this in common, civil, or hybrid legal systems. The geography of the history of industrialization, one quickly discovers, is inextricably linked to patents, suggesting, as many historians have, that patents have served a generative rather than ancillary role in the world's industrialization.[1] But there is also a way to read this geography against the grain, which I do in this book. In the colonial system, national patent regimes were sometimes exported and sometimes not. Comparing the vicissitudes of colonial and imperial patent regimes is itself important in defining what relationship, if any, exists between sovereignty and invention and demonstrating the considerable variegation between patent culture in, for example, the Raj and the Dutch East Indies.

There are also ways to read the axiomatic history of industrialization against the grain *within* national patent systems, an even more central aspect of this volume. For example, patents provided women with unprecedented property rights and inventive influence well before they could matriculate at universities or vote. Even the professional culture of architects and corporations bringing products to market for the construction sector defied the primacy of the heroic cis white male inventor by modeling invention practices that were integrally and expansively collaborative and nonhierarchical. Finally, for the fullest possible story of patents, it is also necessary to look beyond the convention dictating that the inventor (as attributed on an invention's paper trail) is the beginning and end of inventive activity. Other actors—from financiers, lab assistants, and modelmakers to patent attorneys, judges, and consumers—are integral to the story of how patents come to be and why.

Alongside tracking the history of industrialization geographically, this investigation also does so chronologically. There is a rich array of literature across economic, legal, social, and cultural history that analyzes the premodern practices of guilds, privileges, and exclusive rights and how these presage patents. Indeed, all of these institutions supported ownership of intangible property without ever identifying it as a discrete concept per se.[2] The story I narrate in this volume begins at the point—the very late eighteenth century—when patent offices in Britain and the United States formally opened for citizen inventors who had the newfangled right to apply for (rather than simply be granted) a patent, a process that in turn necessitated the brick-and-mortar patent office and a defined and transparent patent bureaucracy. Where to end this story is a bit more complicated. Although some aspects of the material contained in this book bring the discussion of architectural patents up to the present day, most of the subject material does not go beyond 1990. There are several reasons why this date—often deemed the end of the "short twentieth century"—is vital as a bookend for my account.[3] Perhaps the most important reason is that this is roughly when architectural practices and building offices began to go digital. In addition to radically rearranging the relationship between architects and their creations, digital design also fundamentally changed the way in which patentable inventions were conceived, prototyped, manufactured, filed, and disseminated. Digitization also had a radical impact on global patent offices, allowing inventors and patent lawyers to swiftly search for prior art online and eventually to establish an invention's viability as a "new" creation by comparing it to other inventions from across the world, and replacing the draftspeople who generated patent drawings by computers, which dramatically altered the nature of the disclosure process. In the globally influential patent system of the United States, 1990 was also the year that the Architectural Works Copyright Protection Act took effect, which made major strides in legally enshrining architecture as intellectual property (IP) in a way that had hitherto been ambiguous. With the end of the Cold War, which also happened at this time, alternative regimes to capitalist patent systems—such as that of East Germany, which is discussed later in this book—ceased to exist. Finally, as with all good history, an arm's length from the recent past is necessary, and 1990 more or less marks such a length from the time of writing.

Patents as History

The intellectual products that patents represent require definition. They tend to have objectively discernible forms and properties that are expressive, informational, or both, and their "invention" can be traceable to at least one human being.[4] Several scholars have attempted the tricky task of pinning down the origins and contextualizing the prehistory of modern patents, with results that raise more questions than they answer. As Christine MacLeod has pointed out in her history of the British patent system, many of the "standard" histories of patents have been written by "lawyers or economists with present day cases to plead," lending these histories something of an uneven and instrumentalist character.[5] In any event, the most common landing point for the quest for origins is the Renaissance city-state in Italy, which adopted the system of *privilegio*, kingly privileges that afforded artisans exclusive rights to certain practices.[6] Yet there are other contenders, including medieval England and China and even antiquity with its siloed structures of labor. In the end, though, the precise origins of either the patent system or IP as a formal concept are somewhat moot from the point of view of "natural law," which holds that the fruits of one's mental and physical labor at least ought to be one's own. The rise of legal systems and the financial incentivization spurred by capitalism thrust this principle into a domain where it ultimately needed to be explicitly codified and regulated.

This volume adopts a general chronological assessment to mark what makes the patent system "modern." It comes from the historian of science Mario Biagioli, who divides the history of inventive monopolies and negative rights into two broad categories: a period of privileges (roughly 1450–1790) and the period of modern patents that this book covers (roughly 1790 to the present).[7] Biagioli's basic scheme, pared down a bit here, goes as follows:

Privileges	*Modern Patents*
Local protection	Increasingly global protection
Novelty as geographical priority	Novelty as chronological priority
Local novelty requirement	Global novelty requirement
Local and present utility	Temporally deferred and geographically unspecified utility
Training of local users	Textual and pictorial disclosure to anybody, anywhere
Invention as a material object	Invention as an embodied inventive idea
Invention as a practice	Invention as knowledge

Inventor as an artisan and importer	Inventor as an author
Presentation	Representation[8]

As helpful as this division is, it is not intended to suggest that the transition from privilege to patent happened overnight, as it was in fact a gradual process of sedimentation. We also cannot ignore premodern practices regarding property. In Roman law, for example, an individual who painted or drew an image on a canvas belonging to another individual became the owner of that image and of the canvas on which it rested through the principle known as *accesio*.[9] Vitruvius, for his part, insisted in *De architectura* that "changing the titles of other men's books" and inserting his own name would not make their writings his. He cited his gratefulness, instead, to authors preceding him whom he could draw upon like "water from springs . . . converting them to our own purposes, we find our powers of writing rendered more fluent and easy, and, relying upon such authorities we venture to produce new systems of instruction."[10] This influential principle—directly linking a human act to property—was the bedrock on which much of subsequent European property rights came to rest. Records from China's Song dynasty, which extended from 960 to 1279 AD, also testify to a rudimentary system akin to modern copyright in which grants were occasionally extended for the exclusive printing of texts.[11] In addition to exclusive printing and publishing rights, this system included, for example, monopolies in the area of salt distilling and iron smelting. However, a full-fledged privilege system did not materialize in China, because the exploitation of such rights relied on the integrity and goodwill of local authorities tasked with enforcing them, something that the public very often knew did not exist.[12] In England, the Crown used its ability to enjoy special trade privileges to attract skilled foreign workers, such as John Kempe, a wool weaver from Flanders, in 1331.[13] In Elizabethan times, "inventor" could also connote anyone who imported manufactured and technical devices from elsewhere that were hitherto unknown in the country.[14] Moreover, from England to China, the medieval world hosted myriad permutations of "craft secrecy," in which guilds and artisans kept trade secrets, effectuating a culture of monopolies without the need for a patent system.[15]

Whether or not it marks the definitive origin of the patent, as many claim, a 1421 grant extended to Filippo Brunelleschi by the Signoria of Florence is certainly an important moment in the deployment of a monopoly to encourage innovation.[16] Brunelleschi, who was then one year into the construction of the cathedral of Florence's dome, was granted the exclusive rights to manufacture and use a new technique for the transportation of large loads of material on and across rivers for a period of three years.[17] His invention was a barge ("Il Badalone" or "The Monster") that integrated cranes and transported Carrara marble from the Arno river to the construction staging area for the cathedral's massive dome.[18] The Brunelleschi example is noteworthy not merely as a possible origin story for the patent but also for apparent needlessness of the exclusive rights: Florence certainly had other construction projects under way, but none were in direct competition with Brunelleschi's dome, which was by far the biggest undertaking of its day. The purpose of granting the exclusive rights, it would seem, was more to recognize Brunelleschi's ingenuity in designing the system in the first place, which in turn cast the state in an official role as an arbiter of ingenuity. This role was further formalized under a general patent statute issued in Venice in 1474. Here again the state was acting as the arbiter of so-called "genius," but doing this in such a way that would incentivize more such genius, not merely recognize it as it occurred. The Venetian statute noted the existence "of men of great genius, apt to invent and discover ingenious devices," who would be afforded the "honor" of the exclusive right to manufacture such devices for a period of ten years, in turn encouraging others to "apply their genius [to] discover, and . . . build devices of great utility and benefit to our commonwealth."[19] The use of the word "genius" in the Venetian statute reflects a wider trend in the development of IP as a concept. The notion of "genius"—a rare manifestation of skill or talent—transformed from a fringe, quasi-mystical concept into a mainstream concept in the fifteenth century and was the engine behind the desire to recognize exceptional accomplishments at the level of the state.[20]

There is a general assumption that the Venetian system's appeal spread across Europe during the sixteenth century. But this neglects some contemporaneous activities in places like England, which, like Italy, had a strong system of craft patronage. In 1449 King Henry VI bestowed a special 20-year grant to a certain John of Utynam to produce colored glass for the chapel at

Eton college and later other college chapels.[21] A century later, in 1551, King Henry II of France granted a ten-year monopoly to the court mechanic Abel Foullon for his invention of the holometer, an architectural measuring device, on the condition that he publish a detailed description of the invention for the use of the general public upon the expiry of the grant; this is likely the first published patent specification written as a condition for a monopoly.[22] Many of the great Renaissance scientists were folded into a privilege system: Galileo, for example, received a 20-year privilege from the Republic of Venice in 1591 for his design of a water pump.[23] Patents or their equivalents grew slowly through the beginning of the seventeenth century, but grantees still tended to have some inside connection to the courts and the ruling class, so that the system could not yet be described as truly democratic in nature.[24] The English Parliament recognized the profound problems with this system of kingly privileges (and the limited patronage associated with it) and prohibited the Crown from granting exclusive privileges under the 1623 Statute of Monopolies.[25] In the Low Countries, privileges, particularly in the field of hydraulic engineering, not only existed but also became a kind of rivalrous currency between cities and nations.[26]

Just as architecture had been the subject of many of the patent system's progenitors in the fifteenth and sixteenth centuries, textiles became the driving force behind the rapid uptick in patent activity in the eighteenth century, when the patent system became a quasi-democratic institution with procedures for application in England and the United States (where Thomas Jefferson was one of the greatest champions of patents in history) and an entirely orthodox cultural practice.[27] Indeed, the spread of the Industrial Revolution brought the spread of a new zeitgeist, transforming an age of projects into an age of invention.[28] It was at this point that the dialectic of the modern patent system came into focus: there was a spur for progress, to be sure, but also, and often, a force of entrenched conservatism and antidynamism and something of a "parlor game of concealment and obfuscation."[29] Indeed, as this dialectic cemented, so too did resentment of the patent system, which in turn sparked patent abolition movements, which peaked in the middle decades of the nineteenth century.[30] Countries that were industrially advanced at the time, such as Switzerland and the Netherlands, went so far as to reject or abolish patents for a certain period before ultimately acquiescing to them.[31] Patent

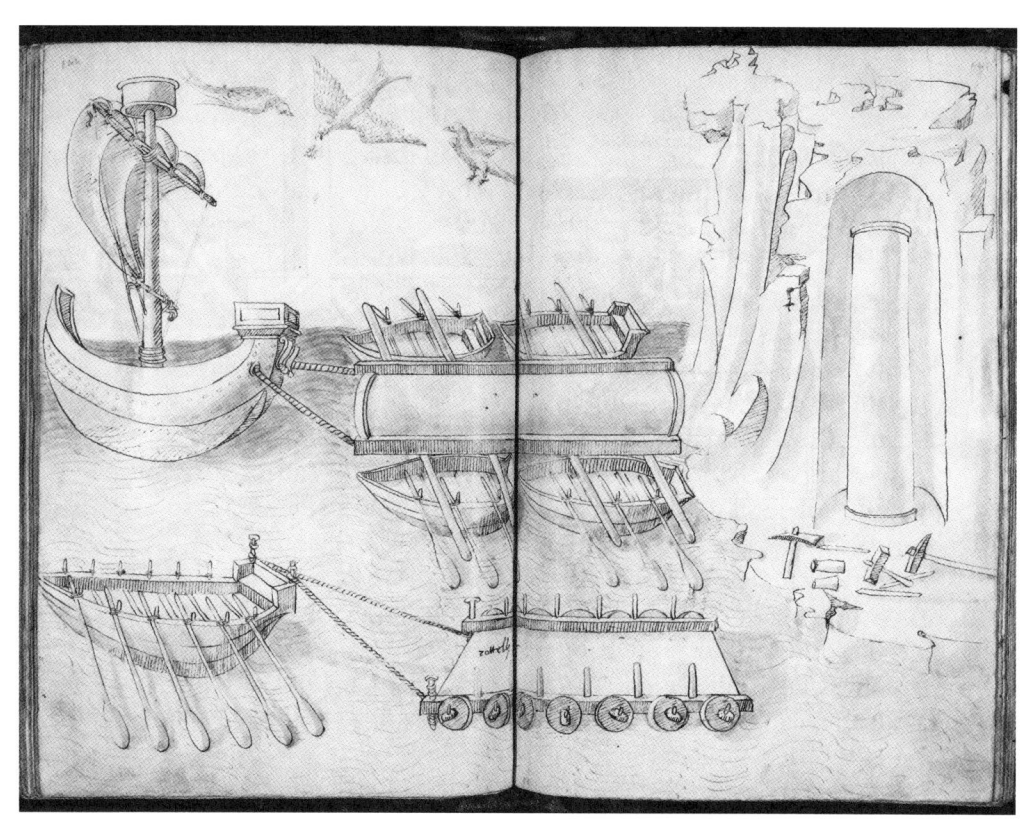

Mariano Taccola, illustration of Brunelleschi's "Il Badalone" from *De ingeneis*, c. 1449, Codex Palatinus 766, p. 40 verso / p. 41 recto. Also color plate 5.

abolitionists surmised that patents perverted the nature of invention and the social role of the inventor as a collaborative figure within the community.[32]

The historian of engineering Henry Petroski has juxtaposed two of the great English inventors of the nineteenth century—Henry Bessemer and Isambard Kingdom Brunel, both men of great consequence to the development of architecture—to highlight the profound differences between inventors at the height of the patent abolition movement. Petroski's essential characterization of Bessemer, who was pro-patent, is that of an opportunist who incidentally furthered the art of engineering by using patents as a way to incentivize competitors to surpass his own inventions in metallurgy. His characterization of Brunel, who was stridently opposed to patents, is that of an ideologue whose commitment to the belief that patents perverted the inventive process also furthered the art of engineering, but in an altogether different way. Brunel's contribution to the field was the well-argued point that true invention came from considered observation and problem solving, not from the anxiety associated with a competitive condition, a position that remains a valuable reason for a healthy public skepticism about patents.[33]

One of the numerous reasons why the pro-patent camp won out is probably that the rhetoric of invention and progress that was baked into the patent system was a sweeping and convincing form of social theory, ultimately more convincing than fears over "grandmotherly government."[34] As Adrian Johns has described it, the pro-patent camp made allies of the "worker-artisan, the public, and the new nature of science and technology, tying them all to the doctrines of political economy."[35] The patent system emerged stronger than ever out of this period of tumult and concretized much of the patent landscape as we know it today. One of the continuities lies in *who* applies for patents, a person Christine MacLeod has put into one of three broad categories: (1) the amateur inventor, (2) the professional inventor, and (3) the businessman (an artisan or manufacturer).[36] Another continuity since the late nineteenth century is the three pillars of IP at the global scale: patents, trademark, and copyright.[37]

Internationalization and ultimately globalization made it clear that atomistic national systems, where patents were only as good as the borders within which they were granted, could not survive as the be-all and end-all of IP regimes—particularly in Europe, where industrialized countries were tightly packed together and conducted constant trade with one another.[38] This can first be seen in 1862, a decade before German political unification, in the formation of a commission of German states trying to unify their patent laws, an effort that was largely thwarted by the most powerful participant: Prussia.[39] By the time of unification in 1871, the German patent system was widely seen as one of the strongest in the world, fulfilling its mission to compete with the British system.[40] The 1883 Paris Convention for the Protection of Industrial Property, a landmark agreement among most of the European countries with patent systems, repressed unfair competition in all forms of IP—signaling, as the World Intellectual Property Organization calls it, "the first major step taken to help creators ensure that their intellectual works were protected in other countries."[41] This was followed by the Berne Convention for the Protection of Literary and Artistic Property in 1886 and the 1893 creation of the United International Bureau for the Protection of Intellectual Property (widely known as BIRPI, after the French name Bureaux Internationaux Réunis pour la Protection de la Propriété Intellectuelle), which elevated the goals of the Paris Convention into an institution that enjoyed broader powers of enforcement and better organization.[42] While Europe and North America advanced their patent systems at the most rapid clip in the nineteenth century, this did not mean changes were not afoot elsewhere. In China, for example, the "comprador system"—a kind of administrative system with its own bureaucracy—functioned as an intermediary in China's trade with states that had patent systems in an effort to stymie the flow of Chinese exports infringing upon local patent laws in trade partner states.[43]

During and between the two world wars, patents took on a new importance as part of a global arms race, with much of the transatlantic patent activity in this period directly related to the machinations of the military-industrial complex. Patent laws, which were ostensibly politically neutral and had provisos against inventions meant to kill or hurt people outside of the context of war, nevertheless proved malleable enough for regimes like that of the National Socialists in Germany to co-opt the laws to their own ends.[44] In the Reich patent law, a vague "good morals" doctrine regarding what was patentable was easily subsumed by Nazi ideology to create internment and killing technologies that were the very opposite of moral inventions.[45]

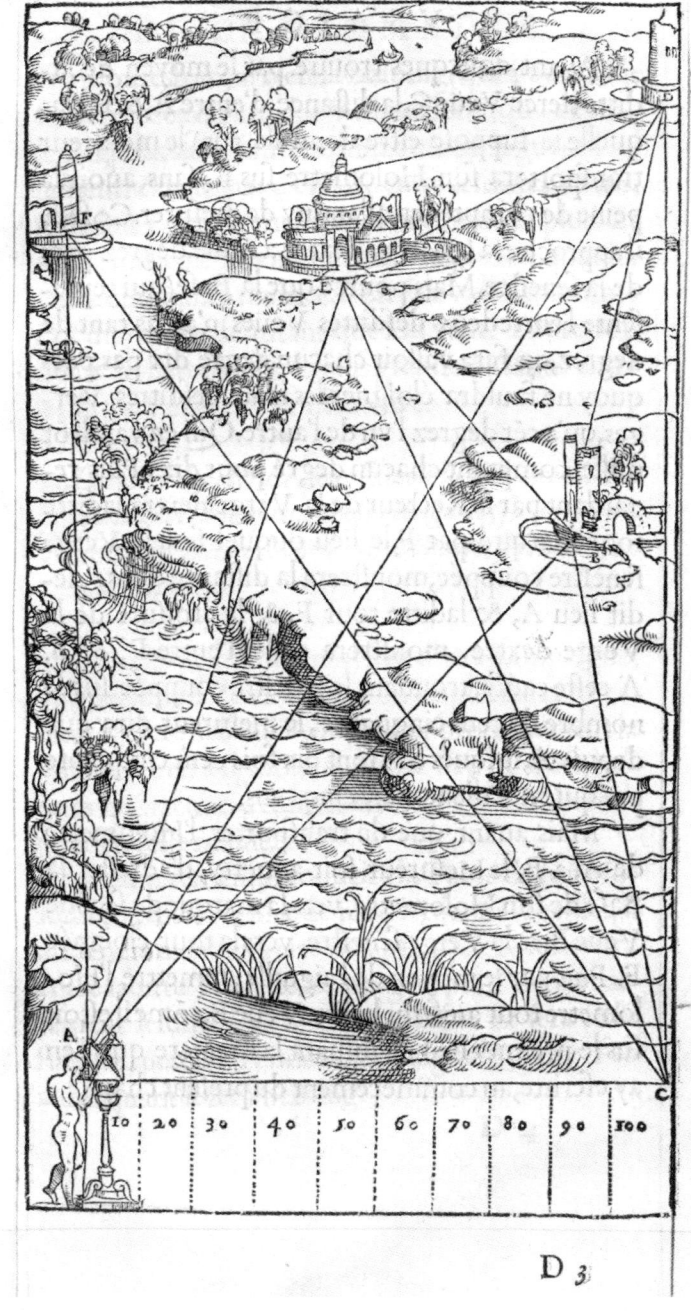

Abel Foullon, page from *Usaige et description de l'holometre: pour scavoir mesurer toutes choses qui sont soubs l'estandüe de l'oeil, tant en longeur & largeur, qu'en hauteur & profondité / inuenté par Abel Foullon* (Paris: chez Pierre Beguin, 1561).

The post-World War II order shook up patent systems across the globe, moving generally in a trend toward international cooperation and streamlining, although the Cold War was, of course, a notable hurdle. BIRPI ultimately relocated to Geneva in 1960 to be closer to the United Nations and transformed into the World Intellectual Property Organization in 1970, with both moves superimposed on the larger landscape of Europe's gradual political integration.[46] A bit earlier than that, in 1949, Europe had also established the International Patent Institute in The Hague.[47] During the Cold War, patents took on new meaning as symbols of the dynamism of capitalist ideology, something latent in the famous Nixon-Khrushchev kitchen debate in which the two men looked over a sea of patented home appliances that perpetuated the argument that a country's economic merits were measurable by its technological bounty.[48] A unified patent system became a reality in Europe under the 1973 Munich Convention, which instituted the European Patent Organisation (EPO) as an autonomous legal body that could grant pan-European patents. The EPO began granting European patents in 1977 from its headquarters in Munich.

The Question of Architecture

Today, patents are manifest in all aspects of our built environment. Our homes are full of patented items, from our floorboards and countertops to our window fittings and plumbing fixtures. So too are our public spaces: in traffic lights, sidewalk pavers, bridges, and tunnels. Ana Miljački has argued that architectural knowledge is presently in the initial stages of regulatory transition, a process that would begin with sweeping changes to the landscapes of copyright and fair use and perhaps result in the widespread adoption of patenting as well.[49] Ever since Brunelleschi, architects have played an auspicious role in the codification of inventiveness in our built environment. Yet architects have faded far into the background in the modern patent system and are presently one of the design professions least likely to participate in patent activity. Figuring out why this is the case is one goal of this book. Is it that the profession has indulged in a belief, a kind of gentleman's understanding, that direct copying stands to be more potentially damaging to one's reputation than any kind of patent infringement

Unknown artist, political cartoon showing the Oliver Patent—Monopoly in Crucible Steel, c. 1882, as a giant steam-driven robot rampaging through a scattering crowd of people with the Tariff Commission riding on its back.

could be? Or is it embedded in the primary characterization of the architectural elite as a profession that is, ultimately, more art than engineering? Or something else entirely?

As muted as patent activity in the field of architecture may be, there is enough of it in the modern history of the profession to make the argument that it has been neither a negligible nor an inconsequential practice. Furthermore, patent activity in architecture points directly to the tension between the view of architecture as a practice of singular vision and "genius" and architecture as an essentially coauthored, collaborative practice where inventions occur slowly. Patent culture telegraphs the kind of hagiography that prefers the former over the latter. This is why the likes of Auguste Perret, Le Corbusier, Wallace Harrison, Buckminster Fuller, Frederick Kiesler, Frank Lloyd Wright, Eduardo Torroja, Shigeru Ban, Rem Koolhaas, Santiago Calatrava, and Frank Gehry hold patents in their names alone when in all of their cases their inventions benefited from the contributions of a vast array of other professionals.

In his authoritative survey of IP, Siva Vaidhyanathan has characterized architecture as a burgeoning field of IP law—indeed, a field that is now covered by copyright law in most parts of the world.[50] This coverage is something that the legal scholar Kevin Emerson Collins has tried to understand as a semantic question within law itself, as the legal profession has very different understandings of what words and concepts like "invention," "innovation," and even "space" mean than the understandings within the architectural profession.[51] After all, the flat roofs of modernism were certainly inventive in the architectural sense, but not at all in a technical and legal sense. That said, someone, somewhere, had to figure out how to facilitate drainage off those flat roofs, and that was patentable in the legal sense in a manner that the flat roof itself was not. Patenting architecture is thus a topic hiding in plain sight.

As Collins has shown, most patents that we might call architectural patents—of both building materials and building systems—are mundane construction technology patents, evolutionary and not revolutionary.[52] Yet many of these patents form the technological backbone of architectural modernism: cast iron, curtain walls, and poured and precast concrete.[53] Indeed, over the course of the nineteenth and twentieth centuries, patent offices across the globe have tended to have an expansive understanding of what is patentable in the fields of architecture and construction, not

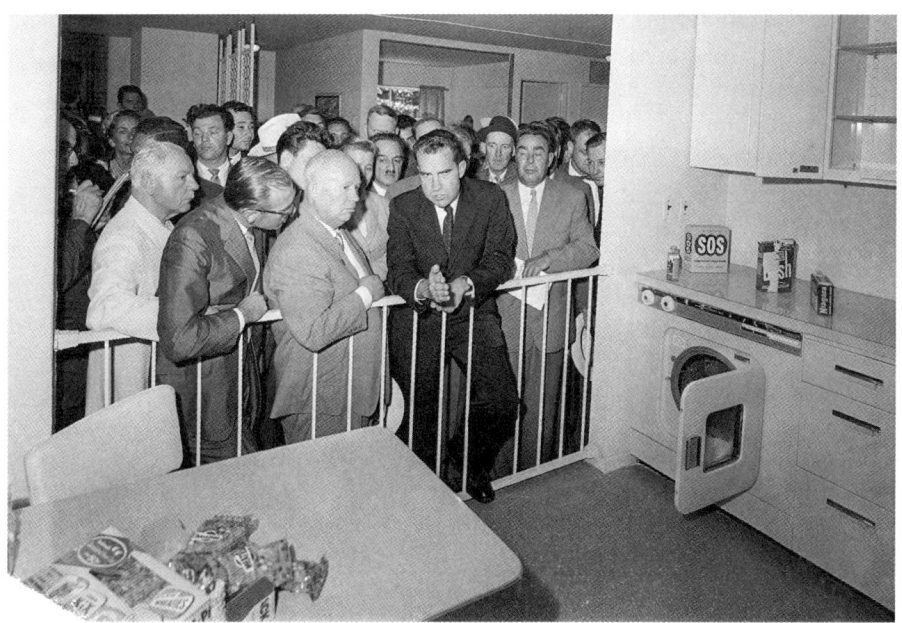

Associated Press, photograph of Soviet Premier Nikita Khrushchev, center left, talking with US Vice President Richard Nixon during their famous "kitchen debate" at the United States exhibit at Sokolniki Park, July 24, 1959. While touring the exhibit, both men kept a running debate on the merits of their respective countries. Standing to the right is Khrushchev's deputy, Leonid Brezhnev.

least the recent trend toward patents codifying spatial configurations as inventions. Despite this, the hard evidence supplied by patent records demonstrates a clear reticence to adopt the practice wholesale. This means, among other things, that many patentable inventions in architecture were not, and never will be, patented. Moreover, those holding these so-called architectural patents are often not architects, but rather engineers and real estate developers.[54] The reality may be that many architects did not know and continue not to know that their inventive activities, including spatial arrangements, are patentable in the first place.

The issue of patents in architecture may also suffer from being overshadowed by a more prominent member of the IP triad: copyright. The more frequent invocation of copyright in architectural work and legal disputes in architecture is in part indebted to the profession's own self-fashioning, wherein copyright is more clearly associated with artistic endeavors such as music, literature, and the visual arts and hence more allied with the cultural cachet that architects tend to desire.[55] The domain of copyright celebrates imagination and original expression, not inventiveness per se. The result, Collins argues, is that "architectural patents languish in copyright's shadow, garnering little to no attention from either [architects or lawyers]."[56]

The historical journey of concrete through the legal system offers an excellent example of the patent system's largely inadvertent and stealthy role in the development of modernism as a cultural phenomenon. The process of integrating concrete into every facet of our built environment was neither a revolutionary nor a heroic undertaking.[57] Sigfried Giedion was perhaps the first architectural historian to subdue the hagiographic strain in architectural history and look at the ways in which architectural innovations have occurred sedimentarily, bureaucratically, and prosaically.[58] Giedion cites the workaday men, like François Hennebique, whose contributions were "not the result of fantastic visions—they were . . . [e]xpressed in the more modest language of our time—patents."[59] This "stepwise" character, as Michael Osman has called it, demonstrates Hennebique's incremental domination of the European concrete industry through the monopolies afforded by IP, and not through moments of inspired genius. It was this stepwise incrementalism that afforded reinforced concrete its radical new formalism, as it was conceived nearly exclusively in the technocratic infrastructure of European patent systems, proving that concrete's "radical" break with history was embedded in prosaic, bureaucratic procedures. This helps us understand to some degree why designs for buildings across the twentieth century often evoked historicizing forms that are at odds with high modernism.[60]

This book explores the ways in which the patent, as a heuristic device, can help us to better understand how three key features of IP are manifest in the domain of architecture: creativity, novelty, and property. As integral as these are to the creation of architecture, they rarely buttress book-length studies. Thus, while this book may be about the history of architectural patents, it is also a work that specifically uses patents as an operative tool for more rigorously teasing out what is meant when we call something in architecture "creative" or "novel" or when we refer to an architectural idea or design as "property."

Creativity, Novelty, Property

It is worth bearing in mind that creativity is not a synonym of inventiveness, although the two are commonly conflated in both the scholarly literature on patents as well as the general bureaucratic language used by patent offices and patent attorneys. Creativity is slippery: it is, on the one hand, entangled with the phenomenon of aesthetic distraction and neoliberalism and, on the other, posited as a means of emancipation.[61] Isolating creativity from the larger context of inventiveness is especially important in architecture, as the latter has long been associated with creative thought and production. It is instructive to consider creativity in light of the work of the German sociologist and cultural theorist Andreas Reckwitz. Reckwitz invokes something he calls the "creativity dispositive" to identify the constitutive elements of creativity, which in turn provide a way to understand the sociological contexts of creative production. This moves creativity toward a more rigorous definition that is harder to conflate with other concepts and into the paradoxical worlds of neoliberalism and emancipation. "A dispositive," Reckwitz contends, "comprises four different social elements: practices and everyday technologies informed by implicit knowledge; forms of discursive truth production, imaginary and collective problematization or thematization; artifacts (instruments, architecture, media technology, accessories, vehicles, etc.); and patterns of subjectivization—that is to say, ways in which people are shaped, and the way people adapt their abilities, identities, sensibilities and desires to the dispositive and so help to carry it."[62] These four pillars, which we could call implicit knowledge, truth production, objects, and subjectivization for short, notably exclude the time-honored design trope of problem solving that is embedded not only in much of architecture's own rhetoric of self-definition but also in the "utility" mandate of most patent systems—which is that an object must solve a problem (known or unknown) in order to be useful.

Disentangling creativity and problem solving is a critical exercise because it demands that we be more precise about what we mean when we designate something as creative. Reckwitz's contention that creativity is socially and culturally inscribed is also an important precept for moving beyond staid notions of creativity formed and projected onto the rest of the world by Western thinkers and institutions. It provides new ways of thinking not only about *what* is a manifestation of creativity but also *who* is a creator, sidelining commonly deployed but outmoded ideas such as genius or creativity as an individuated act. Reckwitz argues that creativity has two major significations. The first is the "potential and the act of producing something dramatically new. Creativity privileges the new over the old, divergence over the standard, otherness over sameness. This production of novelty is thought of not as an act occurring once only but, rather, as something that happens again and again over a longer period of time."[63] Secondly, he argues, the topos of creativity "harks back to the modern figure of the artist, the artistic and the aesthetic in general. In this sense, creativity is more than purely technical innovation."[64]

Creativity is germane to the study of patents in architecture because as a term (in English as in many other European languages) it is, like the patent system, a product of modernity. Adrian Johns has shown how, in Germany, creativity came to be construed as a power that afforded certain regimes of property, powers emanating from a common principle of creativity.[65] In the contemporary world, critics contend, the so-called creativity dispositive has become the primary force behind an accelerating appetite for cultural and aesthetic novelty, an acceleration that can more or less be charted against the proliferation of patents in global legal systems.[66] Creativity, according to Reckwitz, "assumes the guise of some natural potential that was there all along. Yet,

United States Patent [19]
McHarg, Jr.

[11] Des. 245,013
[45] ** July 12, 1977

[54] BUILDING

[75] Inventor: **Joe Albert McHarg, Jr.,** Smyrna, Ga.

[73] Assignee: **Pizza Ring Enterprises,** Chamblee, Ga.

[**] Term: **14 Years**

[21] Appl. No.: **617,644**

[22] Filed: **Sept. 29, 1975**

[51] Int. Cl. .... D25—03
[52] U.S. Cl. D25/17
[58] Field of Search D25/17, 22, 24, 25

[56] **References Cited**
U.S. PATENT DOCUMENTS

D. 210,007	1/1968	Cassano	D25/17 X
D. 228,640	10/1973	Long	D25/22 X
D. 233,143	10/1974	Laughner	D25/17

Primary Examiner—A. Hugo Word
Attorney, Agent, or Firm—Harold D. Jones, Jr.

[57] **CLAIM**

The ornamental design for a building, as shown and described.

DESCRIPTION

FIG. 1 is a perspective view of the building.
FIG. 2 is a front view of the building.
FIG. 3 is a back view of the building.
FIG. 4 is a left side elevational view of the building.
FIG. 5 is a right side elevational view of the building.

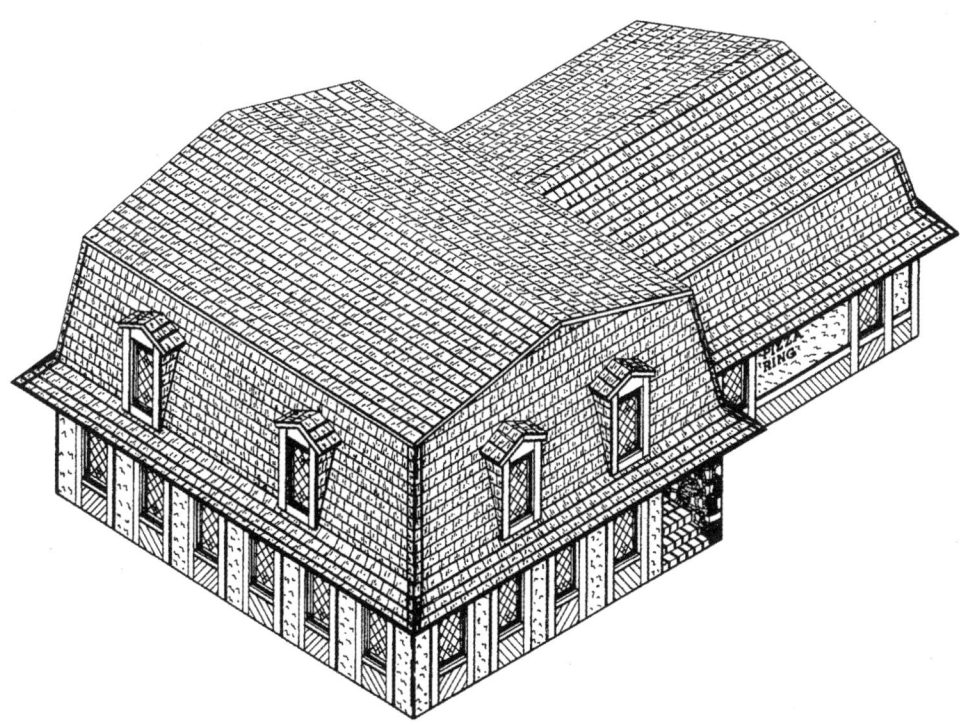

Joe Albert McHarg Jr., "Building" for Pizza Ring Enterprises, US Patent D245,013, filed September 29, 1975.

at the same time, we find ourselves systemically admonished to develop it, and we fervently desire to possess it."[67] This goes a long way toward explaining why, as many legal scholars have argued, global patent systems have become strategy-based and oversaturated, hardly a reflection of the natural course that creativity would take without the market pressure exacted by hypercapitalism. The creativity dispositive's opposite, Reckwitz contends, is "the undifferentiated and thus aesthetically uninteresting repetition of the eternally same" as well as "those divergences that explode the limits of what is sensuously and affectively bearable."[68] All who know architecture and its criticism will know the loathsome status accorded to banality, sameness, and repetition. Architecture too has a creativity imperative.

Creativity as a modern concept has a fascinating history. Pamela Long identifies *Idiota*, a mid-fifteenth-century text by Nicholas of Cusa, the German polymath and one of the first proponents of Renaissance humanism, as a seminal text on the origins of the concept of creativity.[69] The text describes the superiority of the maker of pots and spoons over the painter. The painter, as he sees it, imitates nature, whereas the potter creates something with no model to consult in the natural world, bringing it closer to the divine. This enshrinement of the potter over the painter would not last, ceding ground in the decades to follow to interest in the "manner" of painting, a way of discerning a kind of authorial subconscious in technique and form, no matter how imitative the subject matter.

The idea of a kind of "organic" creativity similar to what Reckwitz describes, a creativity that arises from a creativity dispositive free from market pressure, was explored by Kant and the romantics and was a fundament of the case against patents in the mid-nineteenth century.[70] Robert Andrew MacFie, a Scottish businessman and later a member of the British Parliament, evoked this organic idea of creativity to campaign against what he saw as the artifice of patents and the way they misrepresented creativity: "As in the vegetable kingdom, fit conditions of soil and climate quickly cause the appearance of suitable plants, so in the intellectual world fitness of time and circumstances promptly calls forth appropriate devices. The seeds of invention exist, as it were, in the air, ready to germinate whenever suitable conditions arise; and no legislative interference is needed to ensure their growth in the proper season."[71] Here the main divergence from Kant is that although he too believed in a creativity dispositive that was similarly organic, he also held that such creativity buoyed and should support the modern notion of authorship. This view buttressed the opposite of MacFie's cause, namely, a patent system that reinforced creativity as a regime of authorship (and, in turn, some element of proprietary rights). The legal scholar Robert Merges has characterized the protection of creative works as the "honoring and rewarding" of their creative authors.[72] Authorial recognition was part and parcel of Kant's sharp focus on personal autonomy and brings into view the underlying Enlightenment thinking behind the holding of IP as a right, not a privilege, where third-party interests are reached only through the individual at its origin.[73] Kantian creativity also subtends the primacy of legal conflict that has swamped the modern patent system and its entwinement with artistic creativity.

Just as the disassociation of creativity and problem solving brings into focus a more precise understanding of invention, so too does this disassociation provide an opportunity to pursue questions about the psychology of creativity and what prompts inventors to want to solve problems in the first place. In his influential *The Art and Science of Inventing*, Gilbert Kivenson wove human psychology into his study of the creativity dispositive. Kivenson notes that many inventors conduct their work from a place of deep-seated convictions, often entangled with motivations related to their own personal history, such as a biologist who commits his career to the subject of water metabolism after suffering humiliation for his enuresis at the hands of his father as a child.[74] Kivenson finds that neurosis and anxiety are commonplace among successful inventors. Anxious inventors, in particular, demonstrate a need for constant validation, something the patent system offers, even at very incremental levels.[75]

At the same time, Reckwitz has argued that novelty does not exist as an objective fact in the creative domain. This is important when one considers that novelty is one of the key criteria for patentability in most patent systems.[76] Further obfuscating architecture's place in the patent system, modern patent law construes novelty in primarily temporal, as opposed to spatial, terms.[77] The novelty imperative acts extrinsically on creativity, becoming something of a tyrant demanding "sensuous and affective offerings with the value of surprise" at every turn.[78] Not all inventors, architects included, will find that surprise is the affect they are after when inventing. However, like creativity, the demand for novelty is socially inscribed in addition to being legally inscribed. Ingenuity has been a tacit expectation since at least the nineteenth century in white-collar labor.[79]

The legal scholar Charles Colman is one of the few who have explicitly pointed out the unevenness of the notion of "novelty" *within* the patent system, as it applies to novelty in both utility patents (patents with new "functions") and design patents (patents with new "designs"), a subject that this book explores. Colman describes how the association of utility patents with scientists and design patents with the "ornamental" creations of designers has manifested ontological differences between the two that reveal assumptions, biases, and values regarding the "triviality" of design as a subject of novelty under the law.[80] "The outcome," Colman contends, "if not the purpose, of this normative framework is a potentially invidious division among objects, endeavors, and even people that purportedly have 'value,' and those that do not."[81]

Irrespective of its perceived value, a patent will always manifest a right to property. Like many in the IP field, Merges believes that property (imbued with important limits and constraints) is essential to a fair society.[82] To this end, he views the Kantian conception of property, in which the labor of an individual's mind or body in the absence of a labor agreement is automatically one's own property, as "lend[ing] itself to an even-handed theoretical approach to IP rights, balancing freedom of action for owners with the interests of others in the community."[83] This, by way of natural law, includes the objects of one's labor, a belief echoed by Hume.[84] Kant in fact condemned any law that would render an object ownerless, an anachronistic thought in modern property theory.[85] It is important to note that Kant's thinking on what we would come to call intellectual property was nevertheless centered on works of writing, which leaves some ambiguity about the role of something like architecture, itself very much an object of a community as well as individual authors.[86] Nevertheless, Kant saw boundedness to objects not as a contradiction of personal autonomy and freedom, but in fact as an expansion of these.[87]

John Locke is also essential to any theoretical discussion of property. The idea that the fruits of one's mental and physical labor are automatically one's own de facto property was not so firmly associated with "natural law" until Locke codified it as such.[88] Locke wrote of "mixing" labor with matter as a way of codifying property rights as natural law. The philosopher Robert

Nozick famously ridiculed this idea by pointing out that one could reasonably make a claim to owning the ocean by pouring a can of tomato juice into it.[89] Merges has convincingly suggested that this hang-up is simply semantic and that Locke clearly understood "mixing labor" to simply mean what we would describe today as "applying labor," which sets the absurdity of the tomato juice example aside. What Locke actually meant is that an expenditure of effort acting upon matter confers property rights: one holds property rights whenever one expends an effort to create something. When that thing also solves a problem (and thus is inventive) and does something that nothing else does yet (and thus is novel), it meets the extrinsic threshold of patentability, which provides legally enforceable property rights that do not exist without a government's invention. I suspect that while Locke would have been comfortable with the patent system, he might also have expressed some epistemic discomfort in transferring the role of arbiter of creativity and novelty to the government. In the end, though, what alternative would there be? Hegel was also an ardent supporter of the rights afforded through personhood.[90] "The reasonableness of property consists not in its satisfying our needs," he said, "but in its superseding and replacing the subjective phase of personality. It is in possession first of all that the person becomes rational. The first realization of my freedom in an external object is an imperfect one, it is true, but it is the only realization possible so long as the abstract personality has this firsthand relationship to its object."[91] This book is about seeing architecture as that object.

*

Although *Prior Art* examines the fundamental nature of creativity in architecture, it is not a historiographic project. Rather, it is an examination of a historiographic concern through history and, as such, relies on well-worn methods of architectural history, with extensive use of primary and other archival sources. To this end, the book takes a site-oriented approach, moving sequentially through seven sites where patents in architectural culture provide specific and new insights into the central concerns of creativity, novelty, and property in architecture, the constituent elements of invention, and the patents it produces. These sites are the home, the studio, the laboratory, the repository, the patent office, the courtroom, and the commons.

1 *The Home*

Home is one's birthplace, ratified by memory.
—Henry Anatole Grunwald

Perhaps home is not a place but simply an irrevocable condition.
—James Baldwin, *Giovanni's Room*

Women and Domestic Inventorship

Fans of the cult television program *Twin Peaks*, directed by David Lynch, will undoubtedly be familiar with the story of one the most eccentric of the show's many eccentric characters: Nadine Hurley. Nadine, played by the actress Wendy Robie, is the wife of local go-to guy Ed Hurley, who, despite having deeper feelings for an old flame, is (mostly) loyal to Nadine, in large part due to a debt he believes he owes her. Some years prior, on their honeymoon, Ed had accidentally shot Nadine in the eye while hunting. Nadine, who was briefly in a coma that surged her adrenaline levels and who needed to wear a patch over her damaged left eye, had to return to her normal domestic life despite her newfound, convulsive intensity. Concerns arising from her desire to perfect her surroundings, in particular to fashion drape runners that were 100 percent silent in order to tamp a noise that drove her crazy, grabbed an inordinate amount of her restless energy. We meet Nadine in a handful of situations tinkering with off-the-shelf products such as standard drape runners and cotton balls to no effect—that is, until Ed came home one day and accidentally tripped on Nadine's experimental paraphernalia, dripping motor oil from his car repair shop on her cotton balls. The next day, Nadine discovered that this accident in fact held the secret that would solve her inventive quest for 100 percent silent drape

Still from the television series *Twin Peaks* depicting the character Nadine Hurley, played by Wendy Robie, assembling a 100 percent silent drape runner, 1992.

runners.[1] The coda to this story is offered in the 2016 sequel, when Nadine is reintroduced as the successful proprietor of a store called Run Silent, Run Drapes.

One absurd subplot among many in *Twin Peaks*, Nadine's obsession with 100 percent silent drape runners nevertheless resonates as an archetype of a certain kind of inventor's story that we can find in droves in actual historical records: that of the woman tinkerer, trying to break through the banality of quotidian domestic life and the patriarchy of the patent system writ large to invent something that others will recognize as transformative for women's work. We can also see the acceleration of the patenting of inventions to ease and mechanize domestic labor as having gendered associations even when women were not involved, as was the case with the humble yet revolutionary "lazy Susan," a device whose name suggested an everywoman who was growing lazier as inventions made her life easier.[2] In both cases, the inventor-tinkerer who invents an improvement for the domestic sphere embodies precisely the kind of tactical form of inventorship, backed by neither clients nor corporations, that aligns most closely with what Nicholas Negroponte has described as "frugal innovation."[3] In the world of invention, the home is more than a mere domicile: it is the wellspring for the inventions that represent our most immediate needs and desires.

The crux of the story about women and patents, however, is not what they were patenting or the way in which certain patents became gendered, but rather the very fact that women, in both the United States and the patent powers of Europe, could apply for patents in the first place.[4] Even though the patent statutes of Europe and North America did not draw any kind of distinction between the sexes, the common-law tradition of a married woman's right to property applied to patents post facto.[5] While there were certainly roadblocks well into the twentieth century for women who sought patents—namely, lack of access to the necessary funds and attorneys as well as the general social obstacles to being taken seriously as inventors—women had been, at least on paper, permitted to submit patent applications as long as men had been. This is truly striking when one considers that this right preceded suffrage and the ability to attend the same top universities as men by almost two centuries in certain cases. Indeed, as Deborah Merritt has shown, the hurdles for women in gainng access to the patent system were administrative and cultural, not

The Home

legal, and these hurdles were exacerbated by women's isolation from the world of commerce.[6] Even Voltaire saw an inequitable patent system long before the mainstream would recognize it as such when he said that "[t]here have been very learned women as there have been women warriors, but there have never been women inventors."[7]

There were, in fact, women inventors: in the United States alone, 3,849 women held patents that were filed between 1865 and 1895.[8] Several innovations in the architecture, design, and construction sectors were invented by women: Catherine Mott's Improved Fire Escape (US, 1878), Miriam Benjamin's Gong and Signal Chair (US, 1888; Benjamin was the first African-American woman to hold a patent), Helen Shufelt's Fireproof Doors (US, 1890), Anna Merlien's Tubular Construction System (France, 1906), Astrid Jagdmann's Stone Roofing System (Sweden, 1918), Maria Strowich's Cavity Wall System (Germany, 1919), Lenore Sackur's Wood Joinery System (Germany, 1924), and Eva Zeisel's Resilient Chair (US, 1949), among numerous others.[9] In the next section we focus on four women who stand out as exemplars of the diversity of the conditions under which women who patented were motivated to do so.

Sarah Guppy, Anna Wagner Keichline, Ilse Szegö, Frances Gabe

The story of women claiming IP rights in the arena of the built environment can begin with the tormented story of Sarah Guppy. Born Sarah Beach in 1770, Guppy was 41 when she began patenting inventions developed at her home in Bristol, the city in England with the second-largest concentration of patent holders after London.[10] In her earliest patent, Guppy focused on bridges, a subject of great local interest. As early as 1753, Bristolians had been discussing in newspapers and public hearings the idea of building a bridge that would span the deep and wide Avon Gorge that hampered Bristol's connectivity with Somerset and counties further south and west, as well as the city's ability to expand its prominence in the slave trade and other colonial industries associated with the Triangle of Trade and its position as the major western port of England. A prominent example was a 1793 design by William Bridges, which was to be built entirely out of masonry and inhabitable. As more and more iron bridges were being successfully constructed (albeit over smaller rivers), Guppy, who would later cement her connection to the world of bridges through her friendship with the acclaimed engineer Isambard Kingdom Brunel and his family, submitted a patent for a "hanging bridge" using at least some metal members.[11] It is notable in this connection that Brunel, who was Bristol's most famous inventor, was, as mentioned in the introduction, ardently anti-patent and thus did not himself contribute to the city's patent tally. Guppy's patent, which was site-specific—a fairly unusual practice even today—claimed its pieced-together composition as its innovation: "longitudinally and crosswise fit such pieces of timber, or iron, or other suitable material, as shall and may constitute a platform, which . . . shall afford a proper support for a road or pavement."[12] As Julia Elton has noted, "Guppy may have thought that using iron chains rather than ropes constituted a new patentable element. She certainly understood that a flat, rigid deck was needed to carry the heavy loads of moving vehicles, and to achieve this specified that 'said chains may be drawn tight by secure mechanical means.'"[13]

Guppy promoted her patent with might and main. She published a letter in *The Monthly Magazine* in 1811 describing the advantages of the bridge's design in a narrative prose that the legalese of patent claims could not afford, paying particular attention to credible claims of her design's ability to resist the floods that plagued the Avon and had killed many navigating it. She noted: "The number of lives lost, and the damage and inconvenience sustained, particularly in winter, when torrents of water, and great quantities of ice and snow force their way down rivers . . . carrying away bridges in their courses; having frequently occupied my mind, and believing I have devised a mode by which the same may be prevented, I have obtained a patent for a new method of erecting bridges . . . without arches or sterlings; the advantages to be derived from which are, that they are not subject to be injured or destroyed by floods—no kind of ground is unsuitable for the foundation—they may be erected in the most difficult and almost inaccessible place—roads may be continued over marshy grounds without the danger of being destroyed in winter, and are alike applicable to every situation whether public or private—are erected in a small space of time—and comparatively inconsiderable expence."[14]

Guppy made her efforts to build a scale model of the patent in her parlor known to John Matthew Gutch, editor of *Felix Farley's*

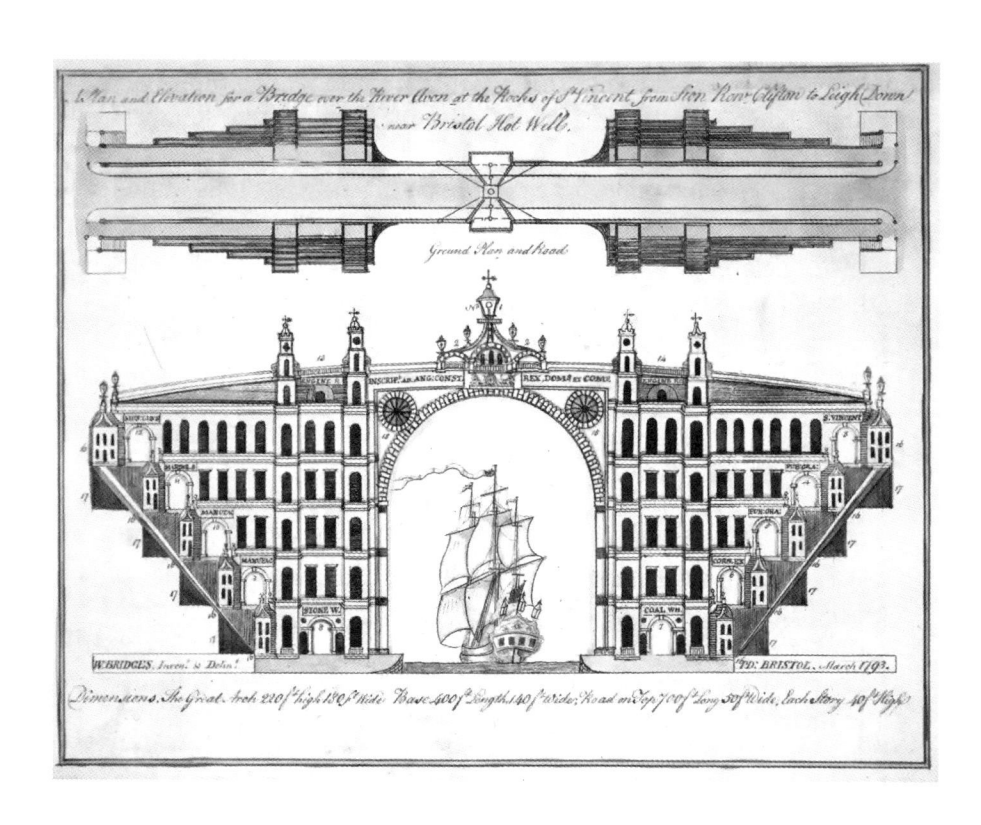

William Bridges, design for a bridge across the Avon River, 1793.

Bristol Journal, who in turn reported on Guppy's "assiduous" efforts to resolve Bristol's Avon River problem.[15] Several days later, Gutch reported on an unnamed "Visitor and Admirer of Clifton" (the southern terminus of the bridge) who was interested in helping Guppy fundraise.[16] Ultimately, however, Guppy's bridge did not gain any more traction than failed proposals that had come before it and, as explained in what follows, she may have had men to thank for this.

At first glance, it might seem that what Guppy was describing was a prefiguration of the modern suspension bridge, a bridge typology that would flourish a few decades later, including in Brunel's own design for the site, the Clifton Suspension Bridge, which was opened five years after his death, in 1864. It would seem, however, that many of Guppy's contemporaries did not want her to receive any kind of credit for this "so-called invention" which, in 1811, did not need to go through an examination process. *The Repertory of Arts, Manufactures and Agriculture* was quick to dismiss her "hanging" design as derivative, noting that "[t]his method of making passages across rivers is of very ancient invention, having been practised in China from remote ages; and also in Peru and Mexico, as may be seen in the many histories and accounts published on these countries."[17] The publication went one step further by directly raising the issue of gender: "Politeness might forbid us to question the ingenuity of a lady, and therefore we presume she must have been ignorant of the well known facts, or she would not have declared this mode of constructing bridges to be her own invention."[18]

Guppy encounters similar hostility in the little existing historiography on her. For example, the historian of engineering Julia Elton accepts that Guppy was the first person to patent a hanging bridge that could take both road and rail traffic, but adamantly rejects the romantic notion that Guppy designed the first suspension bridge as a kind of facile feminist fantasy. As Elton notes, "The patent appeared at the moment when a completely new and far more effective type of structure was emerging. The American, James Finley, in 1801, first came up with the idea of a true suspension bridge, hanging a deck, via a series of stiff vertical hangers, below a suspending system using iron chains running over tall towers to their anchorages; this ensured a horizontal platform able to take wheeled vehicles. . . . Guppy must have realized that her hanging bridge was completely superseded by these developments."[19]

Despite all of her rather bold behavior (when considered in the context of 1811), Guppy is also said to have remarked that "it is unpleasant to speak of oneself—it may seem boastful particularly in a woman."[20] This quip, as well as her subsequent patents, indicate that she may have been neither a charlatan and plagiarist who was trying to somehow self-fashion an image as a pioneering woman engineer, nor a naïve dilettante who was able to draw as much attention as she did because of her high social standing and access to money and the press. For one thing, Guppy could not have cited Chinese or Peruvian precedents, even if they were known to her, because they were not patented and she quite likely believed that her introduction of human-made materials (as opposed to the natural materials of those precedents) was itself a major new development, which it was—even if it might not pass the patent threshold of nonobviousness.

Guppy's next patent, a year later, was a method for boiling eggs in a standard tea or coffee urn that also incorporated a means to keep toast warm, a ladylike invention that unsurprisingly came and went with little fanfare and no opposition from male dissenters.[21] This seems to indicate that her invention aspirations were not as megalomaniacal as the Bristol press intimated. Moreover, when Brunel began his own work on the Clifton Suspension Bridge in 1829, Guppy took an active role in promoting the project among Bristol's society women and through anonymous editorials in the *Bristol Mercury* in the summer of July 1833.[22] Guppy's enthusiastic supporting role in realizing Brunel's vision was clearly not born of narcissism, but one can also imagine their differences on the question of patents: Guppy believed deeply in the patent system, while Brunel was a famous patent skeptic. Yet, as Elton has shown, Guppy herself continued to whisper and insinuate to the press that she was the mother of the suspension bridge, and this was cemented in what was surely a prewritten obituary in 1852 noting that she had "early enunciated the principle of suspension bridges, for which, as long ago as 1811, she took out the first patent ever granted."[23] Guppy's claims may indeed be inflated, but the obsession with determining the "true" origin of the suspension bridge, an obsession shared by both modernists and those in the sphere of patent culture, stands to relegate complicated figures like Guppy unnecessarily to the dustbin of history. Indeed, the pursuit of a

Unknown photographer, photograph of Anna Wagner Keichline (unnumbered individual in the far back right of room hunched over her desk) in the architecture studio at Rand Hall, Cornell University, c. 1909.

"true origin" of the suspension bridge vis-à-vis patents begins to sound like a masculinist prerogative when the target is a woman, a prerogative that will inevitably disqualify the very significant work of women in gradually making the practice of patenting as accessible to women as it was to men.

By the turn of the twentieth century, as women were allowed to study architecture and engineering and choose all manner of allied fields as their professions, patents by women also picked up pace. Anna Wagner Keichline, the first woman to be registered as an architect in the state of Pennsylvania and one of the very first women to practice architecture professionally in the United States, represents a watershed moment, but not in the way one might expect.[24] Keichline benefited from having parents who supported her interest in carpentry and her desire to go to college, first at Pennsylvania State College to study mechanical engineering, and later at Cornell University to study architecture—where, upon graduating in 1911, she became only the fifth woman to obtain an architecture degree from Cornell's prestigious program. A photograph of her among a sea of male colleagues at Cornell suggests some of the challenges faced by intrepid women in architecture schools.

Despite her success at Cornell, the professional landscape awaiting her after graduation was markedly different from that awaiting her male colleagues. Few firms were hiring women, and those that were would typically offer them positions that were not commensurate with the professional skills that came with a professional degree such as hers, positions like working on interior finishes or doing general business work such as accounting for a studio. Although she never said this explicitly, Keichline, who went into her own private practice, seemed to have seen a particular kind of professional horizon for women in the area of patenting, where gender had never been a point of bifurcation the way it had been for architectural practice. As someone with roots in carpentry and industrial design, Keichline would have had a natural affinity for the kind of inventions that a self-run office could actually manage. Between 1912 and 1931, she obtained an impressive seven patents, the very first coming just one year after her graduation from Cornell.[25]

Keichline's design career came of age at the height of the influence of scientific management theory, manifested most famously in Frederick Winslow Taylor's *The Principles of Scientific Management* in 1911. Five of her patents, including Sink for Apartments (1912), Components for Kitchen Construction (1926), Child's Portable Partition (1927), Folding Bed for Apartments (1929), and Air System (1931), were clearly indebted to Taylor and showed Keichline's ability to apply his theories to the domestic realm where they might ease women's domestic responsibilities. But it was her only truly nondomestic patent, for a K Brick (1927), that would bring Keichline as much recognition as any American woman architect had achieved up to that point. The K Brick, a clay building unit in the shape of the roman numeral "II" for hollow wall construction, was immediately recognized as a significantly cheaper and lighter way to build walls, a unit that is a clear marker in the succession of inventions that led to the ubiquitous concrete block we know today.[26]

A 1932 article entitled "Modern Wall Construction" appearing in the journal *Clay-Worker* is the longest single piece of writing that Keichline left behind. It operates as something of a subtle intellectual victory lap in the wake of her influential patent and bears quoting at length:

> There is no doubt that architecture is now passing through a transition period, possibly the most definite and extensive that it has ever had. Changes are being made in materials and methods of construction, but probably the most notable change is in the form of the building itself and the omission of decorative features.... Heretofore we have thought of shelter as something reasonably durable and not always fireproof. In considering the wall itself it should be of durable material both as to installation and upkeep, it should be entirely fireproof, and it should be of high insulating value.... If possible, it should show a saving over the usual methods of construction.... Inherent in burned clay in the form of brick or tile are found many of these properties.[27]

Keichline, who identifies herself as "A. W. Keichline" in the article, perhaps strategically masking her gender, is not the most compelling storyteller, nor does she highlight the transformative nature of her patent. But the argument is clear: ceramic building units such as bricks were the proverbial baby in the bathwater of the "old" being discarded for the "new" machine aesthetic. Keichline's argument is that the brick, while perhaps associated with antique and outmoded forms of architecture, is not in itself anathema to modernism. In fact, an optimized brick, like hers, which reduces mass by half and derives its form through a kind of pure calculus of structural necessity, is as modern as anything else. Masonry, in other words, can simply adapt through systems like patents and need not be discarded as the result of the stylistic dogma of the burgeoning international style. Moreover, hollow bricks, intended as they are to be unseen, are open to the cladding and skins of the modernist material palette and are thus ideal constructive handmaidens of the modernist agenda.

It would be tempting to focus exclusively on patents held by women that had a social or industry impact, as the Guppy and Keichline patents did. But even the most quotidian-seeming patents by women often reveal engrossing stories of provenance. This is where the story of Ilse Szegö (born Ilse Siegel, near Potsdam in 1907) comes in. Szegö was the last woman to be granted a patent in Germany before the beginning of World War II. Szegö's patent for a gridiron (*Balkenrost*), filed in 1935 and granted in 1938, contains a bare-bones description of a relatively modest invention for building foundations.[28] Ilse had no history as an inventor (this was her first and only patent), but anyone versed in the prior art of foundation construction in Germany might have recognized her surname. Her husband, Stefan Szegö, who was part Jewish, had moved to Germany from Hungary as a young engineer and inventor.[29] Stefan received a number of patents in Germany until the National Socialists came to power in 1933 and began the series of legal restrictions on Jewish participation in public life that would ultimately lead to the Holocaust. The writing on the wall was clear to Jewish inventors like Stefan Szegö: he and his patents were no longer welcome in Germany. These restrictions included the removal of Jews from government service (including as patent examiners) in April 1933; the Nuremberg Race Laws in 1935, which revoked and prohibited Jewish citizenship and the ability of Jews to have sexual relations with persons of "German or German-related blood" (such as his wife, Ilse); and then in 1938 the Order for Disclosure of Jewish Assets, which required Jews to report all property (including IP) in excess of 5,000 Reichsmarks, as well as the Decree on the Confiscation of Jewish Property (including patents), which regulated the transfer of assets from Jews to non-Jews in Germany.[30]

Stefan Szegö left his home in Potsdam for the United States sometime around 1935, right around the time his wife applied for the gridiron patent on her husband's behalf, which she was able to do as a German non-Jew despite not having any proof that she had actually participated in any aspect of the invention.[31] Stefan Szegö, who planned to seek new engineering work in the United States and wait for the political tides to turn against the National Socialists before returning to Germany, left Ilse behind to care for the couple's two children as well as to act as his proxy in his continued pursuit of patents in Germany. Sometime shortly after the gridiron patent was granted in 1938, Ilse also emigrated with the couple's two children to be reunited with Stefan in their new home in New York State. Stefan Szegö's name next appears in the public record in the form of a legal memo in the *Federal Register* of August 1944.[32] In this memo, the federal government determined that work executed by Stefan Szegö for a 1944 patent for a Beam Grid Structure for Ships, Airplanes and the Like contracted by the American Diagrid Corporation of Delaware, a structural engineering firm specializing in gridirons, was not to be designated as alien property despite Szegö's (now annulled) status as a German national.[33] While the memo does not cite Szegö's status as a Jewish refugee, it does make clear that at the time of the invention, "the claimant was ... at all times organized in the United States; and ... claimant is not a national of a designated enemy country."[34] Not only had Stefan Szegö established himself as a successful engineering consultant for firms like American Diagrid almost immediately upon his family's arrival; he is also believed to have acted as a covert engineering consultant for the US war effort in Europe.[35]

Although little is known about how Ilse and Stefan Szegö felt about their forced relocation to the United States, their deeds indicate that they embraced their new life as fully as they could. With Stefan's engineering consulting career now viable and most of their assets (except their house in Potsdam) transferred to the United States, the Szegös were able to purchase a property in Cortlandt Manor, New York that had belonged to the Hollow Brook Country Club before it went defunct. It was a generous space with a large meadow, providing Stefan with plenty of room to invent and experiment. In August of 1949, the civil rights activist, football star, and singer Paul Robeson, along with a group of supporters known as Peoples Artists Inc, had planned a concert in the

Lakeland Acres Park in neighboring Peekskill. Many in the area mobilized against Robeson and Peoples Artists Inc, prompted in part by J. Edgar Hoover's claim in 1943 that Robeson was a communist functionary and a threat to national security.[36] Those in the community who were convinced Robeson was a threat showed up at the concert, trying to block it with a demonstration that ultimately turned violent, with one concertgoer stabbed and twelve people sent to the hospital with injuries. The clash gained international attention.[37] Robeson and Peoples Artists Inc were determined to hold the concert and sought a new space that was not on public ground. This is how Paul Robeson met Ilse and Stefan Szegö, who invited Robeson to rent the meadow on their property for the concert. After getting wind of the new venue and the concert planned for September 4, demonstrators fired several shots at the Szegös' property and tried, on four separate occasions, to burn the property down, which resulted in considerable fire damage.[38] When the Szegös tried to collect on the fire damages, their insurance contract was canceled.[39]

A consortium comprising the Szegös, peace-minded residents of Peekskill and Cortlandt Manor, Robeson himself, Peoples Artists Inc, and US Representative Vito Marcantonio (who represented Harlem, where Robeson was based) came together to urge then-governor Thomas Dewey to take violent demonstrators to task and protect the second iteration of the concert on the Szegös' property. The concert did proceed, but despite the presence of 904 police officers, it ended with another melee on the streets around the Szegös' property, resulting in injuries to 145 people.[40] Thus, just ten years after fleeing persecution in Germany, the Szegös were again in the midst of a culture of racially charged antipathy and violence. Although there is little record of either Ilse or Stefan actively taking up political activism, their steadfast support of Robeson and his followers is a testament to their belief in the protection of free speech, political dissent, and equal rights, a set of values certainly strengthened by the trauma they suffered under the Third Reich. The black activism that the Szegös supported is one of many examples of the interrelationship of activism and patent culture.

The feminist savvy of Sarah Guppy was modest compared to that of her twentieth-century American counterpart, Frances Gabe. Dubbed the "original smart home pioneer," Gabe rose to some minor fame in the 1980s and '90s (and again in 2017 when her obituary in the *New York Times* went viral) with her 1984 patent for a self-cleaning home.⁴¹ Gabe, who was a self-identified feminist artist-inventor from Oregon, clad in garb that seemed to deliberately emphasize the image of a matron, was every bit as complicated a figure as Guppy. Gabe lived in the very experiment she patented: her self-cleaning home was a real-life work in progress in the suburbs of Portland.⁴²

As a manifestation of the ideal final version of her own self-cleaning home, Gabe built a model of the patent as part of the application process, despite the fact that models had been expressly prohibited by the US Patent and Trademark Office for almost a century. The model, which was accessioned by the Hagley Museum in Delaware and remains the latest patent model in their large collection of the medium, is indeed instructive regarding what exactly Gabe invented and why this ambiguously ironic project was able to pass formal muster at the patent office as a veritable technological invention.⁴³ On the first floor of the model there is a living and dining space with model furniture, appliances, and amenities. Most of the walls are made of a heavy-duty plastic, while some are in stone. The kitchen fixtures and furniture are also made of plastic or ceramic or another water-safe material. A system of clear tubes running along the perimeter of the house begins to articulate just why the material choices for this house are so unusual: a series of ducts, valves, and nozzles that would be attached to the house's ceiling system are in fact a cleaning system, and at the push of a button, the system begins to drench the house with a soapy solution. Through strategically placed jets and hoses, the house is ostensibly able to clean itself, sparing the inhabitants—particularly the woman typically associated with domestic cleaning—hours of tedious labor. Dirty water leaves the house through a drain system the way it would in a shower.

The point of all this, as Gabe noted in a 1982 profile in *People* magazine, was to use patents to liberate homemakers. "We should be better mothers, wives, neighbors, and spend time improving ourselves instead of saying 'I'm sorry, I have to clean the kitchen.'"⁴⁴ Gabe put it another way in an interview with the *Ottawa Citizen* in 1996: "Housework is a thankless, unending job. It's a nerve-twangling bore. Who wants it? Nobody!"⁴⁵ Then, in an article in the *New York Times* in 2002, Gabe honed her gender critique of American domestic architecture: "The problem with houses is that they are designed by men. . . . They put in far too much space and then you have to take care of it."⁴⁶

Frances Gabe with a model of her self-cleaning house, 1979. The model is now in the permanent collection of the Hagley Museum.

While Gabe's dalliances in media outlets like the *Phil Donahue Show* or *Ripley's Believe It or Not* make the self-cleaning house's status as an art piece rather clear, what may be more interesting are the measured and legal claims and citations used to buoy the validity of her actual patent, US Patent 4,428,085, where gender is only mentioned in the neutral term of "homemaker":

> The present invention relates to a building construction and apparatus characterized by ease of upkeep, and particularly to a building construction or apparatus wherein cleaning is mechanized and substantially automatic. Although the average home or building is provided with various laborsaving devices, very little progress has been made toward the automation of the basic cleaning of the building itself. Thus, cleaning of wall, floor and window surfaces as well as counter tops, table tops, plumbing appliances and the like involves a great deal of hand labor, with the basic cleaning functions consuming a considerable proportion of the average homemaker's time. In addition, appreciable time is expended in the washing of clothing, dishes, and the like even with the aid of conventional apparatus designed for the purpose.[47]

Like all patentees of her era, Gabe also had to provide citations of prior art. Even though her patent was part and parcel of a rather radical artistic statement, the conventions of the patent system required her to place the self-cleaning house within a clear genealogy of recent art. Her selections (or those of her patent attorney at Klarquist, Sparkmann, Campbell, Leigh, Whinston & Dellett) are telling: a sprinkler cleaning system by Dean Whitla of Massachusetts in 1965 and two self-cleaning restrooms, one by Richard Stock of Mississippi and another by Glenwood Garvey of California, both in 1973, among a handful of others, are all domestic cleaning systems designed by men.[48] As such, there is a certain dissonance between Gabe's contention in her artistic and media statements that men are at the root of the cleaning tasks foisted on women and the reality that her patent's very precedents are laborsaving mechanisms designed by men. The truth is probably that Gabe did not pay much mind to the citations and left that to her lawyers, who had the responsibility of proving that Gabe was in fact doing the opposite of what she said she was: creating an *ex nihilo* invention. Others were similarly tasked with another crucial part of the patent: the drawings. The overwhelming complexity of Gabe's design, with its hundreds of elements from dishracks to drains to soap dispensers, based as it is on a meandering narrative of the intricacies of home economics, is also plainly evident in the whopping number of figure callouts—588—in her 23 patent figures. These rote illustrations detail every single element of the house's functionality as a self-cleaning mechanism.

The Home Tinkerers

Indeed, it would be incorrect to represent domestic tinkering as the domain of women alone. In reality, as with patents in general, men have held the vast majority of patents seeking to reorder aspects of domestic life both large and small. This fact can also be read through a feminist lens insofar as it demonstrates that, contrary to stereotypes, men were concerned with the efficiency and advancement of the domestic sphere even if they were not the ones doing the heavy lifting of domestic labor. Male and female inventors alike have clearly recognized the potency of innovating the home, because home, and its material condition, are something that resonates with every person who has a regular place to lay their head at night. Home is a universal concept, and its betterment is a universal concern, to say nothing of the unequaled market that home innovation patents could reach and the revenue that home innovation could generate for savvy inventors.

Unsurprisingly, Scandinavia, with its longstanding and rich tradition of timber frame home production, was the wellspring of many of the most influential European patents related to timber construction.[49] Scandinavian designers also took to patenting entire "house kits"—what would later be labeled as "prefabrication systems"—at a rate that surpassed their European counterparts despite Scandinavia not being a major colonial power (for whom prefabricated kit homes served the clear purpose of quick and easy settler colonialism).[50] Although not prefabricated in the traditional sense, a patent for a "Method of Erecting Residential Houses and Other Buildings Made of Wood Without Timber and Without So-called Cross- or Rafter-Timber" by the Swedish inventor E. Ericsson in 1881 embodies the same kind of ready-made and thrifty logic of prefabrication, and is perhaps one of the most striking home patents to emerge from the Scandinavian countries in the nineteenth century.[51]

Ericsson's patent is distinctive more for its economic inventiveness than for any particular element of its design. Through his experience as both a logger and a carpenter, Ericsson knew of the high amount of waste in the Swedish logging industry, where inferior timber, sawed-off boards, and planks were regularly discarded. To economically minded (and perhaps also ecologically minded) observers, this represented an enormous waste. Ericsson's patent involved using these lesser pieces of wood to cut new housing costs by as much as half. The system involved interlocking irregular pieces of timber to form a beam-like element that could achieve the desired length and mass of regular timber framing. Because the "chaining together" of these irregular pieces was unconventional and foreign to a Swedish vernacular aesthetic, Ericsson knew that the system had to have some kind of aesthetic veil. Here is where he introduced a concomitant system for lime plastering which, in addition to giving the structure a smooth orderly appearance, tightened and secured the joints of the piecemeal wooden members beneath. Other patents by Ericsson's Swedish contemporaries, such as P. J. Ekman's 1882 patent for "Load-Bearing Door and Window Frames" and K. Wildhagen's 1883 patent for "A Compound for Roof Cladding," could easily be recombined to further optimize the cost of new housing.[52] What is striking in all three of these later nineteenth-century patents and countless more, both in Sweden and beyond, is that not a word was spoken about the radical changes these inventions stood to make in the vernacular built environment. Ericsson's plastered house, Ekman's large and bulky windows and doors, and Wildhagen's tarlike roof compound were all entirely at odds with the very longstanding timber frame construction traditions of Sweden. We can see in this kind of tinkering with parts (and less frequently wholes) how the stylistic changes that modernism would begin to dictate a few decades later were already receiving a kind of priming of the pump by inventors who doggedly—and perhaps myopically—obsessed over innovation in home design as a technical and economic issue without ever outwardly framing it as a stylistic one. Patents, of course, suppressed any discourse on style because style had no intrinsic utility within the logic of the patent process.

Whereas wood was the preoccupation of Scandinavian inventors, the material preoccupying home innovation tinkerers in

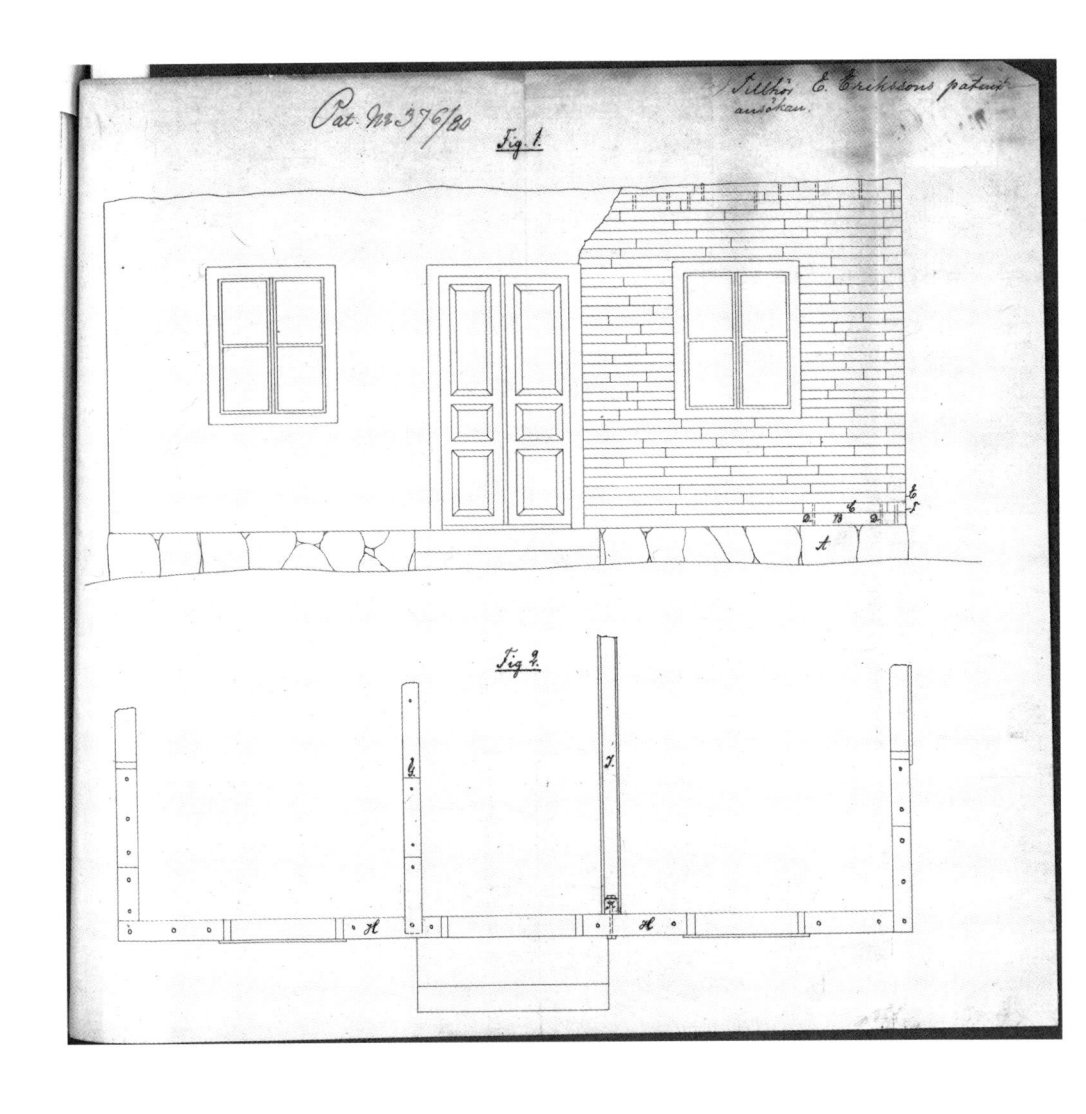

E. Ericsson, "Metod att uppföra boningshus och andra byggander af trä utan timmer och utan så kalladt kors- eller resvirke" (Method of erecting residential houses and other buildings made of wood without timber and without so-called cross- or rafter-timber), Swedish patent 376, filed December 21, 1880.

most of the other industrializing countries in Europe was concrete. Take, for example, William Henry Lascelles, a prominent builder in Croydon, England, who patented an Improved Method of the Construction of Buildings in 1875.[53] Lascelles's patent, comprising simple, thin slabs of precast concrete in a large variety of shapes and sizes, was intended to be paired with ready-made semidetached house designs by the prominent domestic architect R. Norman Shaw, marking a unique patent partnership between architect and builder.[54] Three years later, Lascelles had 28 schematic drawings of Shaw's Queen Anne-style designs published as single sheets or, alternatively as a compendium of all the drawings in a bound copy under the title "Sketches for Cottages and Other Buildings," which he deployed as both a promotional tool for his patent and a published record of his collaboration with Shaw.[55] Lascelles's system for Shaw's designs comprised sundry joinery solutions like inlaid iron rods or hooks that meant the concrete slabs were never themselves load-bearing (an auxiliary wood frame would do that).[56] Lascelles had envisioned the patented system as a product that could span socioeconomic classes, noting in the patent application that certain of Shaw's designs were for houses "of a better sort where the cost is not so great an object."[57] Others, meanwhile, had the object of reducing "the cost of small houses or cottages and to facilitate their construction in such a manner that they may be erected for the most part with unskilled labour in a short space of time."[58] In a letter published in the *Builder* on December 30, 1876, Lascelles boasted that he had managed to produce a four-room house for only 100 pounds and that he was setting up facilities for mass production that would begin manufacturing the following year.[59] A mention of the same system in the *Furniture Gazette* in 1878 notes that the cost of the system had been further reduced to 80 pounds, likely indicating that Lascelles's plans for mass production were successful.[60] At least two examples of the patented house are extant (or one if the joined units are considered as one), located at 226–228 Sydenham Road in Croydon, which is believed to be the original demonstration house.[61] Here, the concrete ornamentation, which was originally in a very rich terra-cotta color, is on full display in cast-concrete moldings, pilasters, cornices, and a swag of flowers on the gable end.

The American grocer Clarence Saunders was not a home tinkerer per se, but his 1917 patent for a Self Serving Grocery Store revolutionized domestic life.[62] Through a series of winding aisles,

Chris Nash, photograph of 226–288 Sydenham Row, Croydon, England, 2022. The design of the semidetached houses, completed c. 1881, is attributed to Norman Shaw, but is more likely to have been developed by his pupil Ernest Newton. The houses were built by the contractor W. H. Lascelles of concrete slabs dyed red, a system he patented in 1875.

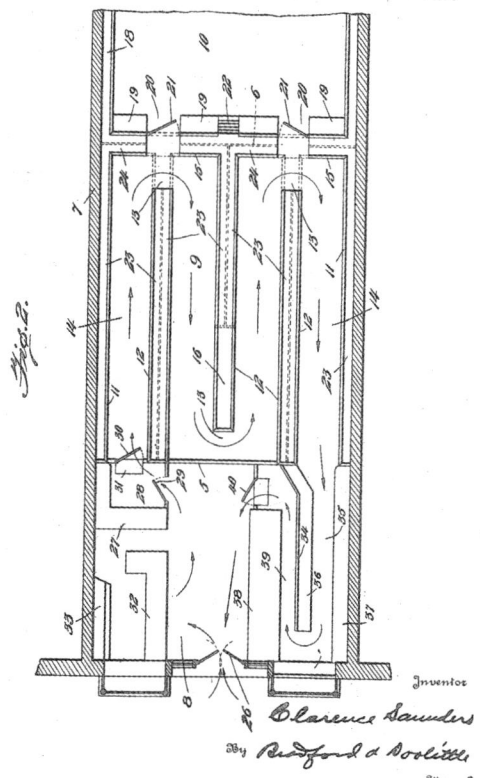

Clarence Saunders, Self Serving Grocery Store, US Patent 1,242,872, filed October 21, 1916.

cages, and turnstiles, Saunders facilitated the grocery store shopping experience that is familiar to most shoppers today. Instead of asking the grocer to retrieve specific items, Saunders's system allowed shoppers to peruse the aisles themselves and fill their own shopping baskets (and later, as the scale of this type of store grew, their carts). This afforded a groundbreaking consumer interface in which shoppers could compare brands and labels before committing to one product or another. This trend in turn changed the food industry, which now needed to pay more attention to its packaging and marketing. It also reduced the burden on grocers who could now divide their laborers between stockers and checkout clerks. As the patent states,

> the object of my said invention is to provide a store equipment by which the customer will be enabled to serve himself and, in so doing, will be required to review the entire assortment of goods carried in stock, conveniently and attractively displayed, and after selecting the list of goods desired, will be required to pass a checking and paying station at which the goods selected may be billed, packed, and settled for before retiring from the store, thus relieving the store of a large proportion of the usual incidental expenses, or overhead charges, required to operate it.[63]

Saunders goes on to note the two primary purposes of the patent:

> One purpose of the invention is to provide an arrangement for distributing the merchandise of a store in such a manner that the goods may be selected and taken by the customers themselves while making a circuitous path through the store; and whereby a large number of customers may be accommodated at the same time without confusion, and in an expeditious manner. Another purpose is to utilize all the available floor space of the room to the best advantage; to dispense with the employment of many clerks who are usually engaged to wait upon the customers; and to insure that the customers become acquainted with the variety of the lines of goods in the store and with the various items in the several lines.[64]

To read such a description of a grocery store today is to read a spatial logic of capitalism that seems completely self-evident. That it appears so self-evident a century later is what, at the risk of some hyperbole, makes Saunders's invention one of the most radical and transformative single patents relating to spatial design. Saunders has rightfully taken his place in the history of commerce; so too can he take a place in the history of architecture when that history is inclusive of patents and the inventive spatial practices to which they have given rise.[65]

If a grocer could be an inventor, so too could a poet. The avant-garde French poet Raymond Roussel took an unexpected career turn in 1922 when he filed a patent for a mechanism to counter heat loss in homes and automobiles, revealing that he was also keen on revolutionizing the home through patents.[66] Roussel's invention comprised the use of a vacuum to prevent heat loss in all aspects of housing and locomotion, and he derived methods for installing the vacuums in new homes as well as through the retrofitting of existing homes. A state official visiting Roussel's mansion in the Parisian neighborhood of Neuilly in 1931 discovered a small masonry building in the back of the property with just one door and one window with the following inscription: "the walls, floors, ceilings, doors and the windows of this house are entirely filled with vacuum-filled glass tubes. In case of demolition, please take all the necessary precautions to avoid an accident."[67] Roussel, like every good tinkerer, knew that his invention stood the best chance of success if it could be tested first, but unlike many of his tinkerer peers, Roussel had the money—and space—to create this full-scale experimental mockup behind his mansion, one that he curiously never publicized.

The third system in the family of vacuum devices, for vehicles, reflected Roussel's interest in the automotive industry and is a close ally of his *automobile roulette*, a kind of mobile home that was capturing the inventive imagination of France at the time.[68] Roussel's version of the mobile home was patented, exhibited at the 1925 Salon de l'Auto in Paris, and used by Roussel in his own trip from Paris to Rome in 1926 to visit Pope Pius XI and Benito Mussolini in an attempt to interest them in the vehicle.[69] Roussel, a darling of the French press, garnered a great deal of attention in the Parisian newspapers for his invention. As one anonymous editorial in *Le Matin* noted, "[T]he great poet of *Locus Solus* is no less an

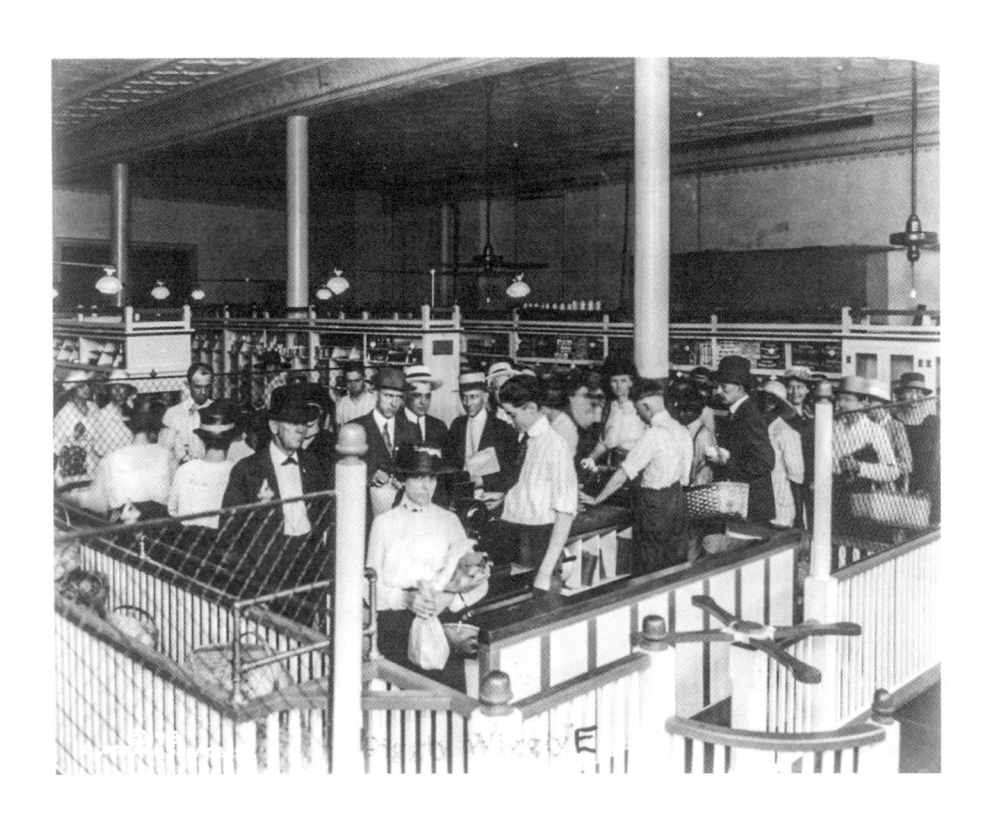

Clifford H. Poland, interior view of a Piggly-Wiggly self-service grocery store in Memphis, Tennessee, c. 1918.

innovator in the domain of reality than in the domain of dreams," drawing a direct parallel between the fantasy of Roussel's armchair travel and the fantasticness of his eccentric tourism machine.[70]

Homes that could travel were one subgenre that the patent system cultivated with great verve. The patent system also cultivated an array of inventions for traveling *within* one's home. A 1956 patent by Myron S. Teller, an architect-tinkerer in Ulster County, New York, brings us back to the topic of the lazy Susan. Teller's invention, dubbed the Lazy Susan Home, comprises rooms arranged around a circular moving platform that doubles as the structure's hallway.[71] To move from a bedroom to the kitchen, for example, one exits the room, stands on the platform, pushes a button, and waits until the platform is aligned with the kitchen. A utility core of sorts—including a fireplace, utility room, closets, and stairs between levels—is embedded in the platform and remains stationary at all times. In any event, patents put the home on the move, and that mobility was often facilitated by a new palette of materials that would articulate spatial innovation by way of matter.

Materials for Making Homes

While tinkerers working out of their basements and living rooms were indeed concerned with innovating the home itself, there were also plenty of home tinkerers who were more interested in the construction industry writ large. Throughout the history of the modern patent system, one finds a consistent interest in the innovation of materials that have already been around for decades if not centuries, from aluminum siding to fake marble to linoleum.[72]

A myriad of patents from the mid-nineteenth century onward, particularly in the United States and Great Britain, were concerned with ways industrial production could make expensive building materials accessible to a wider swath of the consumer market. The reality, as Pamela Simpson has shown in her landmark work on imitative building materials, was that the only real way to bring the look of marble, limestone, cedar, or leather into the homes of the international middle classes was to do so by faking it.[73] Faking it required making it: imitating these materials to the point of even passing believability was not easy, and the recipes and processes for doing so were, and remain, the subject of constant patent innovation, not least because those inventing these materials knew from the outset the scale of profit that a successful imitative material could bring. Unsurprisingly, the architectural elite—high-end architects, architectural historians, preservationists, and critics—found the trend toward artifice repugnant, believing that it went against not only the tenets of modernism (which demanded truth in material form) but also the very quality of new construction, as these materials were most often less durable and had less structural integrity than the ones they imitated. Writing in 1960, Ada Louise Huxtable, then the widely read architecture critic for the *New York Times*, saw the trend toward material verisimilitude proliferating in the American suburbs as a pattern that originated in the nineteenth century with the concrete industry, where the lure of near-infinitely plastic form gave rise to a kind of chronic tastelessness in materiality: "The development of concrete construction in the United States," she quipped, "was fueled by the primary American desire of doing things that were 'cheap, quick, and easy.' Its story is a characteristic mixture of the immediate American recognition of unprecedented technological possibilities and the willingness to do what had never been done before with the tastelessness of a new middle class society that accepted substitute gimcrackery for traditional materials and ideas."[74]

Huxtable's critique will of course resonate with anyone who believes in the integrity of materials and the shame of artifice (or "gimcrackery"). Yet, decades later, her critique also comes across as uppish and suppressive of tactical forms of material expression and the desires that they represent and seek to signify. A rote dismissal of artificial building materials is made more difficult when one considers their rich history of innovation through patents. In fact, artifice as such has a rich history of innovation and design, and should be able to claim at least a small element of dignity in architectural culture.

To be sure, Huxtable was not mindful enough of the long history of artifice in architecture, a history that was in large part bolstered by the patent system. One early patent for an artificial building material came from the Swedish tinkerer Gustav Dahlström, who in 1868 patented a pair of methods for making artificial sandstone and artificial marble.[75] In his patent, Dahlström describes a process in which completely dry sand and chalk are mixed with soda water. Once the resulting pulp is sufficiently mixed, it is cast into molds of the desired shapes and sizes. The cast pulp is then processed with a solution of chlorocalcium, which immediately hardens the pulp. The sodium chloride that is a by-product of the chemical process is then extracted with water, leaving the artificial stone. For artificial marble, the sand is simply replaced with pulverized marble (which could be made from the refuse products of marble mining and processing). Dahlström touts the invention as particularly advantageous in its ability to use existing materials and industrial equipment in Sweden, thus requiring no investment in new facilities or mines.

Meanwhile, the English inventor Frederick Edward Walton, was at work on a substitute for another expensive material: rubber. In his lab in Chiswick, Walton experimented with the effects of oxidization on linseed oil, gaining his first patent for that process in 1860.[76] In later experiments he added bits of cork and dyes to the mix and discovered what he found to be a useful floor covering, which he dubbed linoleum, patented in 1863, and began exporting it to the rest of Europe and the United States by 1869.[77] The product was a runaway success: no other single patent mentioned in this book made as much profit for its inventor as Walton's linoleum patent did. In its highest-quality form, linoleum was able to do things no other material could: because of its flexibility, it could render uneven surfaces where ceramic tile might crack, and it absorbed virtually no liquids, making staining difficult and cleaning easy. The material was extremely durable and, as one advertisement later boasted, possessed "the noiseless characteristics of carpet, without its cost."[78] Linoleum's global popularity continued well into the twentieth century, with linoleum options and variations becoming ever more sophisticated.

The theme of artifice in patents is not limited to faking a cognate material. Some patents retain a desired material but seek out ways to make it do new things. This was the case with Otto Karl Friedrich Hetzer's 1901 Swiss patent for glue-laminated construction.[79] Since at least 1890 Hetzer had been conducting experiments in which he laminated thin sheets of wood with high-grade glue adhesive, allowing wood beams to take on strong and flexible profiles and signaling the birth of clear-span timber arch construction.[80] Clear timber arch construction was particularly useful to architects who wanted to obtain the large clear spans that steel and iron framing afforded without using those materials (whether for stylistic reasons or as a matter of cost). Hetzer, and Hetzerholzbau, the firm he formed through the success of his early patents, was keen on cornering the European market, patenting their

The Home

Unknown artist, advertisement for Nairn's Art Linoleum "for large halls, passages, etc.," c. 1879. Photomechanical print from a sales catalogue.

Unknown artist, advertisement for linoleum, 1928.

laminated beams and forming methods across Europe: in Austria-Hungary, Belgium, Czechoslovakia, Denmark, England, Finland, France, Holland, Italy, Norway, Spain, Sweden, and Switzerland. Famous architects who designed with "glu-lam" wood construction systems, most notably Alvar Aalto, are clearly indebted to Hetzer's desire to create a continental trend in the method. Indeed, the Nordic nations, with their strong traditions of timber construction, were among the most responsive to the process. Hetzer licensed his patents to H. J. Kornerup-Koch in Denmark to promote the system there in 1914, and then to Guttorm N. Brekke in 1918 to do the same in Norway and Sweden.[81] The systems were wildly popular in both places.

The use of structural steel in architecture—and all of the pioneering architecture, most notably the skyscraper, that came from it—were indebted to the engineers who homed in on the particularly dynamic range of uses for steel construction in the form of the I-beam, an "I"-shaped structural unit. Assessing the history of the I-beam and its relationship to the railway gauge, the Scottish railway engineer William Fairbairn recognized a cotton mill built in Salford, Manchester in 1801 as exhibiting the first use of the I-sectioned beam (likely a disused railway gauge) in a building's lateral load distribution. He hailed this as an example of the "intuitive recognition of the most efficient shape in advance of the calculations that would prove it to be such."[82] Intuitive, perhaps, but neither Fairbairn nor his contemporaries ever believed that the reuse of railway gauges as structural members in construction was worth patenting, despite the amazing structural work it could apparently perform. The gradual transition from timber-framed buildings to steel framed-buildings in the modern era, a paradigm shift that would be largely afforded by the I-beam, is also well known in the history of infrastructure, where it proved particularly critical in the construction of bridges and other forms of civil engineering that were essential to the development of industry and denser cities.

Fairbairn presents the work of his colleagues in 1801 as a sea change, one fueled by the innate, intuitive genius of the Industrial Revolution's first protagonists in the United Kingdom. This laid the foundation for a historiographic narrative of technological progress as something that emanated axiomatically from Great Britain. Writing in the 1970s, the architectural historian Sigfried Giedion adopted Fairbairn's assertion for architectural history, showing how the I-beam would become the building block for the most heroic aspects of modern architecture and infrastructure: jaw-dropping cantilevers, a dramatic reduction in the density of columnar configurations and the liberation of the floor plan, the ingenuity of an armature for "skin" cladding such as the glass curtain wall, and so on.[83] These spatial traits would come to be explicitly emblematized through what we could call a fetishization of the I-beam by the great modernist architects, as can be seen in Mies van der Rohe's mullions for the Seagram Building (1958).[84]

Fairbairn's and Giedion's historiography is, at the very least, unduly Anglocentric.[85] What's worse, though, is that it takes no account of the patent system and the work that any half-savvy inventor would have ventured when they realized how railway gauges could be repurposed for architecture. The adage that "necessity is the mother of invention" does not exclude the act of reinvention as well, as illustrated by an astonishing 1874 Swedish patent by the Russian inventor Nikolay Ivanovich Putilov, House Made from Railway Gauges.[86]

Putilov had established his career as a machinist for the Russian navy around the time of the Crimean War.[87] He established a factory and workshop that built steam machinery, boilers, and propellers, among other naval equipment. Putilov later went into private enterprise, establishing four steel factories across the Russian empire, his rise largely propelled by the mass construction associated with the development of the Russian imperial railway network. His 1874 foray into building construction was as unexpected as it was innovative. He describes his patent as one for which the "construction, mainly consisting of old railway rails, is presented in the attached drawing, and, depending on the ground quality, these buildings can be erected on a foundation of stone or wood.... These buildings made of old railway rails are completely solid, light and cheap as well as, to a high degree, spacious."[88] The primary plan submitted as part of the suite of drawings, while not particularly beautiful, reveals that the main invention of Putilov's design was not what he believed it to be. It shows something railway historians (including this author) have long suspected: that the I-beam, the universal building block of steel-framed structures the world over, grew morphologically out of railway gauges. Putilov knew that the I-shape was optimal both structurally and economically *and* articulated that in the form of a patent (something that previous incarnations of this usage of railway ties documented by other scholars did not do). This is a masterful example of technology transfer and how patents contributed to architectural innovation (in this case presaging the mass-market I-beam). The reason Putilov patented this invention in Sweden was because that country produced, as it continues to do to this day, high-grade steel at a high per capita rate.

The focus of material patents for the construction and architecture sectors in the first half of the twentieth century remained largely on enhancing and refining the materials we have already discussed: wood, concrete, and steel. By the second half of the twentieth century, however, other forms of experimentation were necessary, and inventors responded in due course. Some of these new directions entailed radical new methods for assembling familiar materials. Another new direction turned instead to questions of space and spatial configuration. Indeed, the 1960s bore witness to some of the most visionary and experimental—and perhaps also least profitable—patents in the field of buildings. Prominent among these is the Domecrete building system by Haim Heifetz, patented in the United States in 1972 (filed in 1969).[89] The Domecrete system comprised a technique for creating inflatable concrete dome houses in which the semispherical and self-supporting concrete shell is achieved by being sprayed on top of an inflatable structure that is removed after the concrete dries. Heifetz, who touted the design's resistance to earthquakes, deployed the structure across Israel in the form of villas, synagogues, and bungalows. He soon found, however, that the US patent offered him little protection in Israel, where he and his associates regularly found copycat structures, which Heifetz believed was ultimately what made the Domecrete venture unprofitable.[90]

Meanwhile, the "spatial turn" in patents was being borne out in the 1970s, a period that saw some of the most liberal international expansions of the scope of the patent system. Inventors in the construction industry were reckoning with the fact that patenting was as much a game of getting ahead of trends (offensive patenting) as it was claiming inventions (defensive patenting). Increasingly, this meant that patents in the construction sector did not even have to designate materials for inventions whose innovation lay in spatial relationships, not technological advances. This was a trend that has led to the patentability of something as general as a building layout or arrangement, or its "disposition of space," as it has come to be known in patent parlance.[91] An important example of such a patent, which proved influential across the building development

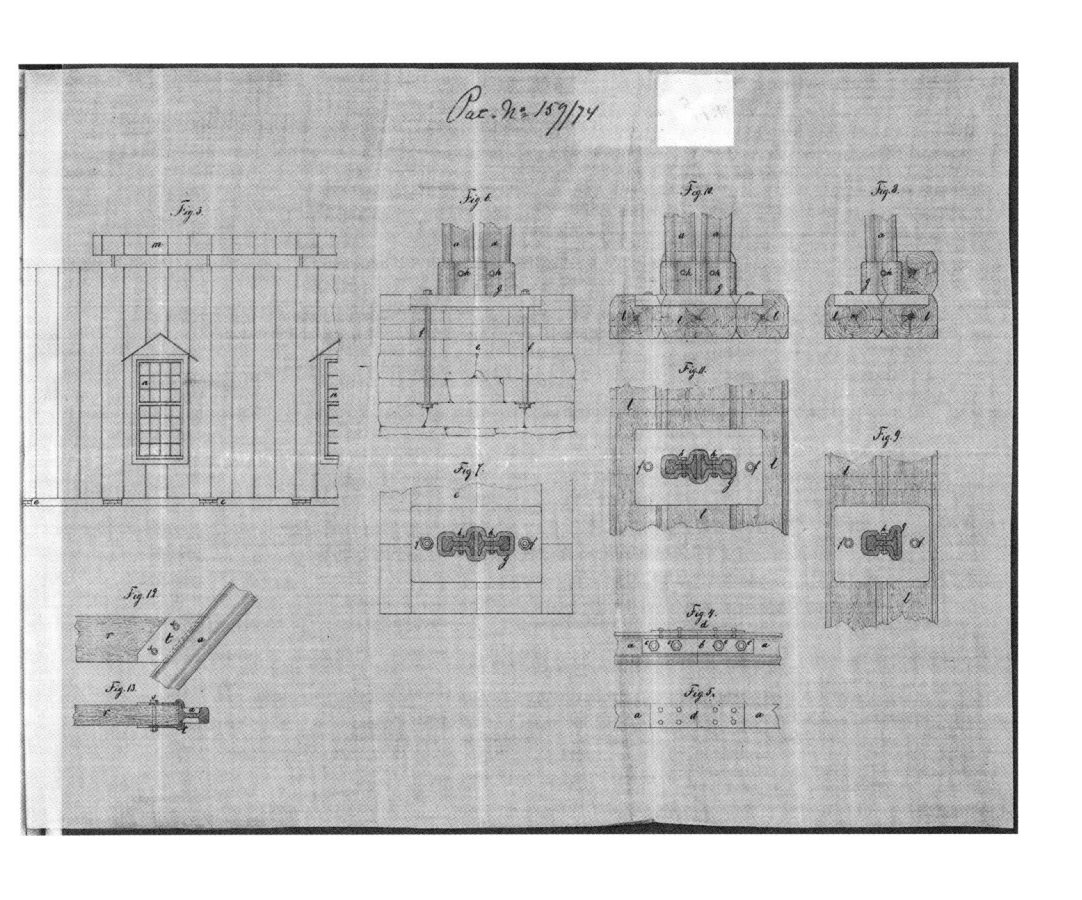

Nikolay Ivanovich Putilov, "Nytt sätt att af jernvägskenor bygga hus" (New way to build houses from railway rails), Swedish patent 159, filed October 2, 1874.

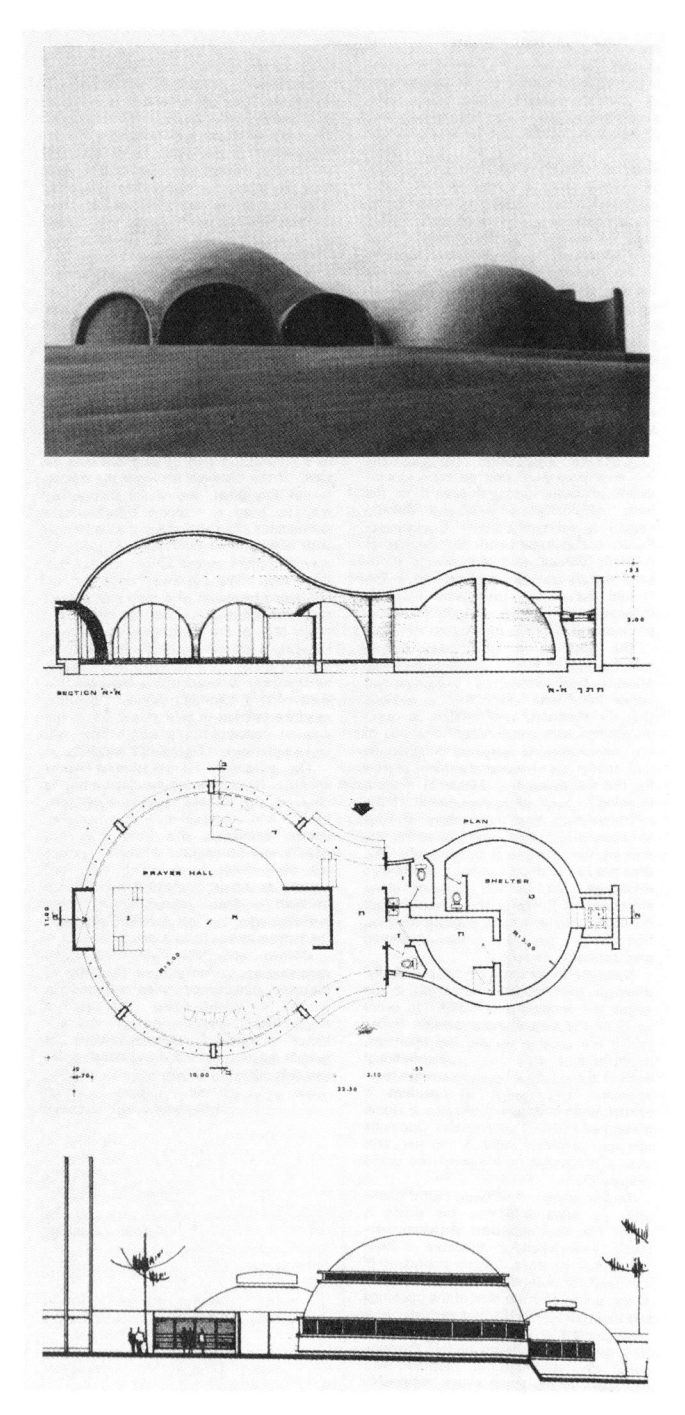

Illustrations of the "Domecrete" system by Haim Heifetz deployed as a synagogue, featured in *Bâtiment International* 3, no. 1/2 (January-February 1970): 38.

sector, was a patent for a House with Unfinished Bonus Space by the Michigan real estate developer William J. Pulte in 1977.[92] Pulte's invention, which appears to be a generic colonial revival suburban home, comprises a house in two parts: the center and one side of the interior, which are "finished with complete living quarters, including all normal accommodations for single family occupancy," and another part "having a built-in garage, adjacent room space, and second floor overlying same providing unfinished room space adapted with roughed in facilities for optional future completion to finished rooms."[93] The "utility" of the patent is summarized as follows: "A common two-story full length rectangular shell for the entire house together with the arrangement of unfinished room space in relation to the built-in garage provide a potential and many options for substantial economic future expansion at minimal initial and ultimate cost."[94]

In other words, the crux of the patent is the inclusion of space that is *not designed* and that is meant to attract the countless homebuyers who covet more space than they can afford at the time of their home purchase. It is a patent representing aspiration, not invention, proving that aspirations too could be engineered. In a bit of social analysis that is rare in patent documents, Pulte explains the patent's social context as follows:

> The need for a more adequate solution to home expansion problems has existed for many years and become increasingly important with accelerating costs of building, high home loan interest rates and depressed new and used housing sales. Home builders have been faced with the dilemma of conventional construction costs exceeding the reach of the market. Growing families with the need for supplemental accommodations have been forced to resort to small initial homes which are rapidly outgrown; and are then confronted with excessive costs in attempting to remodel or in transferring to larger homes in which the total prices reflect the uninterrupted inflation of the intervening years.[95]

Pulte's "bonus space" is ostensibly an all-in-one solution to a socioeconomic problem in housing and thus departs from virtually any kind of prior art, where solutions are developed for problems that are fundamentally technological in nature.

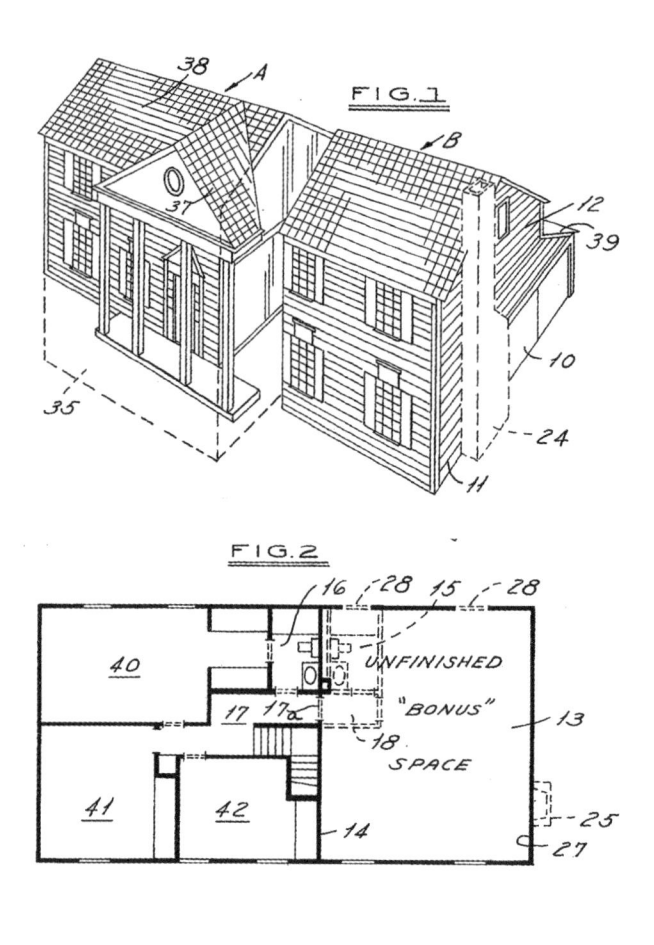

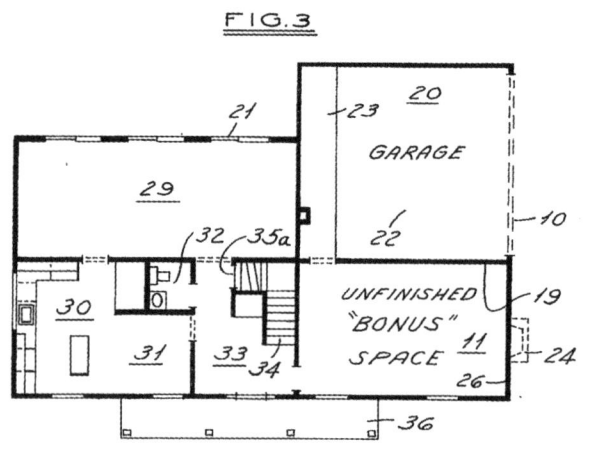

William J. Pulte, House with Unfinished Bonus Space, US Patent 4,015,385, filed June 2, 1975.

From Guppy's hanging bridge to Pulte's bonus space, we are reminded of the potency of the home as both a place from which to draw inspiration and a place to project wishes, as was the case in Nadine Hurley's quest to invent 100 percent silent drape runners. Nadine's eccentric obsession may prove to be the perfect entree into the world of home tinkerers who, over the course of the nineteenth and twentieth centuries, sought to use the patent system's openness to all sorts of people as not only a site for capitalist speculation but also an outlet for creative expression. We began our story in the humble setting of the home because it seems to be the wellspring of so many inventions in the built environment where we reflect on what could make everyday life incrementally better, easier, or smarter.

2 *The Studio*

The basement inventors and architectural tinkerers who have pursued patents have done so with the straightforward objective of introducing incremental improvements in the way the built environment is constituted. They have also done this in the hope that their toil would be paid back with the remuneration that comes with owning (and leveraging) IP. By contrast, for those trained as architects and working in architecture studios, whose day-to-day work revolves around clients, competitions, and commissions, the incentives for patenting are ambiguous at best. This does not mean, however, that there are no good reasons to do so. Indeed, a long list of well-known architects, particularly well-known twentieth-century modernists, have filed patents in patent offices across the world; some have filed as many as 60. This trend has remained oddly invisible. Why do we still know so little about it?

For starters, there is no imperative for architects to patent, as there is for corporate professionals working in fields such as chemistry, pharmaceuticals, materials, technology, and even architecture's close ally of product design. The history of the architecture profession offers two reasons why this might be the case. The first is the profession's deep connection to the arts, another domain where patents are nearly nonexistent (but where copyright reigns supreme).[1] As a form of artistic expression, architecture is an awkward fit for the market logic of patents, and the fact that a stylistic innovation cannot by itself be patented is a testament to architecture's status as a square peg in a patent system that is shaped like a round hole. The second reason is a kind of honor system that has long existed among architects. Architects certainly do compete with one another for work, but ever since the profession took shape, it has been considered indecorous to copy or steal, even from a rival. One reason is that wholesale copying of a design that will be in public view is virtually impossible to pull off without someone noticing (although this is far less the case with smaller elements of architectural design).[2] An architect who incontrovertibly copies and deploys the design of a colleague will quickly earn the status of a pariah. It is hard to say how this honor system emerged, but it is in some part indebted to the nature of architectural education, where architecture students share large open spaces with other students and are free to look at and discuss designs with one another. An architecture student could never copy a design and get away with it the way a math student could copy the answers from another student's exam.

Before the rise of large corporate firms in the twentieth century, the most common types of patents that were granted to architects were for floor plans for small domiciles such as cottages and bungalows. The patenting of floor plans by sole practitioners of architecture, which was particularly common in the United States and the United Kingdom, testifies to a kind of transition period in which architects who were eager to gain some exposure essentially published their floor plans via the patent system as a means to share their spatial schemes with a wide audience for free (excluding, of course, the cost of the patent).[3] Needless to say, it is very difficult to build from a plan alone, something that these architects certainly realized but patent offices and patent examiners presumably didn't—which is another source of disjuncture. What can be inferred from this is that the patent was essentially functioning as an advertisement. If a developer or customer liked the plan, they could contact the architect listed on the patent. If they wanted to proceed, they could then hire the architect for a complete design, including all of the building information that is not contained in a plan, without having to license the patent. A telling point is that this trend was never as effective for apartment and urban dwelling plans, because these were rarely built without an architect having already been lined up to make the plans.[4] For a time, it was a clever tactic, but eventually it petered out, in no small part due to the rise in catalog and mail-order homes, which effectively bypassed the database of home plans in the patent office.

This chapter discusses roughly a dozen of the architects following the demise of the plan patent trend—most of them familiar names, but several not—who patented everything from tiny building details to entire urban configurations. This survey excludes architects like Michael Graves and Norman Foster, who hold one-off design patents (as opposed to utility patents) that are the result of a special collaboration with a manufacturer or fashion house, itself a trend of the late twentieth and early twenty-first centuries.[5] The breadth of examples will give us a large enough body of evidence to draw larger conclusions, however provisional, about when, why, and how architects have chosen to engage the patent system. Such a survey may certainly be instructive for any architect seeking to understand the value (or lack thereof) that patents may hold for their work.

The selection of architects and offices discussed in this chapter does not represent any judgment of value, although such judgments could certainly be made. Rather, the examples reflect several pragmatic methodological choices: the availability of sources beyond the public patent file, the parallels that can be drawn between the examples, and geographic diversity, the latter to avoid the suggestion that global trends can be extrapolated from national examples. The parallels that emerge among the patentees have been used to structure the four sections of this chapter. The first two sections, working as a contrastive pair, bring together architects who approach patenting as a way to articulate macro and Platonic ideas about the nature of architecture and construction ("macro" patenting) and those who instead consider architectural detail to be the humble heuristic for a design logic ("micro" patenting). The following two sections, also working as a contrastive pair, bring together architects who approach patenting as a distinct business practice ("business" patenting) and those who instead consider patenting to be a domain for experimentation ("experimental" patenting). While these four categories are imperfect and exclude many worthy examples, they help bring clarity and focus to the questions of when, why, and how architects have engaged with patent culture and in turn staked a legal position in relation to their peers.

The Studio

"Macro" Patenting

The story of architects patenting under the auspices of their own design practice (as opposed to inventors who happen to be architects filing patents unrelated to corporate research and work) is largely a twentieth-century story. The roots of the practice, however, can be traced through architects' engineering brethren, who began patenting their corporate work in significant numbers in the last third of the nineteenth century. The Spanish emigré engineer Rafael Guastavino and his son, Rafael Jr., who held a total of 24 patents, provide a case in point.

The two Guastavinos, who emigrated to the United States in 1881 (when Rafael Jr. was just nine years old), are known for introducing traditional Catalan timbrel vaulting techniques to the United States and ultimately the world.[6] Their name became synonymous with a herringbone pattern of tile layout known as the Guastavino tile, which was first patented in the United States in 1885. Their system furnished muscular, self-supporting vaults and arches that became immensely popular in the design of large civic and institutional spaces, a structural system whose elegant mix of ancient tradition and modern materials spoke perfectly to the moment and the rhetoric of late Beaux-Arts architecture. As Jonathan Ochsendorf explains, "The Guastavino Company is a story of personal success, of a Spanish immigrant family transforming an ancient form of construction into a modern one. It is also part of an important moment in American building history, when craftsmen and architects collaborated closely together, and when architects trusted craftsmen to successfully design significant components of buildings. The Guastavino Company's work marked, as Guastavino Sr. said before the Congress of American Institute of Architects in 1893, 'a page in the American history of constructive art.'"[7] Ultimately, however, the immense success of the Guastavinos and their company would be cut short with the rise of architectural modernism, which was resistant to the syntax of vaults and arches.[8]

Even so, by the time the company, which was called the Guastavino Fireproof Construction Company, ceased to exist in 1962, it had contributed vaults, domes, and other structural systems to buildings in more than forty US states, four Canadian provinces, and eleven countries.[9] These included well-known spaces at the Boston Public Library, Grand Central Terminal, Carnegie Hall, the American Museum of Natural History, the US Supreme Court Building, Pittsburgh's Union Station, the Nebraska Capitol in Lincoln, and the Astor Tennis Courts, among many other projects. The Guastavino firm's patents, granted between 1885 and 1939, span everything from a kiln for glazing tiles to a fireproof vaulting system to a massive domed architectural structure made of masonry and steel.[10] Yet despite this range, virtually all of the patents are concerned with "macro" patenting in the sense that they seek to support (in the case of the kiln), detail (in the case of the fireproof vaulting), or demonstrate (in the case of the masonry and steel structure) the logic of a replicable structural system that could in theory extend ad infinitum.

The name of the firm touted the word "fireproof," as did the official titles of many of Rafael Sr.'s earliest patents.[11] Upon closer inspection, however, there is little in these patent files that attests to the primary feature being fire safety in a way that would meet the key threshold of patentability: nonobviousness. Masonry bricks and ceramic tiles, forged as they are through the fire of the kiln, are *obviously* fireproof. It would seem, perhaps, that what the Guastavinos were in fact patenting was the formal and even stylistic logic of their structural systems, not their fireproofness. To use fire safety as a veil for something like form or style (say, the herringbone pattern) that is unpatentable as a utility patent (at least under the more rigid standards of the late nineteenth-century US Patent Office) makes perfect sense if it was the form of the timbrel vault that was the central invention for the Guastavinos. There are no written records that indicate that they were trying to game the patent system in this manner, but when one considers the growing number of similar patents both in the US and abroad—such as E. Edwards's 1889 patent for Arches, Floors, and Ceiling Vaults in England—it becomes easier to see the defensive nature of these early fireproof patents and how they were more about laying claim to a tectonic system than about fire safety.[12] To be sure, patents also lent a proprietary aspect to the Guastavinos' work, which may have seemed more important to Rafael Sr. than to most other Americans as he was making his career in the United States as an immigrant. Additionally, patent protection did prove useful to the Guastavinos in 1917 when a former assistant, John Comerma, received a patent that clearly infringed on patents held by his former employers.[13] After a successful court case, the Guastavinos were able to take over a number of Comerma's projects.

The Studio

Stanford White with Rafael Guastavino, Col. J. J. Astor's Tennis House in Rhinebeck, New York, completed 1904. Photograph by Michael Freeman.

One need only look at the claims made in Rafael Sr.'s very first patent in 1885, Construction of Fire-Proof Building, to see how the spatial concerns of the Guastavino patents quickly surpass any explicit commentary on fire safety.[14] He notes:

> My present invention relates to the construction of fire-proof buildings, and has particular reference to the partitions of such buildings. It is applicable to buildings of all descriptions—such as private dwellings, factories, theaters, school-houses, &c. The object of the invention is to produce a more substantial and more economical system for constructing the partitions than any now in use. By my construction of partitions I attain economy, solidity and incombustibility. In addition to the qualities mentioned, partitions constructed according to my invention are light in weight, clean, and entirely free from the usual cavities incident to the common form of partitions, thus insuring the building against the lodgment of pests—such as rats, roaches, and the like—besides they are entirely free from joints, one integral structure without solution of continuity resembling a large stone.[15]

Guastavino's evocation of an "integral structure without solution of continuity resembling a large stone" and his assertion of the patent's applicability to many different building typologies are both key features of the visionary nature of "macro" patentees, seeking as they did to create strikingly Platonic and repetitive structural solutions to all manner of architectural subjects within the legal framework of the patent.

Although the work of the Guastavinos and the subsequent work of the American inventor R. Buckminster Fuller have never been juxtaposed, one can see through the lens of "macro" patenting why they should be brought into conversation. Fuller had a similarly large number of patents: 28 in total, granted over a 55-year period spanning from 1927 to 1981.[16] Most of Fuller's best-known designs are, in fact, also patents. These include the 4D House (1928), the Dymaxion Bathroom (1940), the Dymaxion Deployment Unit (1944), the Dymaxion House (1946), the Geodesic Dome (1954), and the Octet Truss (1961).[17] As with the Guastavinos, there are many files that relate to the preparation of Fuller's patents, but little in the way of written records about his actual motivations for patenting, in Fuller's extensive archive at Stanford University.[18] In the broad sense, his reasons for patenting seem connected to his identity as an inventor, not an architect. We gain at least some insight into Fuller's reasons for patenting in his 1983 book *Inventions*, which looks retrospectively at the 28 patents filed over the course of his career—a book that Fuller uses to clearly position patents as one of the markers of his career, if not the most important.[19] A personal reflection in his discussion of his very first patent, Stockade: Building Structure, offers some insight into how patents might have fit into Fuller's self-fashioning as an inventor:

> My father died on my twelfth birthday. He had not been able to communicate with me since I was ten. My mother, her friends, and my father's, and my relatives, all tried to make me conform to the most acceptable type of reliable employee of whoever might offer me a job after my education was finished. All of them said "Never mind what you think, listen. We are trying to teach you." Because I knew how much they loved me, I did my best to pay no attention to my own thoughts. When I became engaged to Anne Hewlett, her father, a leading architect of New York, was the first grown individual to tell me to pay attention to my own thoughts, which he said he found to be constructive, inventive, and order-seeking.[20]

The most striking trend in Fuller's career as a patentee is his gradual move from objects to systems; in other words, his gravitation toward "macro patenting." His earlier patents, those invented through the end of World War II, are certainly about macro solutions—particularly for housing—but these are evinced through discrete, finite objects: the 4D House and the Dymaxion family of patents. Beginning in the 1950s, Fuller moves toward the more "open-source" inventions of the Geodesic Dome and the Octet Truss which, like the Guastavino vaults, set forth a logic of systems. There are certainly some historical explanations for this. Fuller invented the Dymaxion family of patents, for example, in the long shadow of World War II, and the questions prompting those inventions were classic World War II-era conundrums: how to reappropriate wartime industries (such as the manufacture of aircraft) for times of peace, and how to solve the massive housing

View of R. Buckminster Fuller's Biosphere for Expo '67, St. Helen's Island, Montreal, 1967.

shortages that followed in the wake of the war. As radical as the Dymaxion inventions were, they also unambiguously tackled the zeitgeist. This does not mean that they were not "macro"-minded, for indeed they were. It was just that their "macro"-ness stemmed from the prospects of mass production, not from the flexibility of a system.

It is fairly easy to see why Fuller's patents evolved away from military-industrial issues and in the direction of systems design in the 1950s, most famously with the Geodesic Dome. The Geodesic Dome was unequivocally his most successful invention, with permutations across the globe. The Biosphere that Fuller constructed in conjunction with the 1967 Montreal Expo is one of the most crystalline real-world manifestations of the patent, in fact of any of his patents.[21] Writing in *Inventions* about the Geodesic Dome, Fuller contends that its success was dictated in large part by his own lack of concern about the patent's appearance: "At no time in my last 56 years have I paid any attention to conventional architecture's 'orders' about the superficial appearance of my structures. I never try to anticipate what my structures are 'going to look like.' I am concerned only with providing comprehensive, logical, pleasingly adequate, and most economical solutions to all design problems."[22]

Fuller's philosophical bent toward systems design and "macro" patenting is also borne out through analogies he draws with the natural and scientific world. In his reflection on his 1962 Tensegrity patent, Fuller explains how theoretical physics (as distinct from applied physics) informs the logic of his patent.[23] He argues:

> Nothing in the universe touches anything else. The Greeks misassumed that there was something called a solid.... Today we know that the electron is as remote from its nucleus as is the Earth from the Moon in respect to their diameters. We know that macrocosmically none of the celestial bodies touch each other.... There is no way for structural analysis to analyze a geodesic dome. This still continues to be true. I think it is reprehensible. The only way to analyze it is with pneumatics and hydraulics. At the molecular level, this is the method of quantum mechanics; it could never be done with crystalline continuity, for such continuity does not exist in the Universe.[24]

The technical verbiage of Fuller's patent applications is the verso to his philosophical reflections in *Inventions*. Again patenting as a process seems to be veiling something else, as when the Guastavinos include form-making beneath the veil of fire safety. Here it is an aesthetic philosophy about the primacy of Platonic form that is veiled beneath a discourse of efficiency (in which, of course, Fuller was also interested). One only need compare Fuller's high-minded rhetoric on the Geodesic Dome in *Inventions* with the logic-driven rhetoric in the patent application: "A good index to the performance of any building frame is the structural weight required to shelter a square foot of floor from the weather. In conventional wall and roof designs the figure is often 50 lbs to the sq. ft. I have discovered how to do the job at around .78 lbs per sq. ft. by constructing a frame of generally spherical form in which the main structural elements are interconnected in a geodesic pattern of approximate great circle arcs intersecting to form a three-way grid, and covering or lining the frame with a skin of plastic material."[25]

Just as there are two divergent voices in these writings, so too is there a division in his relationships with industry and the academy. Fuller's patent portfolio, whether deliberately or not, brought him into a very cozy orbit with a number of big companies interested in his work if not also his patents. In his early years, while based in Wilmington, Delaware, Fuller was in close proximity to DuPont. In his later years, companies like Minnesota Mining and Manufacturing (3M) even invited Fuller to their campuses to give motivational talks on the subject of creativity to a group of the firm's scientists involved in patent research, known as the 3M Technical Forum.[26] Reporting on the lively event, 3M correspondents noted how the

> famed inventor of the geodesic dome agreed with his audience that through intelligent use of technology man can do more with less, and raise the world's standard of living. Fuller differed with those who set arbitrary limits to growth, based on obsolete physical accounting methods. Older ways of measuring wealth, resources and potential, he said, offer no adequate way of dealing with the metaphysical, the transcendental and the uniquely human ability to modify tools and control the regeneration of the

The Studio

> physical universe. His four and a half hour discourse without notes earned him a standing ovation from 3M's technical community. He accepted the applause "as a cheer for the truth ... nothing personal."[27]

Fuller's lofty talk at 3M headquarters in Minneapolis, with its calls to explore universal truths and ditch traditional capitalist logic, was completely at odds with his own unwillingness to lease his patents out freely.

Even if Fuller truly believed that knowledge—and patents—should be shared freely to solve the world's problems, this was simply not possible if he wanted to sustain his practice economically. The same kind of dissonance between the universal aims of Fuller's designs and the economic reality of his practice is also on display in his relationship with the academy. By the late 1950s, Fuller was in high demand as a guest lecturer or critic at architecture schools across the world. His assistant, John Dixon, was tasked with fielding the myriad requests that came in from architecture deans and department chairs. Fuller taught at Tulane, the University of North Carolina, Cornell, Princeton, and the University of Minnesota, among other schools. In one letter to the chair of the Architecture Department at the Rhode Island School of Design, Dixon unapologetically lays out how students would be instrumentalized in patent research, research that of course stood to be profitable for Fuller while offering the students nothing in the way of compensation for the ungodly amount of time they were expected to commit:[28]

> In addition to the events on the enclosed schedule, Mr. Fuller and the Fuller Research Foundation are conducting for the US Marine Corps an investigation concerning advance shelter bases. We have been pressed for time. The Fuller Projects at the Department of Architecture, North Carolina State College and Tulane University, are simultaneously engaged in phases of the Marine Corps problem. . . . In order that the students may be properly oriented historically, economically, technically, and psychologically, Mr. Fuller must give approximately fifty hours of lectures during a project, but a project resulting in a new prototype is predicated upon more than lectures: Mr. Fuller and those working with him each dedicate a minimum of 100 hours every week throughout an

undertaking. The result is a high-speed integration of talent, techniques, and nation-wide resources, brought to new foci at the respective universities.[29]

No architect can compete with the patent record of Japanese architect Kiyonori Kikutake.[30] Between 1975 (well after his career started) and 2006, Kikutake filed a whopping 54 patents. A large portion of them reflect his firmly held belief that the marine world would be the future home of architecture. There was a paucity of prior art for any of the inventions needed for large-scale construction on water, so much of Kikutake's strategy with regard to patents appeared to be patenting incremental inventions fairly rapidly and then citing himself as prior art. This goes at least part of the way in explaining why Kikutake held so many patents. A partial list of these aquatic patents, ordered chronologically, includes: Method of Building Floating Structure for Aquatic City Buildings (1977), Floating Structure for Aquatic City Buildings (1977), Artificial Ground (1978), Aquatic Construction (1978), Structure on Water I (1979), Structure on Water II (1979), Floating Structure for Watersurface City (1979), Structure for Artificial Multilayer Land (1980), and Floating Structure for Building (1980); about a dozen more came later.[31] Each of the successive patents related integrally to one of Kikutake's preceding patents; in the first two patents in this list, for example, he first patented a method for building a floating aquatic city, then patented the floating structure of said aquatic city. Methods for making forms and the forms themselves would certainly be separable patents in most patent systems around the globe, but many of Kikutake's patents, aquatic and otherwise, were such minute and interrelated inventions that they would likely have needed to be consolidated with others to justify patentability in non-Japanese patent offices. Indeed, the Japanese patent office was known for holding a different view than others regarding the question of nonobviousness, which increased the number of patents protected by Japanese IP law.[32]

In any event, the collective invention of many of these aquatic patents was Kikutake's Marine City project. It is important to note that the patents mentioned earlier were filed nearly 20 years *after* Kikutake began designing Marine City. This gap is critical for understanding why he was motivated to patent in the first place. Beginning in the 1950s, Kikutake's work came to be recognized as an exemplar of the new Metabolist movement that was taking hold in Japan. Coalescing in the wake of the disbanding of the Congrès Internationaux d'Architecture Moderne (CIAM) in 1959, Kikutake, along with Kisho Kurokawa, Fumihiko Maki, and others, published what came to be known as the Metabolist manifesto (which included four sections: Ocean City, Space City, Towards Group Form, and Material and Man).[33] Rooted in a heady combination of Marxist philosophy and an appreciation for biomimetic form, the Metabolists sought to create a new way of urban living that could incorporate organic growth. This led to a number of expandable and "plug-in" structures like Kurokawa's Nakagin Capsule Tower with its two primary elements: an infrastructural body and a habitable pod that would parasitically latch into the body, theoretically ad infinitum. The Capsule Tower, which was realized in 1972, was a vertical, land-bound articulation of Metabolist principles.[34] Marine City's infrastructural body was its floating deck and towers, while its residential units were its plug-in elements.

The sheer cost and difficulties of building on water, however poetic the idea and integral to Metabolism it may have been, proved to be hurdles that Kikutake was unable to surmount. It was not until 1973, a year after the Nakagin Capsule Tower joined Tokyo's crowded skyline, that he would submit his first patent application related to Marine City. This timeline seems to indicate that Kikutake was trying to revive the Marine City project by filing for patents. How exactly he imagined that his patents would jumpstart the project after so much stagnation is unclear. One possibility is that the suite of patents could lend the project some credibility to potential developers and financiers. Another is that the patents, if implemented in the gradual manner at which the long sequence of patents seems to hint, could unfold slowly, requiring less outright risk and funding. Ultimately, this strategy also failed, and Marine City was never realized. Even though it was executed in many incremental steps, however, Kikutake's "macro" approach to patenting is self-evident.

One commonality of the Guastavinos, Fuller, and Kikutake is that they were extremely influential theoretically, even as their practices faced challenges. The American artist-inventor-engineer Chuck Hoberman is a superb example of this theoretical lineage, with his work having a particularly strong rooting in Fuller's work. At the time of writing, Hoberman, who is still practicing, holds 30 patents, the first of these dating to 1988.[35] The kinship between Fuller and Hoberman is crystal-clear: a love of geometry and how it

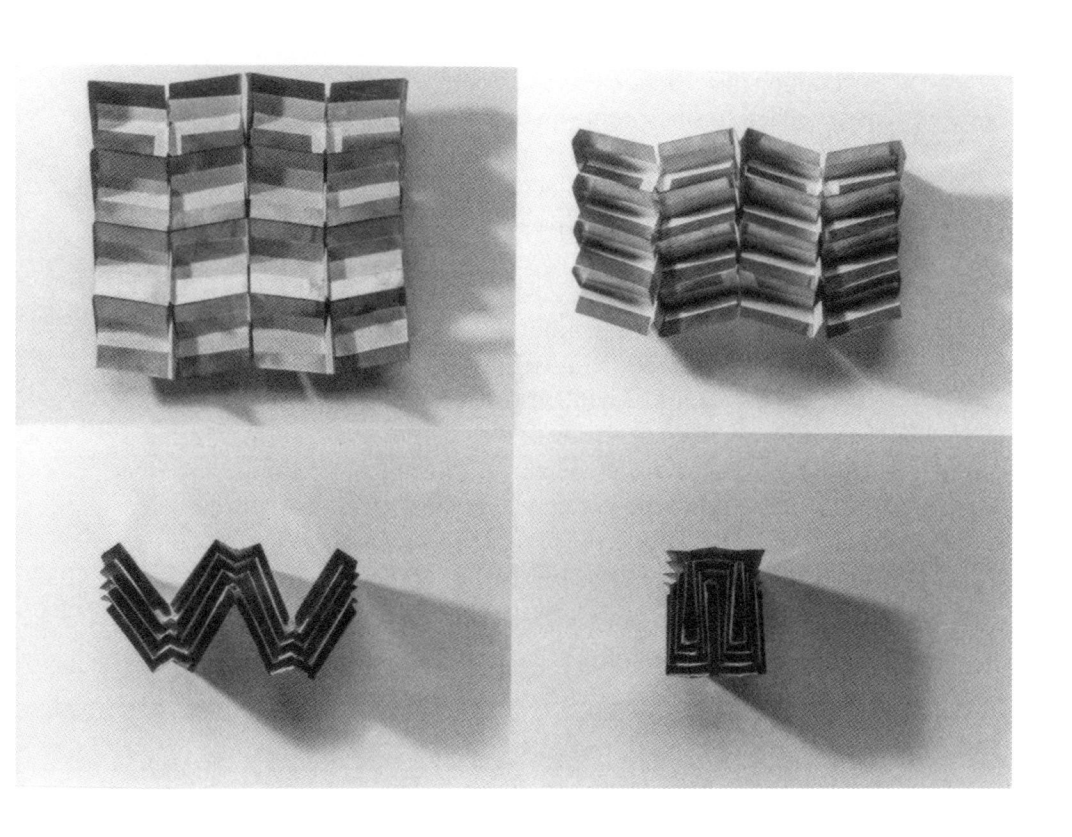

Chuck Hoberman, photograph from patent material for Folding Structure #3, 1986.
Chromogenic color print.

can frame space, the scalelessness of their designs, and a focus on parts and joints and how they perform in repetition. Like Fuller, Hoberman has served as a consultant to NASA.[36] One would expect that Fuller would be a major figure of prior art in Hoberman's patents, but he is not. The reason for this is as straightforward as it is revealing of the blind spots of the patent system's entrenched practices of citations of prior art. Hoberman's primary interest lies in kinetics, which is well demonstrated by his most successful design, the Hoberman sphere, a greatly expandable spherical toy for children.[37] Hoberman's lawyers or patent agents have in turn understood that, to be successful with their patent efforts, they needed to place him squarely in a prior art subcategory of kinetic structures, despite his inventions being much more than that. Precedents, such as Fuller's work, that Hoberman might have cited are essentially subordinate to this strategy.

Hoberman's earlier patents are also instructive for the historical moment they occupy. His Reversibly Expandable Structures patent (1991) was developed as computers and computer modeling were beginning to render the manual drafting of patent drawings obsolete.[38] Records of Hoberman's research and preparatory work reveal that, in the last days of analog design, inventors were still building cardboard maquettes and drawing complex geometrical structures with the assistance only of graph paper.[39] The records also reveal a keen student of IP and the philosophy behind it; among his files Hoberman has copies of early writings, including some that are unpublished, of the influential IP scholar Robert Merges.[40]

"Micro" Patenting

It is not a coincidence that the aforementioned "macro" patentees all shared an interest in the theoretical side of the sciences, particularly theoretical mathematics and physics. In all four cases, the patents are recognizable as physical ruminations on the universal conditions of mathematics and physics, and as such articulate a desire for an infinite, scalable, and universal outlook on the built environment. There is an obvious appeal to the field condition of such productions, but it is also important to recognize a certain level of incommensurability with the patent system, which had long preferred the measurable and incremental. Indeed, most patents in the "macro" vein have tended not to get much traction out of the patent system itself. Fuller's Geodesic Dome is the shining exception, in no small part due to the potency of his personality cult and his uncanny ability to whip up media attention.

The architects I have chosen to exemplify the loose category of "micro" patenting are all European: Alvar Aalto (Finland), Jean Prouvé (France), Konrad Wachsmann (Germany), and Renzo Piano (Italy).[41] Each of these architects had either considerable training in or a professional interlude in artisanal crafts such as wood and metalworking, and their material experiments with these materials are prominent features of their oeuvres. Although these craft industries had certainly come into their own in the United States by the turn of the twentieth century, the artisanal traditions in Europe still retained their higher quality and an integral role in the philosophy of design. Without a craft background, as was the case with Fuller and Hoberman, a patentee seemed much more likely to jump to the theoretical and universal positions undergirding "macro" patents.

The case of Alvar Aalto is instructive in this respect. Aalto was keen on experimenting with materials, particularly wood, and recognized the inventive nature of much of his research, but he had no interest in IP as a concept or as a means to some kind of financial end, nor was he intrigued with the logistics of patenting. This explains why he outsourced the process for filing his seven patents in their entirety to professional patent agents, especially Jal Ant-Wuorinen, the director of a patent law office named Patenttitoimisto Suomea ja Ulkomaita (Patent Office Finland and Abroad), which sought to bring Finnish innovations to broader markets.[42] Of Aalto's seven patents, granted between 1934 and 1965, four were method patents for bending wood or for wooden pieces of furniture.[43] The other three were design objects, including a chair, a stair tread, and an anti-glare shade for light fittings.[44] Ant-Wuorinen successfully brought Aalto's patents abroad as well, to Germany, Sweden, Denmark, Great Britain, Belgium, Italy, France, the Netherlands, Austria, Switzerland, the United States, and Czechoslovakia.

One patent that is highly recognizable today was Finnish patent 18,256, A Method for Making Furniture and Other Objects of That Nature, and Chairs and Other Items of Furniture Made Using the Method (granted 1938), known on the market as the Paimio chair, an armchair with a flexible wooden structure. The primary claim of the patent is a longitudinally U-shaped chair whose back and seat are formed by bending a single sheet of plywood between the seat and the back, on one side, and the upper and lower ends on the other. The U-shaped components were made using Aalto's hallmark method of wood lamination: thin strips of wood glued together and then, once hardened, formed in a press.[45]

Despite Ant-Wuorinen's careful shepherding of Aalto's patent applications in Helsinki and abroad, several of the patents were plagued by challenges that forced Ant-Wuorinen to involve Aalto more integrally in the process. After it was filed, the patent for the lamination method was swiftly faced with counterclaims by the designer Asko Avonius and his manufacturer, the Wilhelm Schaumarin Faneeritehdas Company. Avonius and the manufacturer claimed that the laminar design had already been used for making wooden harness bows for both horses and airplanes.[46] Unfamiliar with the intricacies of laminar design, Ant-Wuorinen needed to press Aalto himself to come up with the counterargument to this claim. In a no-nonsense memo to the patent office, Aalto demonstrated with a brief retort and a diagram how Avonius's lamination method, which would pull apart with friction when Aalto's would not, was simply inferior. It was similar in form, perhaps, but not a durable product in the way Aalto's furniture was: "With reference to the second observation, it is of course open to interpretation whether the reinforcement of the integrity of the strips of wood against each other is a result of 'friction'—however, the fact remains that the bond will open if the bent structure is loaded as in B and will remain intact in Condition A even if the loading is the same in each case."[47] Aalto's simple diagram and explanation sufficed, and the patent was granted.[48]

Perhaps embittered by Aalto's success, Avonius and his manufacturer made a counterclaim to another of Aalto's patents, Finland Patent 18,666, A Bending Method for Wood, and the Articles Produced by This Method (granted belatedly in 1940, well after the patent succeeded in other countries). With an argument that detailed the nature of his lamination process, Avonius had formulated a stronger case. Marianna Heikenheimo has described what happened next as a difficult, protracted, and parsimonious battle with Avonius, mediated by the patent office: "It was the thankless task of the examiners at the Patent Office to place the content and credibility of the statements of opposing witnesses side by side. The debate developed some unpleasant features as the patent examiners held a joint session on the issue because of the claims

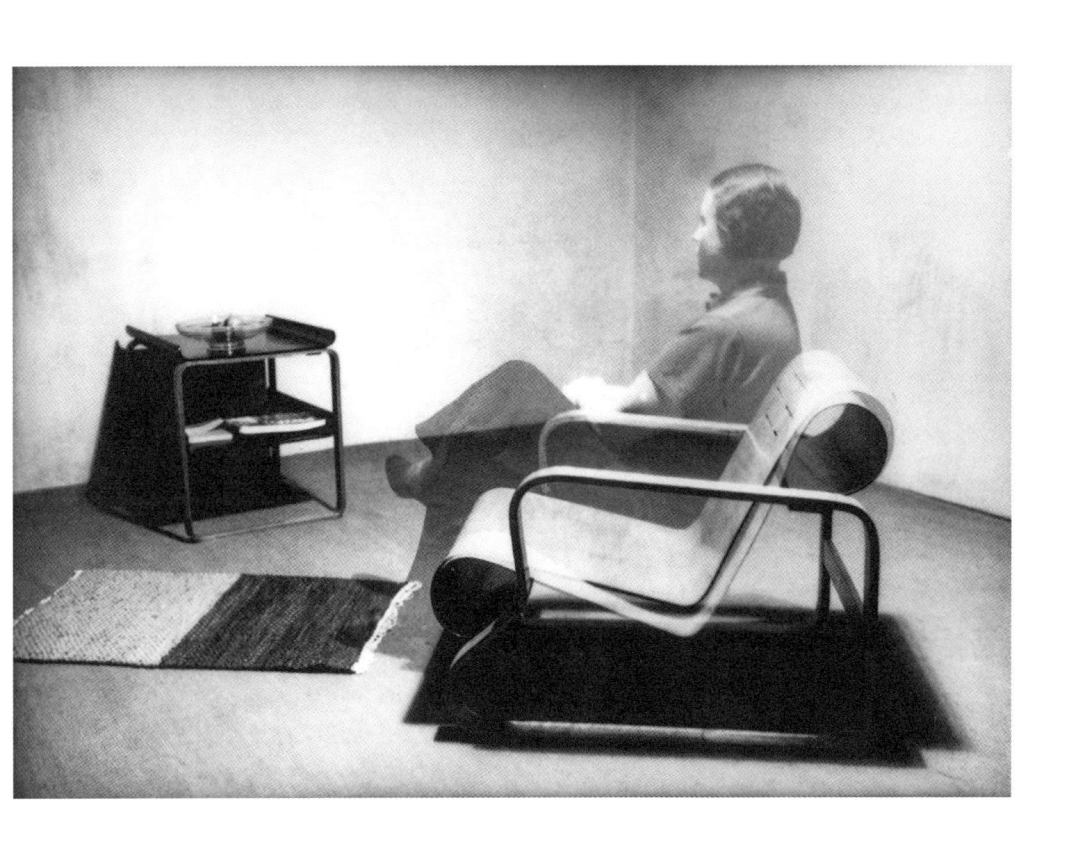

Photorendering of Aino Aalto sitting on Alvar Aalto's Paimio chair, c. 1931.

18256

esitetyistä muutosehdotuksista, olen valmis saattamaan hakemuksen lopulliseen kuntoon, niin että se voitaisiin hyväksyä.

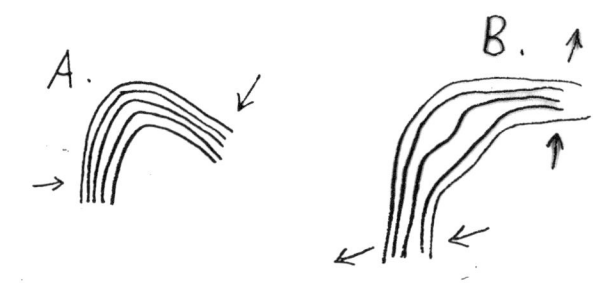

A. B.

Helsingissä syyskuun 21 p:nä 1936.

Alvar Aalto

psta.
Patenttitoimisto
Suomea ja Ulkomaita varten

FRIKTION!

Alvar Aalto, response to the Finnish patent office's interim ruling on patent no. 18,256 (filed September 21, 1936) involving a process for laminated wood. "[I]t is of course open to interpretation whether the reinforcement of the integrity of the bond derived from the compression of the strips of wood against each other is a result of 'friction'—however, the fact remains that the bond will open if the bent structure is loaded as in B and will remain intact in condition A even if the loading is the same in each case."

for the patent's rejection. It seemed impossible to do anything except grant the patent, since by that stage the invention had been granted patents in eleven other countries."[49] In this case Aalto had won not because he was the most credible party (though he very well may have been) but rather because other European patent offices had granted him the privilege that Avonius, with his vendetta, was trying to block back home. Ant-Wuorinen's strategy of multinational patenting had saved the day.

In 1936, Aalto's willingness to participate in the patent process came into direct conflict with his own ignorance about how the patent process in Finland traditionally worked. Although there was no formal mandate to the effect, inventors typically needed to demonstrate some competence and experience within the field where they sought patents. This was easily achieved for his furniture patents, but was a death knell for a patent he submitted for a "method for allowing uninterrupted highway traffic on streets": the patent was swiftly rejected, not only because prior art existed but also because, as Heikenheimo details, the invention "did not come within the scope of the architect's profession," despite Aalto's longstanding interest in town planning.[50] Another invention, for a window, similarly failed.[51]

Like Aalto, Jean Prouvé was a designer who moved between the scales of industrial, furniture, and architectural design and who, like Aalto, patented his seminal furniture designs.[52] Unlike his fellow "micro" patentees, Prouvé was not formally trained in architecture, but rather gained his experience as a blacksmith's apprentice in and around Paris. In the city of Nancy in 1923, Prouvé established a string of workshops and studios under his own name where he produced domestic metalware, including lamps, chandeliers, grills, and handrails. He also began experimenting with furniture design and quickly garnered the interest of local architects who commissioned numerous projects from him.

Around the same time, Prouvé became associated with a group of progressive artists and designers who were joining forces to advocate for more progressive design in France. The group, led by the French architect and designer Robert Mallet-Stevens and known as the Union des Artistes Modernes (UAM), recruited Prouvé as a founding member in 1929.[53] Other founding members included Eileen Gray and Charlotte Perriand, and later members included Pierre Chareau, Le Corbusier, and Pierre Jeanneret, among many others.[54] As a founding member, Prouvé participated in the writing of UAM's founding manifesto in 1930.

An aspect of Prouvé's participation in the UAM that has hitherto gone unexamined is that his joining the group coincided with the launch of his patent submissions, 32 in total filed between 1929 and 1965.[55] Prouvé's patents emblematize the "micro" patent in their sharp focus on constructive details and reveal why Le Corbusier once referred to Prouvé (in flattering terms!) as a *constructeur*.[56] A sampling of the patent titles reveals Prouvé's focus: Improvements in the Construction of Soundproof Panels (1931), Double-Hung Window Whose Sashes Open by Rotating Around a Vertical Axis (1939), Improvements to Buildings with Box-Girder Structures and Walls in Double-Wall Panels (1946), and Sun-Louvre Stretchers (1964), to name just a few.[57]

Prouvé in fact exhibited his very first patent, Process and System for Building and Installing Opening Panels, Metal Doors, etc. and Their Frames (1929), at the UAM's inaugural exhibition in Paris in 1930, making it clear to both his UAM colleagues and the exhibition's visitors that Prouvé believed a new, modern design ethic was consonant with the patent system (ironically, patenting was not particularly popular with his UAM colleagues, with the exception of Le Corbusier, who will be discussed later).[58] Prouvé's early patents, such as this one, demonstrate a particular affinity for the technical parlance of patent applications. His claims are direct, unambiguous, and short on adjectives. Unlike the veils often placed over "macro" patents, Prouvé's patents appear to be nothing but forthright and economic in nature; after all, this young designer was trying to harness IP law to buoy his new business, the Ateliers Jean Prouvé. In addition, Prouvé's technocratic approach to design was a perfect match to the nature of the patent system, where he could readily argue the case for the elements of his various constructive systems by focusing on their technological inventiveness alone (which he knew very well how to articulate), while never needing to rely on the philosophical position of UAM to pass muster.

As Prouvé continued to patent into the postwar period and as the potency of the UAM as an organization began to fade (they would disband in 1959), there were growing indications that Prouvé's approach to patenting was broadening in scope. This period brought many of his important "open-source" inventions, such as the Metal Construction System (1946), Construction System

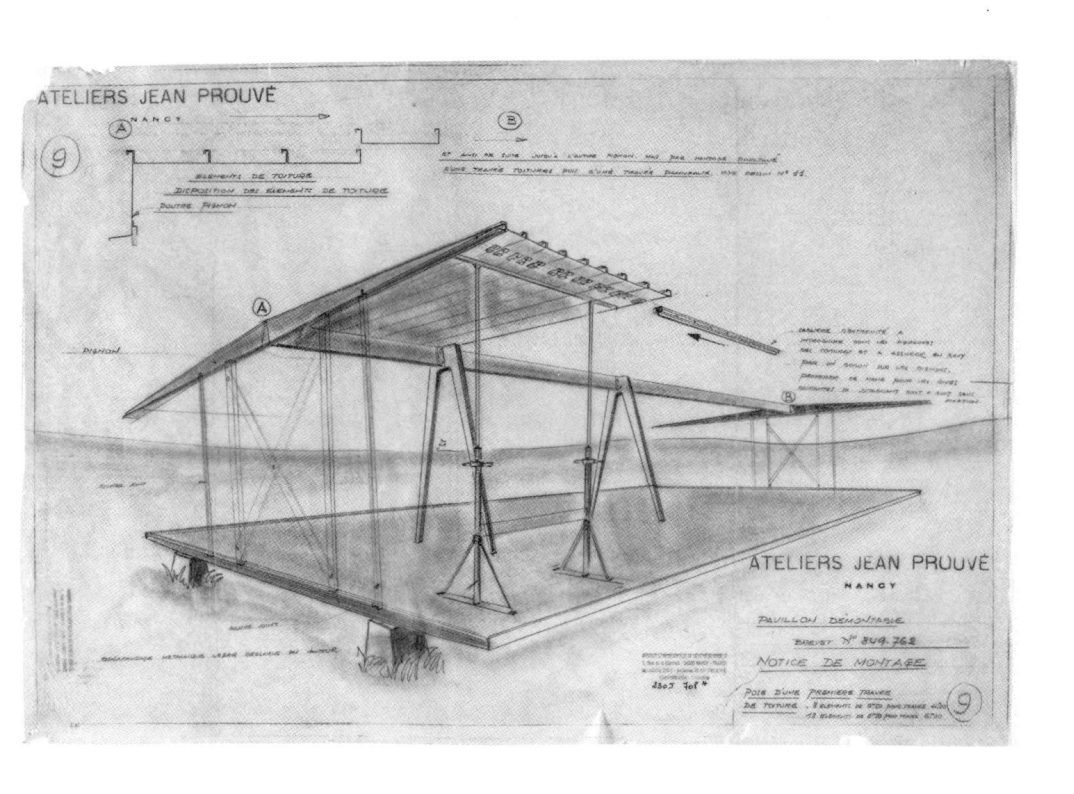

Jean Prouvé, Demountable House, perspective drawing showing the installation of the first roof span, 1947. Pencil on tracing paper, 48.5 × 71 cm. Musée National d'Art Moderne / Centre Georges Pompidou, Paris, France. Inv. AM 2007-2-342. Photo: Georges Megeurditchian. Also color plate 8.

for Single-Story Buildings (1947), Elements for Erecting Buildings and Their Installation (patent with extensions spanning 1955–1957), and Structural Building Method and Systems Applying Said Method (1964), all of which indicate a migration to a more "macro" philosophy about the relationship between design and patents.[59] Indeed, these inventions supplied the constitutive logic for a number of his best-known architectural designs, including the Maisons à Meudon (1949) and the Maison Tropicale (1949), demonstrating that he knew of the potency and effectiveness of a more systems-based approach to patenting.[60] Prouvé did not need to patent these works as they existed in their closed form because they were conjugated from his already patented methods and products. The creative work of the designer thereby became the assemblage of parts, neither more nor less. Even in this shift toward systems, Prouvé never patented finished (finite) products; his concern was the patenting of methods and components, signaling an embrace of an ostensibly democratic architecture that mirrored the ostensibly democratic nature of the patent system.[61]

Konrad Wachsmann, who was the same age as Prouvé, also had a background in a skilled craft—cabinetry—before studying architecture in Berlin under Hans Poelzig. Upon graduating, he took a job as the chief architect of a manufacturer of timber buildings, Christoph & Unmack.[62] Wachsmann, who was Jewish, fled Germany for Paris in 1938 (he was eventually interned in Grenoble) and then emigrated to the United States in 1941. Abruptly deprived of his burgeoning career, the accomplished architect and construction expert arrived in the United States virtually penniless and with little idea of how to start over. He was greatly assisted by his friend Walter Gropius, who was then teaching at Harvard, and Wachsmann moved in with Gropius and his wife, Ise Gropius, at their home in Lincoln, Massachusetts, taking a basement room and promising to move out once he had landed on his feet.[63] The historian Gilbert Herbert describes the situation best:

> When Konrad Wachsmann arrived at the Gropius house in Lincoln, Massachusetts, in September 1941, a destitute refugee, among his few possessions were two precious rolls of drawings, which he believed would one day make his fortune. One of these was the design of a tubular steel structural system.... The second roll of drawings contained ten small sheets, unannotated, unsigned, and undated,

which delineated with exquisite precision a modular universal building system, consisting of load-bearing panels, weatherboarded externally, flush-paneled internally, thermally insulated, and combining freely (as indicated by the plans, sections, elevations, and details) to generate a house plan adhering to a rectilinear three-dimensional modular grid. The edge [sic] of the wall panels were beveled at 45 degrees, and were secured to each other by elaborate Y-shaped metal connectors. This proposal for a universal housing system lies at the heart of the subject.[64]

These drawings, which Wachsmann had drafted during his traumatizing internment in Grenoble, were the basis for a packaged house proposal that suited the wartime concerns of the day, but Wachsmann did not initially disclose them to Gropius. Gropius had long been dazzled by Wachsmann's technical drawing skills and, seeking to help Wachsmann financially, professionally, and psychologically, offered him work under the auspices of his own firm.[65] But the bombing of Pearl Harbor abruptly stopped these projects, and it was at this point that Wachsmann disclosed his ideas for the housing system to Gropius, who was most enthusiastic.[66] Wachsmann, who now had Gropius's blessing to pursue work on the so-called packaged house system conceived in his roll of drawings, appears to have been a classic perfectionist. He was not satisfied with some of the details of the system he had drafted in Grenoble and returned to the drawing board to focus on the design of the Y-shaped connectors of the system. His compulsive energy devoted to perfecting the system may also have been colored by the extremely difficult conditions under which he was working. As Herbert explains: "He was an uprooted and displaced person, living on the kindness and hospitality of others; his wife Anna was ill in New York, and he was torn by anguish over the news from Europe that his mother and sister had been transported by the Nazis to Poland. In his case work was not only a means of fulfilling long-held ambitions, and a way of regaining independence and self-respect, it was also an anodyne to pain."[67]

Eager to launch Wachsmann into some professional success, Gropius was in fact the impetus for Wachsmann's patent efforts, the first of which, for a Prefabricated Building, was filed jointly by the two men (with Wachsmann in the primary position) in 1942.[68] Outside of the patent parlance, the system was known as the

Packaged House.⁶⁹ Research into prior art, which was likely conducted by Gropius, would have revealed a relatively extensive array of timber prefabricated housing systems. This is why the narrative of the patent application stresses the connector pieces, with which Wachsmann had been so obsessed, as the primary inventive element. Yet the technical focus on the connector was not unlike the fireproofing focus of the Guastavinos insofar as it veiled a larger objective, in this case a building embedded in systems theory. In Wachsmann and Gropius's patent application, that aim is made clear in the narrative: "The invention aims to transfer most of the labor involved in the construction of a building from the site of the building itself to a factory and to make the erection of the building primarily one of assembly."⁷⁰ It was at this point that Wachsmann also assisted the US Army Air Forces in the construction of the "German Village" (directly adjacent to a "Japanese Village") in the wilderness of Utah, where he advised on the construction of simulacra German vernacular dwellings to help perfect the military's ability to bomb real German villages into oblivion.⁷¹

The nature of the collaboration between Wachsmann and Gropius on the patent is also worth mentioning. As Herbert notes:

> There is no doubt, from the historical evidence, that the Packaged House was initially the brain child of Konrad Wachsmann. Gropius confirmed this with characteristic generosity, ascribing to Wachsmann a "decisive part in the scheme." However, in saying this, he nevertheless went on to claim that, in the development of the Packaged House, he and Wachsmann had "pooled our experiences." This is literally correct: in the evenings they mulled over the principles and the evolving details of the scheme together. But it is perhaps also true in a more general, and much more significant sense. The original Wachsmann proposal, the French scheme, which is the prototype of all subsequent mutations and developments, is the product not only of Wachsmann's ingenuity but of a whole decade of experience of prefabrication to which Gropius in Germany had given the prime theoretical direction and a great deal of practical impetus.⁷²

By 1943, Wachsmann's fortunes had turned, and he met Wall Street investors and other enthusiasts to start a prefabricated housing company named the General Panel Corporation in Burbank, California.[73] Gropius, lending his name and connections to the venture, helped Wachsmann in the publicity blitz to get the company going. Now under the auspices of his own company, Wachsmann filed an additional four patents that furthered many of the original aspects of the packaged house.[74] By 1948, with only 15 houses realized along with a number of problems with the company's administration, its board called for a complete reorganization.[75] Wachsmann left the company sometime in 1948 or 1949, and the company's assets were liquidated in 1951.[76] Wachsmann would pursue teaching for the rest of his career, first at the Illinois Institute of Technology in Chicago and later at the University of Southern California, where he led a graduate program in industrialization. During this period he filed one final patent, a Locking Device for Building Panels and the Like.[77]

Wachsmann's engagement of the patent system is notable in several ways. Its position in this chapter's organizational scheme as "micro" patenting might at first seem incompatible with his very universal ideas of systems architecture. Yet Wachsmann's patent contributions, like those of Prouvé, were limited in their scope in that they were, even when focused on details, about the construction of finite, often modest structures like the house. His patent career can also very much be charted against the trajectory of his business aspirations, and not his business successes, with the patents marking the systems he sought to develop as turnkey technology that would afford him a place (and an income) in the American construction sector. Ironically, apart from a design patent for a teapot in 1970, Walter Gropius never patented anything else of his own.[78] So eager to help his friend amidst the horrors of war, Gropius seems to have guided Wachsmann to the patent system as a way to establish himself in the United States, but it would appear that the patents themselves were never necessary for this endeavor. This mirrors the situation of the Guastavinos, who engaged the patent system more as a form of signposting their presence on the scene than as something that was unequivocally necessary for entrepreneurship in construction and architecture.

Like Wachsmann, other architects inclined toward patenting were not forever so inclined. Renzo Piano, by way of example, filed five patents in a span of only four years (1965–1969) and then stopped.[79] Piano himself came from a family of builders: His

Konrad Wachsmann, connection details of the Packaged House System, 1942. Published in Konrad Wachsmann, *Wendepunkt im Bauen* (Wiesbaden: Krausskopf, 1959), 142, fig. 209.

grandfather had founded a Genoese masonry firm that his father and uncles expanded into the firm Fratelli Piano. Piano's patent activity originated in his dissertation on modular coordination in construction at the Politecnico di Milano.[80] A star student, he was asked to teach immediately upon graduation, and this teaching engagement, which spanned 1965 to 1968, coincided precisely with Piano's period of patenting. In this time, Piano became a devotee of the work of Prouvé, studying his patents in depth. Piano was so enamored by Prouvé's work that in 1965, having just started both his eponymous firm and his teaching career, Piano made a pilgrimage to Paris to visit the Conservatoire Nationale des Arts et Métiers.[81] The Conservatoire's astounding collection of models (including patent models) and drawings related to French construction history resonated deeply with Piano, seemingly suggesting to him that Prouvé was the contemporary embodiment of the grand tradition of the French *constructeur*.

Piano's five patents include a system for modular walls (which emerged from his dissertation), a shelter made from inflatable elements of polyethylene, an interlocking system for precast reinforced concrete beams and piers, a machine for producing shell structures, and a reinforced polyester roof shed panel.[82] Piano's ability to execute these patents by making prototypes and testing materials benefited from being able to use his family's construction business facilities at no cost.

Piano went on to work for Louis Kahn and the Polish engineer Zygmunt Stanislaw Makowski, received his first international commission for the Pavilion of Italian Industry at the Expo '70 in Osaka, and teamed up with the British architect Richard Rogers in 1971 to found a new firm. Notably, they collaborated on the design of the Centre Pompidou in Paris (completed in 1977), which masterfully combined their shared interest in technological expressionism.[83] Here, nodes, joints, ductwork, and circulation systems are all highly visible on the exterior, as if the subject of a fetish or an anatomical model. The Centre Pompidou is so tectonically demonstrative that it almost looks like a patent drawing itself. Piano was, in many ways, the quintessential architect-patentee. So why did he stop so suddenly in 1969? While it is hard to say with certainty, Piano's abrupt abandonment of patents coincides neatly with the stratospheric rise of his career over the ensuing years, and he may simply have felt too busy to bother with them. It is also telling, however, that his patenting efforts ended at the same time

View of Renzo Piano's study for a shelter made with inflatable polyethylene in the garden of his house in Genova Pegli, 1966, registered as an Italian patent later that year. The design comprises a basic prefabricated module composed of a square surface on which a flexible hemispherical cup is glued. At the top of the threaded sleeves forming the module, a valve from a bicycle tire crosses an aluminum disk, penetrating into the cup. The sleeves are then joined to metal rods in series and in turn linked to a source of forced air, ensuring the inflation of the chambers within. It was the second of five patents filed by the architect.

as his teaching career, indicating that Piano may have seen patents as a kind of worthy academic exercise but nothing more.

Business Patenting

Buckminster Fuller, Jean Prouvé, Konrad Wachsmann, and Renzo Piano all clearly understood that patenting was a particular way to approach their various business goals. There is, however, an important distinction to be drawn between patents that sought to spur business either through a monopoly or as a form of publicity and those that were already part and parcel of a well-oiled business operation that had a kind of assembly line in place for the production of patentable goods or designs. Such operations could certainly include architecture offices, like those of Frank Lloyd Wright, Henri Sauvage, I. M. Pei, Skidmore, Owings & Merrill, and Santiago Calatrava;[84] but classic invention factories like that of Thomas Edison could also turn their attention to architecture.[85]

Over the course of his lifetime (1847–1931), the US Patent Office granted Edison a staggering total of 1,093 patents, the most ever granted to a single person.[86] Many of his inventions are well known to us today, such as the light bulb and the phonograph. Many more are not well known but nevertheless proffered vital, incremental improvements to existing technology, including to the telegraph, the telephone, the typewriter, the microphone, the motion picture camera and film, the storage battery, and the electric railway, among others. Edison also achieved major breakthrough patents in synthetic chemicals, rubber, mining, wax paper, the dynamo, and Portland cement.[87]

When asked about his philosophy on what kind of claims to make in a patent application, Edison is noted for having said "Claim the Earth!"[88] Claims needed to be as broad as they possibly could be for a given invention, so as to sustain and extend his monopolies on so much of the era's IP. It was purely a business strategy. Edison arrived at this approach to patenting over time as he learned, by trial and error, how patents could be leveraged, deployed, and monetized to good effect. In the early part of his career, Edison understood patents as a tool that was essentially offensive in nature, bait for enticing investors and corporate managers, not unlike how Wachsmann would come to see them. It did not take long for him to see the value of defensive patenting also, used, as Paul Israel has noted, "to protect or enhance a company's competitive position."[89] But neither approach alone would build the kind of empire that Edison sought. As Israel notes: "These two understandings came together during his work on electric lighting as he transformed himself from a contract inventor working to improve existing technologies into an innovator seeking to build a new industry. As a consequence, he became as concerned with producing patents as he was with creating new technology. In fact, he could not do one without the other."[90] Edison further understood that filing patents was critical given the dating principles of "first to file," and he utilized the medium of caveats, an official notice of intention to file a patent application at a later date, to this end. Edison's approach to patents also evolved as it related to the patents of others. He had been obstinate in his early career about not pooling or cross-licensing patents, but this changed as he sought to make inroads in industries with formidable competitors, saving him from having to engage in expensive and chancy patent litigation.[91] Edison made certain that patents played a role in all of the myriad fields in which he was working, but he also understood that patent practices in different industries, for example in sound recording and cement, were extremely variegated. His patent strategies—offensive, defensive, or otherwise—needed to respond to the conditions of the specific industry within which he was seeking an inventive footprint.[92]

Needless to say, no single person could do what Edison did without assistance. An army of engineers, assistants, secretaries, and patent lawyers and agents backed his operations at his invention factory in Menlo Park and stood by (for whatever reason) as Edison filed these patents solely under his own name. "Genius" is an easy word to use when achievements are so staggering, but it is interesting to understand how such achievements are conditioned and coalesce. Edison conveyed his broad ideas on an invention to his researchers and machinists while also imploring them to take the idea further without his help. Edison himself described the day-to-day conditions of the Menlo Park studio as such: "I generally instructed them on the general idea of what I wanted carried out and when I came across an assistant who was in any way ingenious, I sometimes refused to help him out in his experiments, telling him to see if he could not work it out himself, so as to encourage him."[93]

Edison's work on what would come to be known as the Single-Pour Concrete House illustrates the dynamics of both patent strategy and patent authorship/attribution.[94] It is also Edison's most bona fide architectural invention, despite his not being an architect and having no background in construction. Edison's notebooks reveal his earliest schematic sketches for the system, dating to 1907. The team of inventors at work on his various inventive enterprises in cement technology would likely have seen these sketches and adapted them as they worked out the details of the construction of the Single-Pour Concrete House and drew the eventual patent drawings submitted to the patent office, which bore a remarkable resemblance to Edison's original sketches. The team also constructed a model of the house, although it was unnecessary at this point, as a didactic element for the patent office.

Edison filed US Patent 1,219,272 in August of 1908, but it would not be patented until a full nine years later due to a number of challenges. As stated in the patent claim, Edison had as the object of his invention the construction

> of a building of a cement mixture by a single molding operation—all its parts, including the sides, roofs, partitions, bath tubs, floors, etc., being formed of an integral mass of a cement mixture. This invention is applicable to buildings of any sort but I contemplate its use particularly for the construction of dwellings, in which the stairs, mantels, ornamental ceilings and other interior decorations and fixtures may all be formed in the same molding operation and integral with the house itself. The house thus made is practically indestructible and is perfectly sanitary. The cost of its construction is low and it is feasible to beautify such a house far beyond anything now possible in so cheap a manner.[95]

Before the patent was even granted, Edison realized a handful of full-scale examples in and around Phillipsburg, New Jersey, a few of which still exist.[96]

Edison, already something of a media darling, received considerable attention for the invention. The *Boston Daily Globe* foresaw that a "large army of wage-earners may yet hail the concrete house as a blessing if all that is claimed for it comes to pass."[97] The house

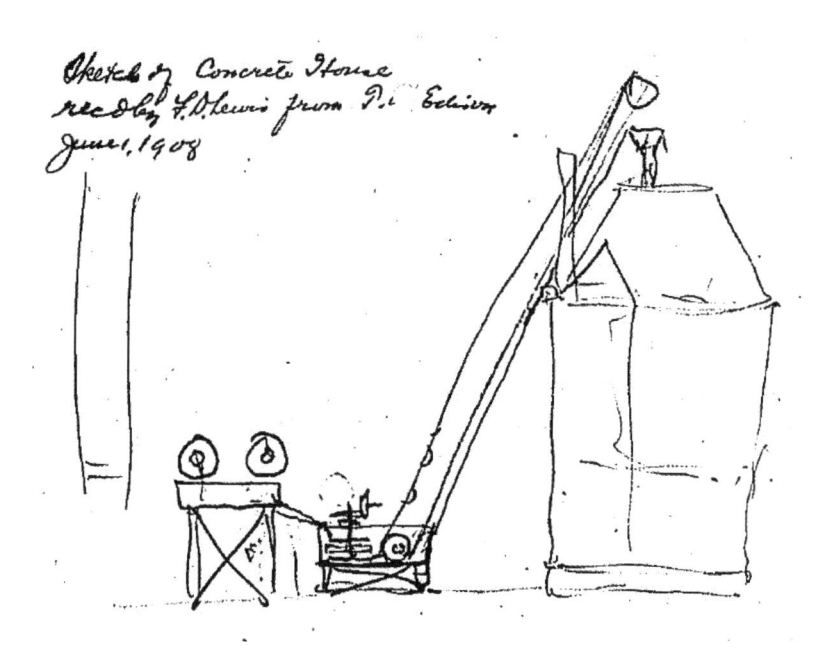

Thomas Edison, laboratory notebook sketch of the Single-Pour Concrete House, 1907.
Thomas A. Edison Papers, NA144-F.

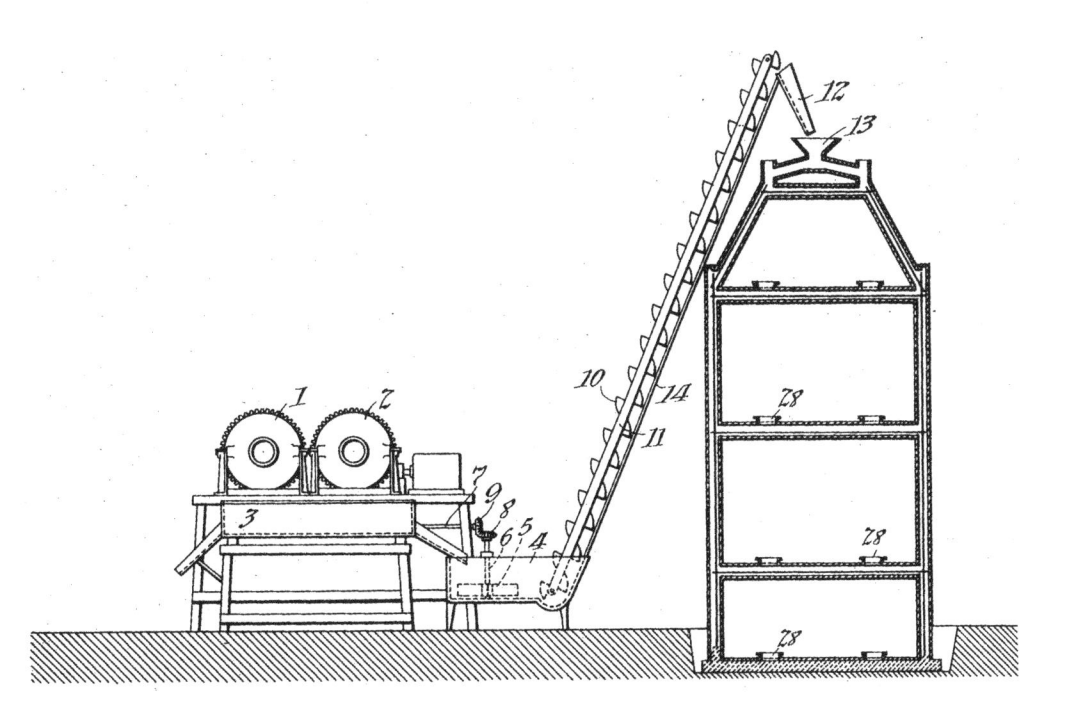

Thomas Edison, Process of Constructing Concrete Building Structure, US Patent 1,219,272, filed August 13, 1908, figure 9.

Thomas Edison with model of the Single-Pour Concrete House, c. 1910.

appeared alongside the dirigible airship, wireless transatlantic communication, and the *Lusitania* in a *Chicago Tribune* article entitled "Seven Great Wonders of Science and Industry Perfected in 1907."[98] A slightly more cautious article in *Architectural Record* thought that Edison had not yet perfected the system (which was true) but nevertheless praised the project for its vision, noting that "in a few generations from now the majority of urban and suburban residents may well be living in concrete houses of one kind or another—without any fear of fire or of vermin, and without paying for these substantial living accommodations any more than they are paying for their more or less flimsy dwellings."[99] Edison's image took a hit, however, in 1911 when, awaiting the approval of the patent, he also touted his forthcoming inventions in concrete furniture that would supposedly be half the cost of regular furniture. Concrete bedroom sets seemed less appealing than concrete walls, and the move opened Edison to public mockery.[100] The venture, due to its ultimately lukewarm public response and difficulties at the patent office, would never fully take off, which was a significant but withstandable blow to Edison's business empire.

The patents of Edison's American contemporary, Frank Lloyd Wright, are instructive regarding the ways in which architects endeavored to engage in business patenting. This practice began with Wright's alliance with the Luxfer company, for which he designed over 40 glass prismatic tiles, each filed as a design patent in 1897.[101] By way of patenting the panoply of prisms, the Luxfer company was making an effort to mass-produce blocks that would contribute to the stylistic development of modern architecture and render light shafts obsolete (in turn providing new floor space).[102] The seemingly straightforward patent drawings submitted for the patent applications warrant further consideration. In accordance with the requirements of the patent office, Wright (or the draftsperson chosen for the task) needed to supply the patent office with drawings of all relevant elevations of the invention, which in the case of the prism were one frontal and one side elevation. Annotations on the two figures would show how they corresponded to one another. Yet the drawing requirements did not stop there. The drawings had to carefully follow the patent office's protocol for representing depth and reflectivity when light was projected from the upper left-hand corner of the elevation (the standard direction from which light was supposed to be rendered in patent drawings for the US Patent Office). The streaky lines in many of Wright's frontal elevations reveal the anticipated intensity and size of shadows and prismatic reflection. Needless to say, these formulaic drawings stood in sharp contrast to the type of drawings that he typically would have produced to illustrate his designs, and they are likely the only example of Wright following the graphic standards of anyone other than himself over the course of his career as the head of his own firm.

Probably in consultation with Luxfer, Wright had to do something else that he was not used to: publicly citing the work that provided a precedent for his work and demonstrating that his invention was incremental rather than ex nihilo. Wright's primary citation for Luxfer was the work of John Pennycuick, whose window glass patents preceded Wright's by about 12 years.[103] A particularly striking characteristic of Wright's patents is the brevity of their explanations, which were far shorter than in the patents of even some of the simplest inventions. Instead of three pages of explanation, which was the average for a utility patent (it was somewhat less for a design patent), Wright got away with justifying his Luxfer patents in fewer than 200 words.[104]

Wright's patent activity next surfaced in 1911 and was again tied to an affiliation with an industrial partner. Collaborating with the Arthur L. Richards factory in Milwaukee, Wisconsin, Wright developed a series of standardized houses made up of elements that were precut and shipped for assembly, which reduced waste and labor costs.[105] Wright produced over 960 drawings for the project, more than for any other project in his archives, detailing over 30 unit variations. In 1917, more than a dozen licensed dealers of American Systems-Built Homes opened for business, offering small units at a cost between $2,750 and $3,500, with large units ranging from $5,000 to $10,000.[106] "I would rather solve the small house problem than build anything else I can think of," Wright once said.[107] The United States' entry into World War I would soon divert materials to efforts abroad, dooming Wright's early attempt to provide a range of modern housing options to Americans of any income. The "Patent applied for" notation can be seen in the corner of the presentation drawing, although successful patents for the system do not appear in the US patent database.

Wright did receive a patent for a dwelling in 1938, a dwelling that appears to develop certain ideas originating in the System Built scheme.[108] This patent was simply titled Design for a Dwelling and had a justification even shorter than those for the humble

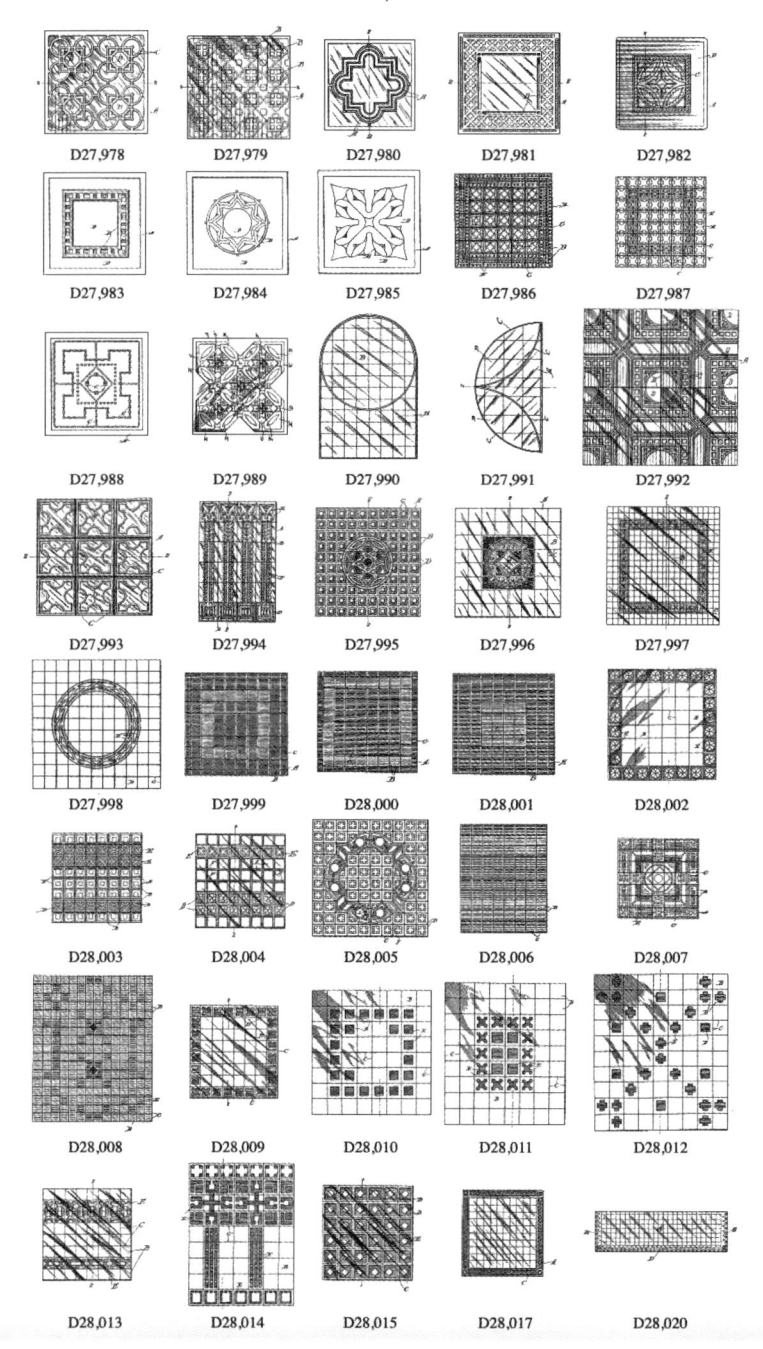

Composite view of Frank Lloyd Wright's Luxfer Prism block design patents.

glass prisms for Luxfer: Wright needed only to state that the design was "new, original, and ornamental" to be granted the patent—an unbelievably low bar, even for a design patent—around 1938.[109] Why was Wright granted this kind of agency, the ability to write such scanty claims? He does not even mention what is clear from the drawings: that the house comprised four equal units rotated on the center, which intimated mass production and spatial flexibility. It is possible that by the 1930s, Wright and his firm were so well known to those in the design field (including patent examiners specializing in the design categories in which Wright was patenting) that he did not have to prove much to anybody—which, in theory, should not be a factor for a patent examiner. But it is also important that he filed architectural (not just industrial) designs as design patents, using the word "ornamental" to describe something that was also ostensibly architectural (something Santiago Calatrava would also do several decades later, as we will shortly see).

One of the reasons to hold a patent is to be able to have a monopoly on the production of owned ideas, presumably to protect one's property and to assure remuneration for inventive work done well. In Wright's archive, there is no record of his pursuing an infringement case against anyone else, but others did pursue infringement cases against him. A key example of this centers on the Ennis house, a sprawling private residence in Los Angeles that was completed in 1924.[110] In January of 1933, Wright received a letter from a Los Angeles attorney by the name of Donald Barker, who indicated that Mabel Ennis, the wife of the late Charles Ennis and client for his well-known 1924 design of the house bearing the Ennis name, had received a claim of infringement from the lawyers of a certain William E. Nelson.[111] Nelson believed that his 1919 patent for a modular concrete building structure was an uncited precedent for Wright's 1921 patented textile block system, which was used in the Ennis House.[112] Particularly noteworthy is the fact that an assistant for Wright wrote in pencil below the lawyer's note: "This letter should be answered"—something that we do not see on similar letters and that is generally only seen on correspondence with a very high level of importance that Wright's staff felt begged his personal attention. While we do not have Wright's response, we do know that there was never any litigation between him and Nelson, which indicates that the conflict was somehow resolved, as is often the case with patents in the realm of design, out of court. One explanation for the claims to infringement by third parties might be their desire simply to be associated with Wright or to make him personally aware of their own work through a legal claim, as opposed to a friendly solicitation, however perverse this may seem. Regardless of the full explanation, we know that Wright shared a willingness to engage the patent system with his son John Lloyd Wright, who also holds a number of patents, including those for the design of the wildly popular microarchitectural Lincoln Logs toy set in 1920.[113] It is noteworthy that the elder Wright's patent activities, apart from those for Luxfer, were simultaneous with those of his son, indicating that Frank Lloyd Wright's strongest encouragement for the practice of patenting may have come from John, geared as he clearly was toward mass production and replicability.

Mass production and mass housing in patents were not an exclusively American affair, and Thomas Edison's experimentations in concrete in the United States enjoyed strong parallels in France. As early as the 1860s, builders like François Coignet sought to simplify the use of reinforced concrete by standardizing elements of the wooden formwork and streamlining the associated work methods.[114] Paris, which had expanded rapidly prior to World War I, was in even more dire need of low-cost, mass housing after World War I. The architect Henri Sauvage and his partner Charles Sazarin had been hard at work on this problem since 1903, addressing the imperatives of the economy through rapid construction. Sauvage linked most of his experiments in rapid construction to the legal procedure of filing a patent, demonstrating his desire, as Jean-Baptiste Minnaert describes it, to present himself as an "architect-inventor" or, perhaps, an "architectural industrialist."[115] From around 1910 onward, the French government was actively encouraging inventors across the fields of science, technology, and engineering to pay greater attention to the patent system and to utilize it as much as they could, with the belief that this would enhance the dynamism of France's economy as well as its long-term standing in international trade.[116] Between 1912 and 1931, the French patent office granted Sauvage an impressive 15 patents, comprising inventions related to various construction systems and processes of manufacturing and assembly. Sauvage also leased rights to other patents, something few of the inventors discussed thus far actually did.[117]

One of Sauvage's most notable buildings, the stepped apartment complex at 26 rue Vavin in Paris, is a prominent example

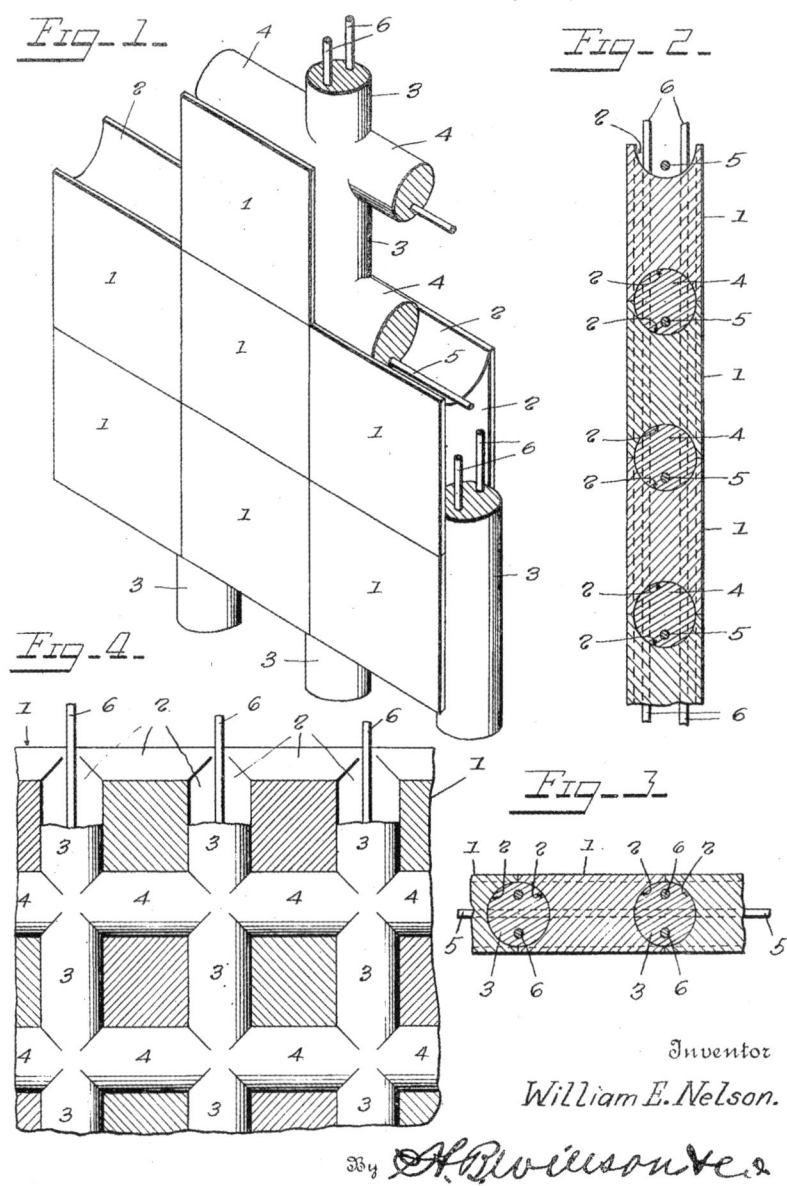

W. E. Nelson, Concrete Building Structure, US Patent 1,535,030, filed April 21, 1925.

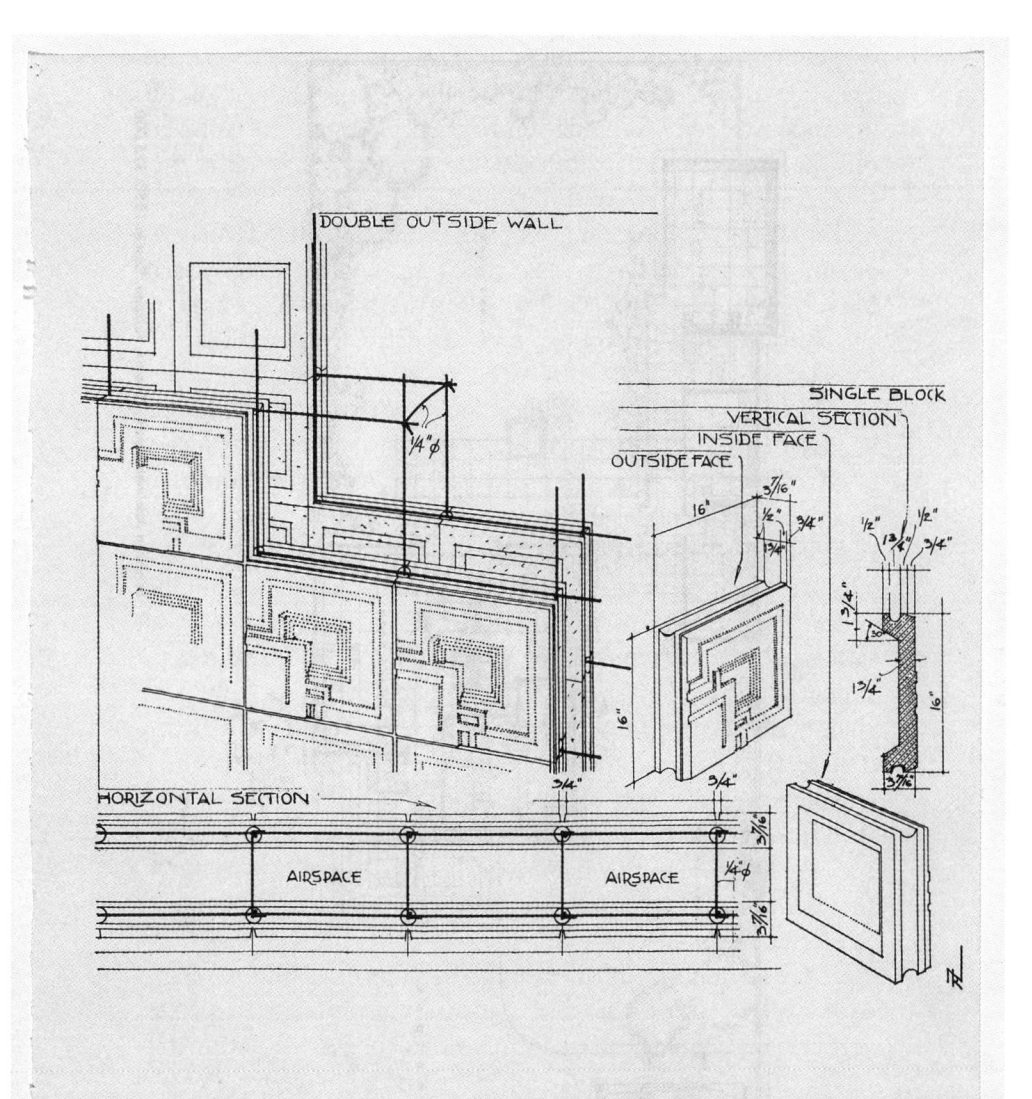

Frank Lloyd Wright, "Textile block design, patent study," 1921. Avery Architectural and Fine Arts Library, Columbia University, drawing no. 2111.004.

of his rapid construction research.[118] It is also the subject of his first patent.[119] The Stepped Building patent came to pass under very complicated circumstances that Sauvage navigated skillfully. The client for the development was the Société des Maisons à Gradins, whose deed of foundation recognized architects as inventors through an unusual profit-sharing arrangement in which they received capital returns on a site's profitability as housing in return for exploiting the invention, which in this case included Sauvage's patent. Sauvage was clever to file the design not as a design patent but rather as a utility patent. The utility patent protected not only the building's stepped appearance, but also its intrinsic logic.[120]

The French government's encouragement of inventors to patent did, however, come into direct conflict with the 1895 code of the Société Centrale des Architectes and the 1899 code of the Société des Architectes Diplômés par le Gouvernement (SADG), the main and rather influential administrative and organizational bodies of professional architects and builders in France. The SADG code stipulated that "if an architect has taken out a patent for a product concerning industry, then he ought not to personally exploit it but rather sell it to an industrialist, giving him all property rights for exploitation of the invention."[121] For Julien Guadet, the influential chair in the theory of architecture at the École des Beaux-Arts who drafted the code, the filing of a patent fell within the technical and commercial sphere of the entrepreneur, and he believed that patents served to distance architects from the artistic (rather than commercial) field with which they should concern themselves.[122] In other words, Guadet wanted to do everything within his power to discourage architects from engaging the commercial sector as directly as patents would dictate and to retain architecture's status in France as that of an art form above all else. In his defense, Sauvage may well have been unaware of this code (which ultimately was neither legal nor enforceable) when he patented the stepped building scheme used at 26 rue Vavin. Another patent, one for an unrolled parquet flooring system, was purchased immediately upon its patenting by an entrepreneur who would then exploit it, indicating that even if Sauvage was not cognizant of Guadet's code, he nevertheless occasionally followed its logic.[123]

In 1925 Sauvage founded his own company, the Société des Constructions Rapides (SCR), where he continued to develop his patent portfolio. One prominent patent that emerged from SCR is a system for constructing a building by using telescopic jacks

Henri Sauvage and Charles Sarazin, terraced residential building clad in glazed white ceramic tiles, 26 rue Vavin, Paris, 1912–1913. Photograph by Chevojon.

as vertical posts, patented in 1929.[124] This invention functions by raising the ceilings of each successive level of a multilevel construction and locking them at their desired height, after which the lathes are lifted by the jacks of the lower level. The true innovation of this patent was in its ability to perform construction primarily from above and to limit structural work to the ground level, nearly eliminating scaffolding and the hoisting of heavy materials to the difficult-to-reach upper levels. Sauvage, like many of his progressive contemporaries, was enamored by the technological innovations of the automobile, railway, and aviation industries, and his appreciation for design that was mobile was at the core of SCR's later work.

Sauvage, in fact, played a quiet part in leading the French architecture profession out from under the aegis of the École and its very influential dictates on the profession into a realm where patents were not only acceptable but logical and desirable. This is one of many hints regarding the dwindling of the influence of the École in the first third of the twentieth century. Sauvage's approach was problem-oriented, concerned with housing shortages, inefficient building methods, and wasted manpower. Under the auspices of his construction firm, he was able to combat the institutional stodginess and reluctance to patent that characterized the architectural profession from the outside in. His brilliance lay not only in his designs but also in his ability to manufacture important architectural change by means of his business and his patents.

Following the earlier exponential rise of utility patent filings, the latter half of the twentieth century witnessed a decline in patents filed by high-profile architects and architectural firms. Architectural patents that arose from the 1950s onward often did so under very specific circumstances. One such example is the patent for a Multistory Building Structure, known commonly as the "helix plan," issued jointly to the real estate developer William Zeckendorf and the architect I. M. Pei in 1955 and described as "a structure type that is characterized by a novel arrangement of the floors of the adjacent units of space in such structure by which maximum flexibility in the utilization of the entire space with the structure is provided, and further characterized by the centralization of the utilities with respect to the units to be served thereby, thus effecting economies in construction."[125] The drive to patent in this case certainly came from Zeckendorf, who had also filed a design for a parking lot a few months earlier and, for a brief moment in 1949, saw patents as a means to monopolies on innovative design solutions for urban real estate.[126]

The Spanish architect Santiago Calatrava offers another example. Calatrava holds 14 unique patents in four countries—Switzerland, Germany, Austria, and the United States—as well as in the European Union. Most are for works of industrial design or auxiliary structural elements.[127] Calatrava's two truly architectural patents—Building (2007) and Bridge and Gondola System (2010)—are both filed in the United States and, curiously, registered as design patents rather than utility patents.[128] In the claims made for both of these, Calatrava, like Wright, refers to the design as "ornamental," tacitly implying that they have no utility, although one would suspect that an "ornamental" bridge still connects two things and that an "ornamental" building still provides shelter. Indeed, the patent for the "ornamental" building, filed in 2006, has a strikingly similar appearance in its overall formal gesture to Calatrava's Turning Torso building in the Swedish city of Malmö, completed one year prior to the patent. When asked, the architect's office contended that there was no relationship between the two and refused to offer further comment.[129] One could be forgiven for suspecting that Calatrava was attempting a novel (and very sneaky) patent strategy wherein generic formal motifs are submitted as design patents that do not require the inventor to articulate prior art or explicate utility, while still benefiting from patent protection. This is most certainly a place where the patent system needs to be reexamined to determine whether it is indeed doing what it is intended to do, or whether architects like Calatrava are merely gaming the system to generate absurd formalistic monopolies that will undoubtedly stymie creativity in the field.

Although rare and normally possible only after an architecture firm has reached international renown, beginning with the last few decades of the twentieth century, a select few firms have developed internal research and development outfits where patenting is a distinct aspect of business activity. This is the case with Skidmore, Owings & Merrill (SOM), which has long considered itself to be an incubator of innovation in construction and design, a prominent example being historic innovations in the curtain wall.[130] From the nature of how the firm builds—internationally and often in the same typology of the tall building—SOM is in a position to sell the same solution for problems that repeat from

Santiago Calatrava's Turning Torso skyscraper in Malmö, Sweden, 2019 (completed 2005). Photograph by Amjad Sheikh.

one project to the next.[131] The firm researches patents on assembly and building systems.[132] For several decades, SOM has utilized inside counsel for initial legal consultations and then employs outside counsel when they decide to go all the way to the patent office. The firm's upper management regularly discusses critical decisions about whether to patent defensively or offensively, and it measures the cost benefits of stamping out derivative designs against the legal costs associated with doing so. Innovation, as the firm's mantra goes, can come from any individual or group of individuals in the office who have a "proprietary solution to a universal problem."[133] Patenting in the United States, with its broadly visible patent system, is always SOM's first stop, followed by the European Union, within which they must decide which countries they want to claim patents in. The firm holds its personnel who contribute to patents in high esteem.

Experimental Patenting

All patents are experimental in nature. Inventions necessitate experiments, and more often than not these experiments fail. Inventors profit epistemologically when these failures bring them one step closer to an invention that is patentable and, crucially, nonobvious. Patents do not, however, need to be infallible, nor is there any particular mandate that they be economically profitable. Of course, from the market's perspective, patented inventions that have the ability to meet the promises made in the patents' claims and that proffer their inventors some form of remuneration are what the system is made for. However, the patent system allows room for patenting for all sorts of unusual reasons apart from the mainstays of property protection and financial profit. Indeed, we have already seen patents deployed by architects that also function as veils for unpatentable formal innovations, as publicity and marketing efforts, and as manifestations of the ego. In the large reservoir of patents related to architecture and construction, there is also a certain kind that extend the mode of experimental thinking endemic to the research phase into the publication phase. These are patents that seem to have no particular motive, it would seem, apart from being thought-provoking.

No discussion of architectural patents can be complete without noting the patent activity of the Swiss-French architect known as Le Corbusier.¹³⁴ Le Corbusier held 16 patents, issued between 1919 and 1961 under his birth name, Charles-Édouard Jeanneret. He submitted a first batch of six patents for a mixture of inventions to the French patent office in October of 1918.¹³⁵ This was followed by two formwork patents in 1919, a window frame patent in 1926, and five patents for prefabricated construction systems submitted between 1945 and 1960.¹³⁶ Le Corbusier holds only one patent outside of France, the patent for Design Rules, issued in Canada in 1950.¹³⁷

The "patents" with which Le Corbusier is most associated, however, are two that were never ultimately approved by the patent office. This is, in fact, no coincidence: these two patents were by far Le Corbusier's most polemical and therefore difficult to patent: the so-called Maison Dom-ino and the so-called Modulor Man, an anthropometric scale of proportions, that Le Corbusier considered patenting when he began working on it in 1945 but ultimately never did.¹³⁸ However, the Modulor serves as the basis for the Canadian patent Design Rules.¹³⁹

The complete genesis story for the Maison Dom-ino does not bear repeating here.¹⁴⁰ However, it has some salient features that bring Le Corbusier into the larger context of architectural patentees and help us understand his patent practice, successful or not, as one that was experimental at its core. The Maison Dom-ino project arose out of Le Corbusier's deep interest in the issue of mass housing and ways that "picturesque groupings," as Tim Benton describes them, could come to pass with a limited array of three or four plan types. The patent itself was to serve as a modular system dictating the dimensions of furniture, windows, walls, and other apertures.¹⁴¹ Le Corbusier worked on the specifics of the Dom-ino as a patent scheme in 1915, and possibly earlier in 1914, alongside a more commercially ready housing scheme to pitch to private investors and developers in the form of brochures.¹⁴² Max Du Bois, an engineer who consulted Le Corbusier on these schemes, also advised him on the patent and submitted it on his behalf on January 29, 1916.¹⁴³ Le Corbusier, who had never before filed a patent, may not have fully understood the limitations of the process, and it is also not entirely impossible that, like Sauvage, he was seeking to patent architecture as an indirect form of confrontation with Guadet and the philosophical tyranny of the École des Beaux-Arts over the French architecture scene. Le Corbusier's adage "Rien n'est transmissible que la pensée [Nothing

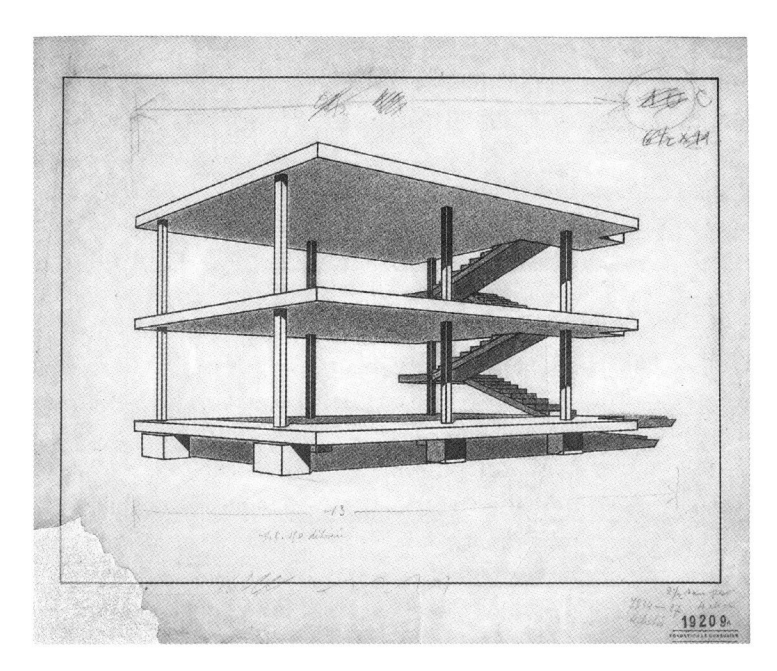

Le Corbusier, "Maison Dom-ino," patent drawing study, c. 1915.
Fondation Le Corbusier inv. 19209.

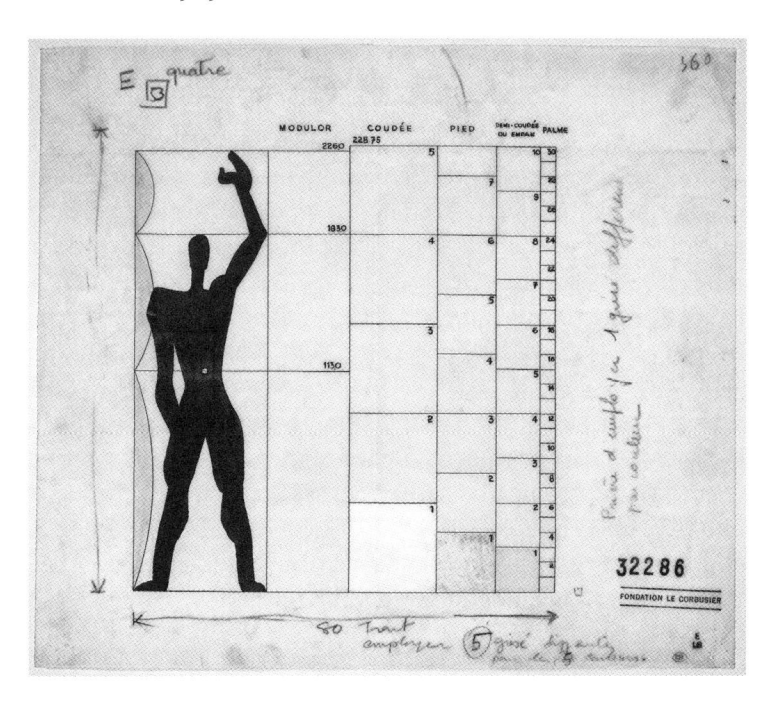

Le Corbusier, "Modulor Man with golden section," patent drawing study, c. 1945.
Fondation Le Corbusier inv. 32286.

can be transmitted except thought]" factors into the equation of why he was motivated to patent the Maison Dom-ino in the first place, in a culture that rarely patented architecture in this way.[144] A patent was indeed a way of transmitting an idea, whether practicable or not and whether presumptive or not. Pride too may have come into play, but the impulse to patent probably hinges most on Le Corbusier's need for recognition. Moreover, Le Corbusier truly admired the Fordist model of manufacturing in the United States, where one could design something (and patent it) and the market would do the rest, promising a (perhaps welcome) divorce between architecture as a design practice and architecture as a business practice. Le Corbusier may have been hoping to jumpstart a more robust Fordist model for France when he submitted his patent, but he clearly and profoundly misunderstood how the market works.[145] His disastrous management of the Briquetterie d'Alfortville is evidence of this, as was his serious consideration of patenting the Modulor in France.[146] After all, he did not design the structure of the Dom-ino; he found it, and one cannot patent things one finds "in nature."

Despite its failure, Benton rightly argues that there is a clear recognition of Le Corbusier's "Dom-ino structure as a founding icon of modern architecture. It is an entirely legible aporia. The charm of the Dom-ino—in its celebrated form as a perspective sketch—is that it is a tangible but unrealizable 'thing,' feasible neither as a single house nor as a combinable unit of terrace housing, and the 'thing' demonstrated to perfection is the theoretical principle of the separation of structure from enclosure all the way round. And to reiterate, what makes the Dom-ino project interesting is precisely this tension between unrealizable ideal and practical project."[147] It was the pioneering Dutch architect Mart Stam who took Le Corbusier's patent and theorized it into the authoritative, polemical leitmotif it remains today.[148] More specifically, Stam lionized the perspectival patent drawing of the structural skeleton as the "completely autonomous art of architecture, one that refers only to its own means."[149] In his reification of the patent drawing, Stam was also unconsciously implicating patents as a system promoting, not demoting, the autonomy of architecture as an art form.

Le Corbusier's other patents are not without their own revelations, even if they were more traditional than the Dom-ino and Modulor. His patents of building systems in the prewar period transmute some of the essential concerns of the Dom-ino into innovative systems for reinforced concrete, for example the project known as the Masion Monolith (or Eternit).[150] In the postwar period, in 1949, he patented another building system that is the fundamental scheme of the Unité d'Habitation, which was completed in Marseille three years later.[151] Dario Matteoni has carefully examined these patents and shown how the variety among them and the Dom-ino (and Modulor) reveal an approach to patenting that was certainly experimental but also opportunistic and not born of a rigid ideology: "Le Corbusier's attitude to patenting was never univocal. At least when he initially engaged with the process, one can recognize its almost instrumental use: the patent serves as a legitimating device for his professional role and as a calling card in the world of industry, affirming his experience and ability to speak the same technical language."[152] Using Le Corbusier's 1926 window frame patent as an example, Matteoni notes that the patent is essentially a legal index of the architect's programmatic statements from the same period.[153]

It would be incorrect to depict experimental patenting as the exclusive domain of architects as broadly renowned as Le Corbusier. A number of small architectural design outfits as well as individuals have sought patents for reasons that are primarily about neither property protection nor economic gain. This includes a 1967 patent by the Polish-Canadian architect Victor Prus entitled Building Structure, a design recalling the well-known Giant's Causeway geological formation in Northern Ireland in its clustered, geometrical appearance.[154] According to Prus, the design offers "an improved building structure, an improved roof construction comprising a plurality of integrally connected, mutually parallel tubes having a polygonal horizontal cross section, in which the tubes are vertically staggered to produce different configurations, and the roof construction is suspended between peripheral supports and the roof construction is trussless."[155] Prus's invention is just as scaleless as any by Buckminster Fuller (who was building for the Expo in Montreal at the same time as Prus was preparing his patent in the same city), but it is also far more polemical, relaying no real aspiration to implement the invention. Prus was known in the field for his belief that ambiance was the ultimate object of architecture.[156] The adventurous, futuristic pyramid of hexagonal clusters surely achieves that, and this may be all that he set out to put into the world when filing his patent. After all, he had

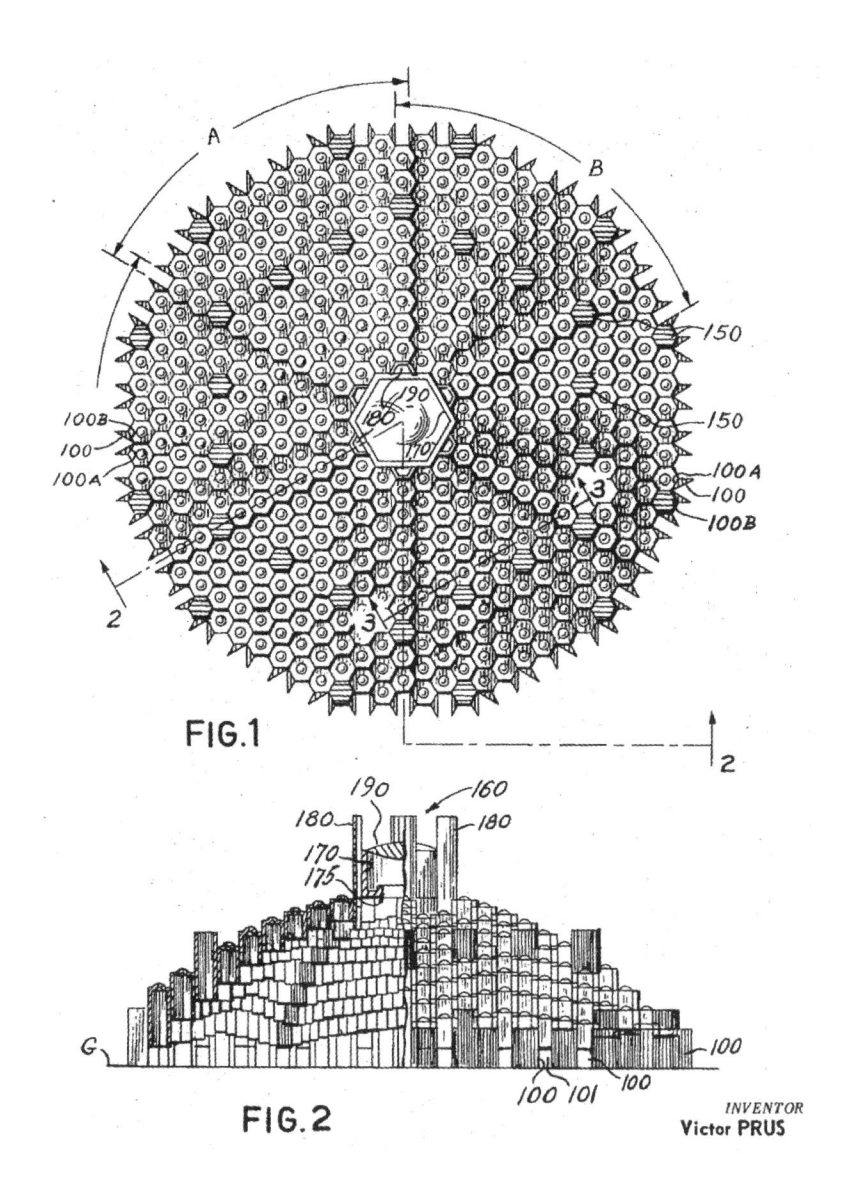

Victor Prus, Building Structure, US Patent 3,337,999, filed March 15, 1965.

a successful practice to fall back on, and this patent would appear unrelated to anything else he was designing at the time.

Another common point of origin for experimental patents is the academy. Professional researchers are, in theory at least, less driven by the financial incentives of those in industry or the architectural profession because they already have a steady base income. Researchers at major research institutions are nevertheless incentivized to patent, as this can (particularly in schools of engineering and architecture) be one of the schools' major markers of research productivity that need to be detailed in annual reports. Mark Goulthorpe, a professor of architecture at the Massachusetts Institute of Technology who specializes in digital design and fabrication, holds a patent that came directly out of his research as a professor.[157] Goulthorpe explains that he was only dimly aware of the history of patents in architecture and that his patent efforts were conducted alone and with some reluctance. His motivations were not tied to any tradition in the field, but rather "to establish some protection to a prospective commercial activity that allows investors to see it as lower risk than it would be if it were able to be reverse engineered and copied."[158] Putting it another way, he notes that he "tended to devise IP strategy that seeks both patent protection (patent, copyright, trademark) AND secrecy (when know-how is not divulged), the latter in how a given software 'works,' for instance—keeping the algorithm and programming private."[159]

Whether macro or micro, a business or an experiment, patenting has occupied a regular if somewhat distended role in the profession of architecture and in the studio culture of architects. The vicissitudes of patent practices in the architect's studio may ultimately reveal more about the architects themselves than about patent practice. This is important, because it reveals how patenting acts as a kind of vessel of validity, one that offers everything from the promise of investment to the signposting of one's architectural ideology, publicity, and good, old-fashioned IP rights protection. As we move next into the corporate arena, where so much of what architects need to build is researched, manufactured, and disseminated, we will encounter yet another landscape where patents play a primary role in defining the nature of invention in architecture.

3 *The Corporate Lab*

The incorporation of research and development (R&D, as it would come to be popularly known) into the corporate environment was at once slow and foreseeable. As global markets for goods in the construction and consumer sectors grew, so too did corporate specialization. Specialized products required the production of specialized scientific know-how that was increasingly difficult to find in the commons of university-driven knowledge. Driven by the incentives of the market, corporations were tasked with establishing that their niche products would both sell *and* work. This imperative steadily drew chemists, physicists, engineers, and specialist scientists of all stripes into the private sector. Patents, in this context, emerged as a regime for standardizing aesthetic qualities (such as paint colors or brick sizes) and establishing consumer viability.

German and American corporations were at the forefront of these new regimes, the two countries marked by certain characteristic differences. For the first several decades of German corporate patent activity, corporations abided largely by the conventions of their day, loath to accommodate the large-scale participatory models of authorship and invention that were increasingly common in the United States. In Germany, the chemist Heinrich Caro was emblematic of this new type of corporate figure as it crystallized in that country.[1] Caro differed from most of his leading chemist peers in having studied chemistry at a trade institute (*Hochschule*), not a university. Intent on applying his knowledge rather than publishing research papers, Caro began his career printing calico in Mühlheim before he was recruited by the Manchester chemical firm Roberts Dale. In 1861 Caro returned to Germany, where he took a senior position advising the Ludwigshafen chemical firm, soon to be known as BASF (Badische Anilin- & Sodafabrik).[2] Caro is most widely known for his patents of the first indigo dye, which allowed the widely popular blue hue to be deployed in all sorts of products without being dependent on the rare genus of *Indigofera* plants. Caro's most important invention, however, is his invention of a rigorous laboratory-based corporate research culture.[3] Caro petitioned BASF's management for exponentially greater funds committed to internal research and effectively argued that the corporation would be best positioned to advance in the German market with a more robust and experimental scientific armature driving its momentum from the inside. Caro successfully disabused BASF's management and shareholders of the long-held tenet that revenue came only from the maximization of production. He emphasized instead a model of continual discovery, one that also brought the risk of failure. This dynamic, laboratory-based research environment quickly caught on across Germany,[4] and it was in this system that the corporate obsession with innovation was born.

Caro's ambitious vision for a culture of corporate research hinged on a key epistemological break with the precedents of attribution established by Enlightenment science. The new province of corporate research and experimentation necessitated the subordination of the individual to the corporate entity. Corporate research was conducted by corporate subjects, and intellectual attribution, most clearly evinced in the parlance of patents, was split between designation of the "inventor" (the scientist) and the "assignee" (the corporation). This merging of strategy and science in the corporate research lab is commonly associated with an increasing concentration of creative activities within the corporate sector in large companies following the Industrial Revolution. The legal scholar Robert Merges has convincingly argued that corporate-sponsored creativity did not anonymize the human sources of innovation as much as it naturalized innovation into the DNA of corporate culture.[5] So too did the law. Merges demonstrates the intrinsic interdependence of the IP regimes of individuals and those of corporations: "[C]orporate ownership is an important feature of the ecosystem that supports individual creative professionals; ... incremental change in IP rules makes sense in some cases.... [A]s long as ... a certain industry dynamism is possible there is no reason to suppose that industry structure is so anathema to the individual creator that the basic premise of IP law has become irrelevant."[6] The naturalization of innovation into corporate behavior was also spurred by the rise of antitrust law, such as the Sherman Antitrust Act of 1890 in the United States.[7]

Not surprisingly, corporations producing goods for the construction sector (along with the chemical, pharmaceutical, and mechanical sectors) were foremost among those defining the contours of the corporate research laboratory in the industrial West. They were, for example, some of the most prominent voices at the 1874 International Patent Congress in Vienna, where corporate managers from across Europe and North America agreed on the importance of international publications in which corporations would share their patent research developments.[8] Two years later, after imperial consolidation, the German government would recalibrate its patent laws to encourage collaboration between private inventors (*Einzelfindern*), such as those we have seen in chapter 1, and big business.[9]

BASF was just the beginning of a long line of corporations that created laboratories for research benefiting the building sector throughout the twentieth century. These included chemical firm peers in the United States, such as DuPont and 3M, firms in other sectors like aviation that had to recalibrate their industries through the vicissitudes of war and sanctions, such as Messerschmitt and Junkers, construction companies such as Dyckerhoff & Widman AG, and countless smaller corporations. In each case, the in-house R&D labs adapted to the wider context of mergers, antitrust law, global competition, and anti-big-business sentiment.[10] This new corporate research culture absorbed a new generation of university-trained (and sometimes not) scientists ready and willing to venture into more lucrative careers outside of academia. The single greatest evidence of this new corporate culture lay in patents, which simultaneously fortified and validated corporate research culture.[11] Indeed, it was not until the advent of the corporate lab that it was even possible to think about patenting as an act that could manifest itself "offensively" or "defensively," or as "thickets" or "pools."[12] Patenting had become big business, with all its trappings of competition and maximizing profit.

Corporate Lab Culture

In the earlier decades of the Industrial Revolution, many industrialists were bewildered by the patent system. Henry Bessemer, the visionary inventor of the refined steelmaking process that would lead to the use of structural steel, admitted in 1834 that he "knew nothing of patents or patent law" even after moving operations to what was at the time the patent capital of the world, London.[13] Whether it was through ignorance or indifference, he sold off his early unprotected inventions for paltry sums. By the end of his career, however, Bessemer and his company had initiated many patents in the arena of iron smelting and steelmaking. Yet even with certain inventions, such as bronze powder, Bessemer deferred to the preindustrial tradition of craft secrecy, something that was only possible because of what he described as a "secure factory": one that employed only trusted relatives in key positions.[14]

The British-American inventor Elihu Thomson, founder of the Thomson-Houston Electric Company, a progenitor of General Electric, was an archetype of the successful nineteenth-century inventor. Thomson combined fierce intelligence (in electrical engineering) with notable business acumen. Unlike Caro, however, he did not hold the belief that corporate research had to maintain a critical autonomy from business practice. Like Alfred Nobel and his electrical colleagues Thomas Edison and Nikola Tesla, Thomson saw research and business as intrinsically intertwined and gladly let his inventions be shaped by the input of his financial backers and managers.[15] Thomson's skepticism about the corporate laboratory model likely also hinged on his firm belief that technological invention was less about incubating innovation qua innovation and more about the marriage of scientific values and craft knowledge.[16] Even later, at GE, it was common for the firm to simply buy outside patents and develop products from those.[17]

The patent system—which many early corporate scientists found to have, at best, a kind of symbolic value—was truly mastered by Heinrich Caro. He taught himself, and in turn his subordinates, how to exploit and manipulate the system, and he did so without ever becoming what one might today call a patent troll. He knew that a patent needed a sound scientific basis and that an invention needed to have robust argumentation around it. What made him the grandfather of corporate patent scientists was that he did not think of patenting as a game that needed to be genteel. His 1869 patent for a dye for the color known as palatine orange was a kind of ex nihilo invention in that it patented something as theoretically ephemeral as a color, which defied all of the day's expectations about the very nature of a patent. One could not produce, for example, line drawings or models of palatine orange; only the chemical formula could be written down.[18] Caro had the kind of verve for patenting that management knew would carry them forward, and, as one author has put it, "it was therefore only too understandable that in the decades that followed [the palatine orange patent], the BASF management did everything to keep this man in their business as a guarantor of scientific progress."[19]

BASF's world-class dye research, and the products that resulted from it, had a multitude of real-world applications. One of the most prominent was paint. BASF's tar paints, derived from their patents, began reshaping the chromatic experience of the city and of architecture. Colors became more vivid, more consistent, and more colorfast. Precise tones and hues could be produced with greater accuracy. City dwellers in Berlin, Cologne, Frankfurt, Düsseldorf, and all over Prussia surely could not have missed the prismatic new environment that emerged in the 1870s and 1880s because of BASF's color patents. This new kaleidoscope of urban and architectural experience was not, however, without ramifications for the industry. Up to the 1860s, BASF and other German paint companies had freely borrowed the chemical compositions of foreign products with little concern about any kind of legal infringement. However, as Caro led the way toward actually creating a regime of ownership around certain colors, BASF could no longer play fast and loose with the colors of the rainbow. In the initial years of dye patenting, BASF adopted the risky policy of patenting their dyes abroad, especially in England and France and later the United States, but not in the newly unified Germany, for fear of making their previous infringements plainly obvious.[20] They also actively courted foreign markets that did not have patent systems, demonstrating dyes in trips to large markets like China as early as the 1880s, where indanthrone dyes, which had been patented by René Bohn, became very popular.[21]

The Corporate Lab

Trade and special interest groups were making this kind of obfuscation more difficult, however, particularly for large and powerful companies like BASF. The German Chemical Society and the German Patent Protection Association were just two of the organizations that came to be seen as unofficial (and sometimes official) watchdogs of German big business, buoyed as they were by the regulatory progress made by the German Patent Act of 1877.[22] These regulatory reforms, at first probably somewhat threatening to BASF, eventually only buttressed industrial monopolies. As BASF's corporate biographers recount: "[T]he competition was spurred on by the legal requirements to develop other processes or substitute goods (circumvention patents) with which they could break through the monopoly that had emerged. In any case, in contrast to France and England, the new patent legislation formed one of the essential institutional prerequisites for the long-term intensive research work in the German chemical industry and contributed to the establishment of the de facto world monopoly of the German tar paint companies."[23] Switzerland, which at the time eschewed having a national patent system, threw another kind of wrench into the equation as they could compete with German corporations with little hindrance. That was, until Swiss jurists decided it behooved them to institute a patent system in 1888.[24]

The aggregate effect of these new laws with their international ramifications and the rise of interest groups raised the stakes of everything the dye world did in the decades that followed. On the one hand, a well-strategized and well-executed patent strategy for a given color, for example, could generate economic bounty by way of a monopoly for years to come. But the risk of not carefully following the letter of the law was also formidable and could result in severe financial losses. Small armies of in-house lawyers and administrators eventually joined the ranks of research scientists at companies like BASF to concertedly navigate these challenges.[25] BASF's patent activity peaked in 1913 at 142 patents, twice as many as it held 13 years prior.[26]

By the turn of the twentieth century, despite the avowed success of corporate research, many of BASF's scientists had grown uneasy with their role in German corporate culture. The increased routinization of industrial research, the growth of big industry, and the rise of trade unions coalesced to create an environment that was hostile to their standing as scientists and their opportunities for advancement.[27] Beginning around 1904, in Berlin, a hitherto sleepy consortium of technical workers, the League of Technical and Industrial Officials (Bund der technisch-industriellen Beamten), began to publicly advocate for reforms to the commercial code and to patent law. Their advocacy rested on a litany of complaints from German research scientists, including unusually low salaries and claims of "intellectual sterility."[28] Addressing this directly as a patent problem and not merely as a labor problem carried their concerns into the domain of the public good, and, alarmed by the political and cultural scope of their grievances, companies like BASF were pressured into rapid change. This included increasing wages for research scientists and an ethos of intellectual research freedom that edged closer to the model of the academy. The patent frenzy and the labor restlessness at BASF came to a head on September 21, 1921, when approximately four and a half thousand tons of ammonium nitrate fertilizer and ammonium sulfate in Oppau (now Ludwigshafen) exploded in a silo tower. The blast killed approximately 600 BASF employees and injured over three times as many. The blast could be heard almost 200 miles away in France.[29] Many feared that the voraciousness with which BASF pursued its own development had come at the price of its employees' safety and contentment.

By the 1920s, companies working in the construction sector were increasingly marketing their patented products directly to architects. Keramik, a product of the A. C. Horn Company, is one example. Keramik was a color penetrant for concrete surfaces, intended to make rough concrete surfaces appear warmer and also more customizable, a stunning example of how one type of patented product buoyed the success of another (concrete), in this case working together to make the latter more palatable to the mass market. Keramik's marketing materials stressed the company's direct relationships with architects by featuring promotional blurbs from well-known architects voicing support for the product and long lists of important buildings where the product had been deployed to rave reviews, including buildings at Barnard College, the Carnegie libraries, the Museum of Natural History in New York, and the United States Capitol building.[30]

In order to accelerate patent activity, as BASF had done to good effect, corporations first needed to clarify protocols and best practices for doing so. Through World War II, corporate patent research on both sides of the Atlantic, while certainly common, was not "designed" in any particular way. Corporations tended

View of the aftermath of the ammonium nitrate explosion at BASF's facilities at Oppau, 1920.

to view their laboratories as experimental places guided by the research convictions of their leading scientists, such as Heinrich Caro. The specter of nuclear war, however, changed all of that, and corporate patent research became a de facto duty of Cold War corporate culture. The popularity of the most common alternative practice—firms buying turnkey technology patents developed by independent inventors—began to dwindle.[31] By the 1960s, most major corporate research labs had organized themselves into well-oiled research machines that were less driven by the research agendas of their lead scientists than by corporate quotas, annual bonuses tied to patent activity, and the ethos of a corporate "patent strategy" that led to the practice of offensive and defensive patenting. Nowhere is this transformation more vivid than in the patent protocols of DuPont, which famously sought to make "better living through chemistry" for the masses.

The 1960s, in particular, were a time of aggressive and ambitious patenting for elements in the construction sector. DuPont articulated a standalone project called the Building Products Venture which adopted a business strategy of developing a broad array of products for the construction sector that were both cheap and required little maintenance.[32] These included everything from roofing and siding to shutters and trim, along with a line of interior products. The goal was to have somewhere between $500 and $2,000 worth of DuPont products in every new house in the United States. The large-scale production goals of the Building Products Venture were, however, only realized for one product: Corian.

DuPont's patent regime was well summarized in a 1964 employee handbook entitled "Patent Policies and Procedures," authored by G. B. Malone of the Central Research Department at DuPont's Experimental Station: "Those choosing the career of an industrial research scientist," Malone opens, "should recognize that patents are an integral element of practically all new or expanding industries. A working knowledge of the nature of a patent, of how patents are obtained, and of how they serve the Company's interests is, therefore, an essential and inescapable part of our chosen profession."[33] Malone asserted that patents were not merely a goal of research, but also an obligation that is systemic to research: DuPont's research personnel had to be "relied upon primarily to initiate patent action promptly on their discoveries."[34] They needed to remain "technically aggressive" until a patent was obtained. Malone even imbued the lofty rhetoric of his opening

View of the indigo laboratory at BASF headquarters in Ludwigshafen, c. 1914.

argument with a nationalist tone: "In the modern sense, a patent may be looked upon as a contract between the inventor or his assignee (in our case, the DuPont Company) and the Federal Government, representing the people."[35] Another booster at DuPont exhorted corporate research to pave the path to national autonomy: "Thanks to research, America is no longer wholly dependent upon Chile for nitrates, upon Europe for synthetic dyes, upon the East Indies for rubber, upon Siberia for bristles, upon Formosa for camphor, upon Japan for silk."[36]

In practice, department heads were tasked with the general responsibility of shepherding all potential patents under their jurisdiction to DuPont's legal department. The department heads were typically so consumed with patent traffic that they themselves did barely any laboratory research. The department heads were also in charge of ensuring that the research scientists were aware of the "Employees Agreement," which prescribed that all IP generated through the course of one's employment was both the property of the company and secret until published.[37]

Over the course of 34 pages in the employee handbook, Malone discharged a list of key clusters of patenting expertise that all DuPont scientific staff needed to have. The list functioned as a sort of crib sheet with which department heads could evaluate a staff member's patent competency, including a working knowledge of the purpose and history of patents, a strong familiarity with prior art in their subspecialty—which, of course, included remaining up to date vis-à-vis patent bulletins or other publications—an understanding of the importance of teamwork, an understanding of the kinds of experimental procedures that support patent applications, a need to be both quick and flexible, the importance of documentation, and a "judicious allocation of effort," among other strengths.[38] Once the department head (or heads in the case of a cross-departmental invention) and the legal team arrived at a consensus regarding an invention's patentability, a tentative application was drafted and presented to an internal patent committee. The committee often included the department head involved in the application proposal, which would necessitate his or her recusal. If the patent committee approved an invention, the application docket would fall back to the desk of the original scientist, who would then be responsible for finalizing all elements of the disclosure: data, phraseology, and the requisite array of descriptive patent drawings if the invention was a physical object.[39]

The Corporate Lab

Patents may have been at the core of corporate R&D objectives, but companies like DuPont also had to resolve a number of slippery questions about proprietorship and governance that came with building research environments tied to corporations.[40] In what manner, besides patents, would the company publish its scientific findings? How would the company manage the protection of hard-won and expensive IP while also providing the public recognition that corporate scientists desired in spite of the Employees Agreement? As research scientists proved themselves, what retention and remuneration strategies could the company deploy to keep them? What level of research freedom was appropriate? Would corporate research have any correlation to research occurring in the academy? DuPont, like BASF and 3M, had come to believe that the answers to such questions could not be monolithic and were largely contingent on the nature of the work being done and the personalities involved. Chemical laboratories differed greatly from materials laboratories, for example, and patent protocols needed to be tailored. From the late nineteenth century onward, this belief led to the steady decentralization of corporate research. At DuPont, this decentralization had a direct impact on manufacturing. The company divided its manufacturing operations along the same lines that divided research departments, tethering the framework for research directly to that of manufacturing and, in turn, sales.[41]

Despite the corporation's proprietary regime, the rhetoric around centralization centered on a narrative of sanctioning the individual intellectual freedom of research scientists. Hugh G. Bryce, vice president of research at 3M in the 1970s, quipped: "Management of research at 3M is not a very difficult thing, because it is mostly taking good people, stimulating them to get involved with something that challenges them and giving some guidance as to which way is the most likely to meet their goal of achieving whatever success means to them."[42] Underneath the breeziness of Bryce's sentiment, presaging the rhetoric of freewheeling flexibility of the twenty-first-century tech world, there is an anxiety about productivity. With research labs now firmly part of big corporate culture, managers needed to be certain that corporate jobs were as attractive to scientists as jobs in the academy (if not more so), where scientists had *always* had autonomy over their own research agendas. This explains why, like many other corporate research labs, 3M allowed its research scientists to commit a certain percentage of their time to research of their own choice (in the case of 3M it was 15 percent).[43] This kind of openness buoyed the rhetoric of autonomy, despite being somewhat misleading. Research scientists constantly had their "free" research audited by one of the firm's several dozen patent attorneys.[44] As Hounshell and Smith note, "In response to requests to publish a paper, [DuPont] research managers frequently asserted that award of a patent should fulfill a ... desire for recognition ... to the field. The problem with this argument ... was that academic chemists did not follow patent literature. In addition, the names that appeared on a patent reflected lawyers' decisions based on patent law rather than the scientists who had contributed most of the work."[45] At 3M, the rhetoric of autonomy stood in direct conflict with the rhetoric of collectivity, with the firm's technologies commonly touted as being the collective property of the firm's employees and the fruit of their friendly collaborations.[46] In fact, real autonomy was driven in large part by the firm's geographical expansion. With labs and factories in England, Italy, Germany, Brazil, and Japan by the close of the 1970s, such autonomy was a kind of de facto necessity in response to different cultural contexts.[47]

Products

For most of the nineteenth century, patenting had mainly been the province of individual inventors. Successful firms tended to emerge through the manufacture of goods utilizing the technology of the most revolutionary patents, not by originating the patents themselves. Indeed, for the most part, invention and manufacturing remained largely distinct from one another. Patents represented a kind of reservoir of opportunity in which companies could fish, making what they would with the catch. The line between the fisherman and the reservoir first began to seriously blur in two corners of the industrial sector. Counterintuitively, the first was the segment of industry where manufacturing was not the core operation. The second was the niche sector of threshold devices: doors, windows, exhaust systems, and chimneys.

Laurent Stalder has drawn attention to a superb example of the latter: the Van Kannel Revolving Door Company.[48] This is the story of a Philadelphia mechanical engineer, Theophilius van Kannel, who had set his eyes on resolving the problem of drafts

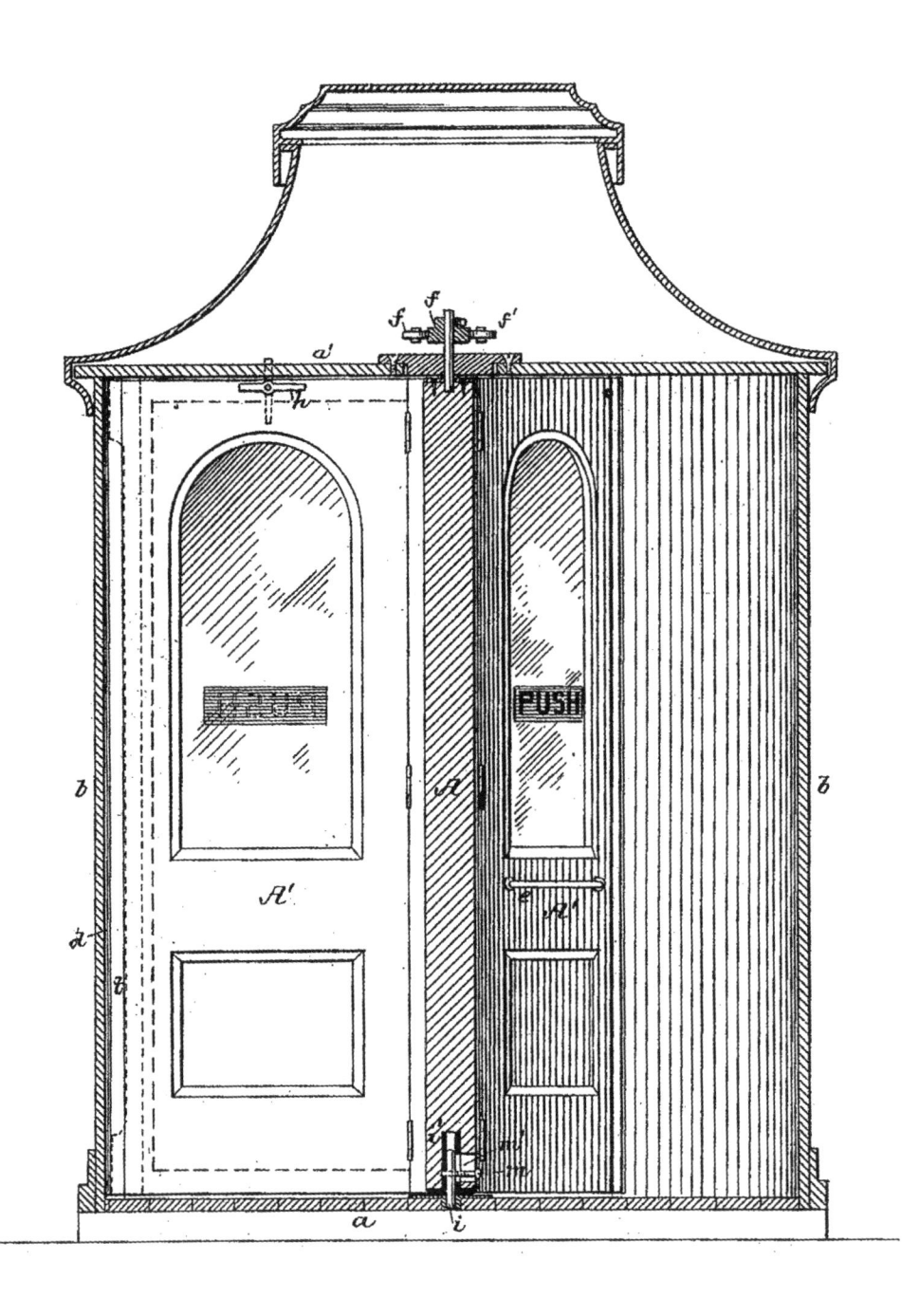

Theophilius van Kannel, Storm-Door Structure, US Patent 387,571, filed February 10, 1888.

entering buildings when doors were opened. This was an especially acute concern for the proprietors of high-traffic businesses like restaurants and department stores, where the constant opening of doors could create a kind of permanent draft at the primary points of ingress and egress, detracting from the interior spaces that were intended to make a good first impression on the consumer. Double-doored vestibules were, theoretically, an option, but they created pedestrian logjams and didn't necessarily even fully prevent drafts because the two doors would often be open at the same time. Van Kannel was able to locate only one example of prior art that he considered to be highly significant (an 1881 German patent for a draft-free door by H. Bockhaker).[49] As Stalder suggests, the essential inventiveness of the circular revolving door, which van Kannel would patent under the aegis of his namesake company, was that it could be both always open and always closed. Indeed, this was the company's motto, one which today would likely also be trademarked. Van Kannel went on to patent the door in the United Kingdom as well.

Van Kannel's revolving door patent is the rather rare example of an invention that is patented for one reason but reveals another efficiency later in its life. That second efficiency is the environmental benefit of the revolving door in preventing regular heat loss, mitigating the environmental and economic inefficiencies of the new building typologies that deployed it, particularly the high-rise building.[50] Van Kannel was also extremely clever in prolonging the power of his initial patent as long as possible, a business move that afforded him a large monopoly over the breathtaking demand for this new type of door. Van Kannel did this by introducing, over long periods of time, significant enough advances to the design of the door to warrant patenting that nevertheless kept the revolving door's essential identity intact, as with his 1912 patent for a revolving door assisted by an electric motor, which made it less difficult—indeed effortless—to rotate doors of heavy wood and brass. Van Kannel's invention of this architectural component was also one that, like the elevator or escalator, fundamentally dictated the architectural form that followed, not the other way around. This positioned the patent as a kind of anticipative spur for further formal change, an active stimulus to, rather than from, the architectural profession.

Construction firms had always been in the business of problem solving, to be sure, but not in the business of invention per se. Yet as innovations coming from firms like Dyckerhoff & Widmann AG (DYWIDAG) would prove, the twentieth century was ushering in an entirely new type of construction firm. DYWIDAG had its origins in the Karlsruhe-based Lan & Co., founded in 1865 by the cement pioneer Wilhelm Gustav Dyckerhoff.[51] Wilhelm's son, Eugen, and Eugen's father-in-law, Gottlieb Widmann, expanded the firm over the ensuing decades and are credited with some of the most impressive concrete structures from late imperial Germany, such as Max Berg's Centennial Hall in Breslau.[52] DYWIDAG was also one of the earliest German companies to go truly global, with factories for precast concrete elements in the Ivory Coast, Hungary, Qatar, Togo, Senegal, Guinea, Gabon, Burkina Faso, and Cameroon. By World War I, the firm's monobloc prestressed concrete sleepers could be found in most European countries as well as in Egypt, South Africa, Mexico, Venezuela, Japan, Pakistan, Turkey, and India.[53] DYWIDAG was the German exemplar of the international trend toward intensified managerial influence on concrete construction, mirrored by Frank Gilbreth and Ernest Ransome in the United States, the former through using a system of flexible troops of workers known as "gangs" and the latter through the patenting and standardization of concrete building parts.[54]

The company's most important contribution to construction technology, however, was not the creation of its own patents but the deployment of an existing patent that DYWIDAG obtained through a licensing agreement. This was a patent for a seemingly simple concrete pile named "Strauss" after its original inventor, Anton Strauss, a Ukrainian engineer who patented the system in United States in 1909. The name of the patent, Betonpfähle Strauss (Strauss concrete pile), is something of a misnomer, as the innovation of the design rests more in its method of construction than its actual form.[55] The innovative method of making the Strauss piles involves a multistep process. A thin-walled iron guide tube is lowered into the earth through a conventional drilling method, with the drill cuts made either to the depth of the load-bearing soil or, if necessary, to a certain depth within the load-bearing soil. Concrete is then introduced in layers by means of a specially designed bucket and tamped vigorously with a heavy rammer that fills the pipe interior cross-sectionally while slowly pulling up the pipe at the same time. This causes the concrete to overflow the lower edges of the casing and displace the soil surrounding the mouth, which compresses the soil firmly under the tamped matter. The

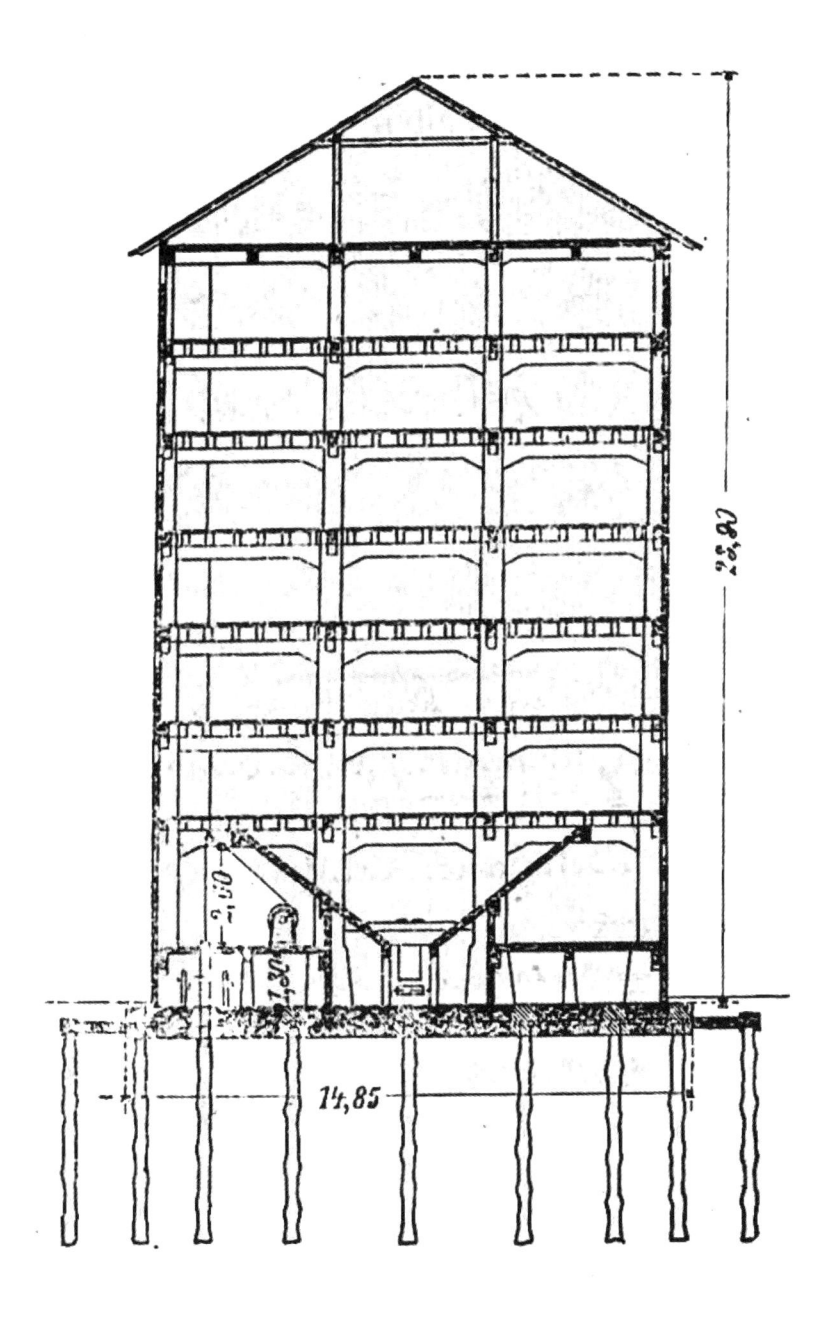

Unknown, drawing depicting the use of the Betonpfähle Strauss patent system of concrete piles by the construction firm Dyckerhoff & Widmann in an informational brochure produced by the company, 1908. The construction depicted is a silo for a mill in the town of Rüningen.

gradual winding up of the pipe according to the amount of concrete pressed out is done in such a way that the pipe is never completely emptied of concrete. A kind of differential pile is created, the cross-section of which changes from layer to layer depending on the different soil conditions, with the surface consequently showing a great deal of unevenness and roughness. This differentiation of the pile's cross-sections also has the advantage of generating an even compaction of subsoil. The concrete is effectively tamped down the entire depth of the pile in an intensive way until a compression of the concrete can no longer be measured. The pile heads are usually joined together by iron-reinforced plates whose inserts reach down into the upper parts of the piles, thereby forming the actual foundation body for the masonry which it is now prepared to support. By 1920, DYWIDAG had laid nearly a half million meters of the system globally in numerous prominent projects, including the Leipzig central railway station, AEG's power station in Concepción, Chile, and the Luitpold bridge in Augsburg, among others.[56]

Although DYWIDAG had not invented the Strauss system, they were the most successful company in popularizing it, and by 1908, just months before Strauss's patent was approved in the United States, DYWIDAG was proclaiming its success in the system's implementation to the construction trade.[57] In a speech delivered to the Society of Saxon Engineers and Architects in Dresden in March of that year, Willy Gehler, DYWIDAG's chief engineer, went to great lengths to place his firm's importance in the history of construction, with verbiage that evoked a grand narrative of progress in architecture and engineering from antiquity through to the present:

> As archaeologists tell us, even the primitive peoples of the Stone Age built pile dwellings in a similar way to the peoples of a lower cultural level, such as the aborigines of the South Sea Islands. In order to find protection from hostile neighboring tribes and wild animals, shallow areas of ponds and inland lakes were preferred, although the unfavorable building ground caused considerable additional work when building the dwelling. Pile dwellings can also often be found in historical antiquity, especially in the age of Roman traffic, which took over the technical knowledge from earlier cultural periods and gained a wealth of experience through

numerous applications. Pile foundations were chosen as a foundation method, especially for bridge construction, with stone blocks placed on top of them and wooden gutters. When the Cestius Bridge in Rome was demolished, the foundations were found to be of a kind of concrete, with four rows of oak piles. The river piles of the Roman stone bridges near Mainz, some of the most important constructions from the heyday of Roman engineering, were founded on 12-meter-long oak posts, as later finds confirm. The execution of a Roman pile foundation is described by Vitruvius in his work *De architectura* as follows: "One cannot get solid ground alone, and if the ground underneath is loose and muddy, one digs it out and empties it, drives in burnt alder, olive tree or oak piles, connects it to sleepers that have been laid close together by machines, fills the space between the piles with coal, and then firmly builds up the foundation." Finally, the use of piles for the Romans' wooden bridges is generally known, as in the case of Caesar's famous Rhine bridge, which was completed in 10 days in 55 BC.[58]

The pile was not the only familiarly modest thing in the world of construction to undergo a transformation through IP. 3M developed a wide array of auxiliary patents for the construction sector: an algae block copper roofing granule system, air filtration systems, urethane foam for wall protection, and cellophane tape for mending ceiling plaster.[59] 3M's most significant inventive contribution to the world of building, however, is undoubtedly the humble piece of sandpaper, crucial to virtually every constructive endeavor involving wood in the modern period and the basis of several 3M patents. Sandpaper is also critical to 3M's own corporate mythology, not least because it was the product that most clearly made the company profitable beginning in about 1914.[60] Paul Carpenter, a Chicago patent attorney, was key in encouraging and prosecuting 3M's early patents for its "WetDry" sandpaper, which, as the name implies, works when either wet or dry. Sandpaper has thus been credited with alerting management to the importance of patenting,[61] and with jumpstarting the creation of 3M's own in-house research laboratory.

The economic success that sandpaper afforded 3M had nevertheless taken a long time to arrive. In the years leading up to 1914, the company's garnet sandpaper was being returned to 3M

The Corporate Lab   145

View of packaging for a set of 3M sandpaper sheet packets, c. 1905.

en masse, the prevailing complaint being that the sandpaper's garnet, the abrasive, would peel off of the backing as soon as customers began using the product in their shops. As corporate legend goes, one evening a janitor spotted an oily film on the surface of a mop bucket. Resting at the bottom of the pail was a small amount of crushed garnet that had been mopped off the floor of one of 3M's production facilities. Eventually, word got back to 3M's managers, who were stumped by the presence of the oil. They did know that a piece of sandpaper contaminated with oil would force the garnet to separate from the glue that attached it to its backing, even after only light use, but no one had any idea how the garnet, which had been imported from Spain, had come to be contaminated with oil.

3M's managers ordered an investigation into the provenance of the Spanish garnet, which led them to the news that months earlier, the cargo ship transporting the garnet across the Atlantic ran into a storm that caused the principal cargo, several thousand bottles of olive oil, to pitch. The bottles broke and contaminated the sacks of garnet stowed next to them.[62] No one reported the incident because no one thought it was of any consequence, and by the time the garnet had reached Minnesota, it had no outward signs of oiliness. This left 3M with 200 sacks of oil-spoiled garnet, a shipment for which they could not collect losses, and they decided that they had no choice but to keep using the garnet. Orson Hull, the factory superintendent, took it upon himself to figure out a way to continue making sandpaper with the oily garnet, making good on the adage that necessity breeds invention. Hull determined that when washed and then heated in thin layers, the garnet could be restored to a state of usefulness. Although the process was tedious and time-consuming, it worked. Hull insisted that the company should continue using the small room he had used to clean and heat the garnet as a dedicated space for internal experiments. From this glorified closet, 3M's research lab was born, and 3M hired its first full-time technical employee, Bill Vievering, in 1916.

Another sector that made unexpected contributions to the building industry through patents was aviation. Junkers Flugzeug- und Motorenwerke AG, based in Dessau, had a short-lived metal house production operation whose fabrication strategy piggybacked on the company's storied aircraft production lines.[63] The latter had to cease operations as a result of World War I: in the ten years following the war, the occupying powers prohibited German aircraft manufacturers from resuming the production of planes. Another manufacturer, Messerschmitt AG based in Augsburg, knew it needed to adapt its facilities to other industry demands to remain solvent.⁶⁴ As would have been clear from the images of flattened cities, Germany, like so many other places torn apart by war, was in desperate need of housing. In the case of Messerschmitt, this meant starting a prefabricated home production line, like that of Junkers, from scratch. Unlike DYWIDAG and the construction industry, patenting had been an august tradition in the aviation industry, due in part to the fact that the industry advanced so rapidly and was extremely competitive. Both firms carried this penchant for patenting over into their house production period, albeit in very different ways.

The Junkers company's connection to construction and architecture hinged in large part on its location in Dessau. In 1925, Hugo Junkers, the company's founder, played a pivotal role in the relocation of the Bauhaus from Weimar to Dessau (indeed the Bauhaus architecture department planned the company's workers' housing estate in 1932).⁶⁵ Junkers embodied the same beliefs as the school's progressive designers: that design and industry both benefited from proximity and mutual exchange. Junkers and his associates in the firm's dedicated house construction department filed patents for individual elements forming the Metal House Construction (*Metallhausbau*) system.⁶⁶ Corporate records reveal a distinct strategy of patenting parts of the house rather than the whole system, with many parts, such as a "wall with installation element," designated as "optional." From a design perspective, this made the house more customizable. From a legal perspective, this functioned to reinforce the protection of the system in its form as a collection of parts and as a system. Elements of furniture, like chairs, desks, and beds, had equal footing with walls, doors, and windows—a parity that was as much conceptual as it was legally strategic.⁶⁷

The very mobile nature of prefabricated housing necessitated multinational patent protection to make good on the promise of architectural mobility. Junkers conducted remarkable international prior art research that allowed them to patent the system in Germany, the United States, France, and Switzerland.⁶⁸ Indeed, Junkers extended their standard procedure of what we might call "matrix patenting" into housing, a system where patentable

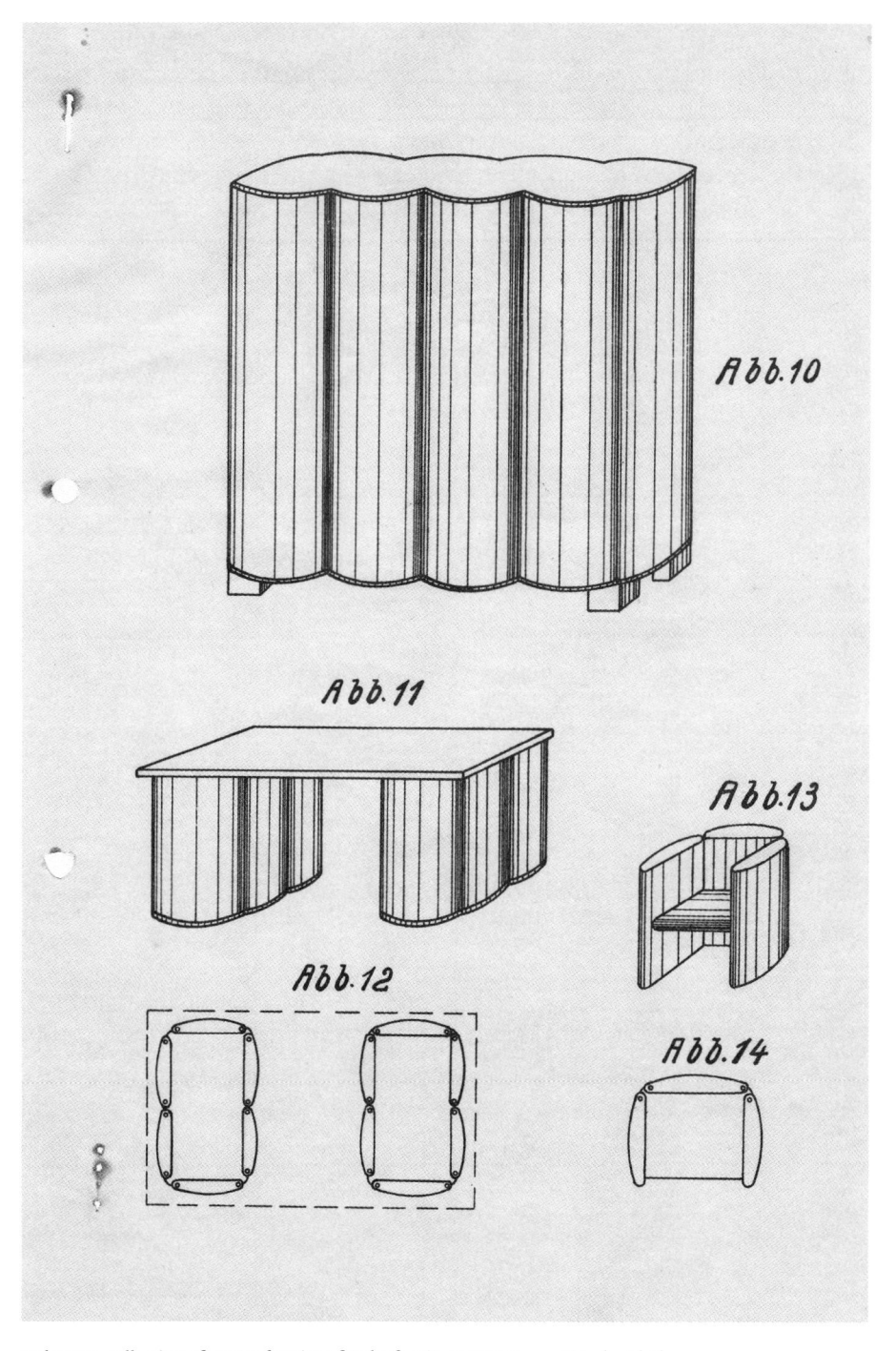

Unknown, collection of patent drawings for the furniture system associated with the Junkers *Metallhaus* system, c. 1931. Also color plate 1.

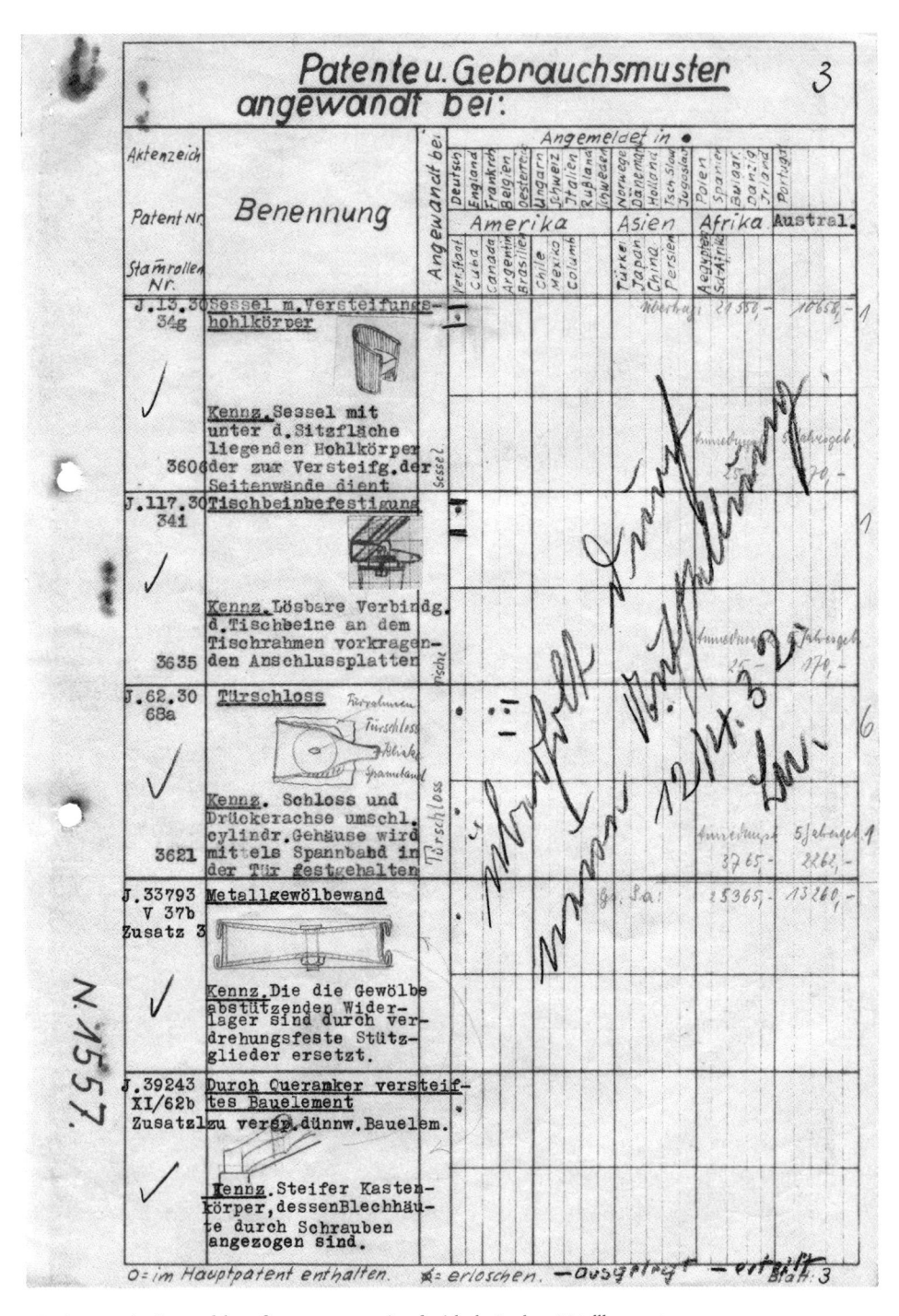

"Matrix patenting" spreadsheet for patents associated with the Junkers *Metallhaus* system, c. 1931.

elements of invention were subdivided and charted along the *y*-axis of a table, while the wide range of countries for potential patents were charted along the *x*-axis. For the Metal House Construction system, Junkers seemed particularly concerned about impinging on the IP rights of two precedents identified in the firm's attorney's prior art search.[69] The first was the Copper House System developed by the Hirsch Kupfer- und Messingwerke in Eberswalde in the state of Brandenburg, developed predominantly for German-Jewish settlers in Palestine by the architect Robert Krafft and the engineer Friedrich Förster and patented a few years earlier.[70] The other precedent was a design by an architect named Reichel, who had offered the design as a freelancer for the Berlin construction firm Kletzin GmbH.[71] Junkers's consultant on the prior art search, Hans Drescher, a bureaucrat for the Berlin city government who apparently moonlighted as a corporate consultant, related why the Kletzin case made assessing infringement risks in the construction sector so difficult:

> As can be seen from the investigations, the property rights of the companies are often not registered in the names of the companies themselves, but in the names of the managing directors or the subsidiaries, which in turn have different names. For further research orders it is therefore advisable, in order not to make the research unnecessarily time-consuming and expensive, to give me as much detailed information as possible about the companies to be investigated and the probable names of the patent holders. . . . If there are prospectuses or company catalogues, it is advisable to send these with them, along with any information that is of any interest, because then I can find more information and use this information for research.[72]

The standard array of countries in the matrix of patents were the countries where the current state of patent laws were monitored and where the benefit of patents would, at the very least, be explored: Belgium, Spain, Austria, England, France, the United States, Russia, Italy, China, Brazil, the Netherlands, Czechoslovakia, Yugoslavia, Turkey, Greece, Romania, and Japan.[73] This necessitated legal personnel capable of reading all of the languages spoken in these countries and familiar with how to conduct, or at least commission, the necessary prior art research. A 1932 report explains the three main criteria to be evaluated in determining the relative worth of seeking IP rights for the metal house system abroad: "(1) To form an opinion about the protection actually achieved for what has been created so far; (2) to gain clues as to what further protection options exist for what has been created so far; (3) to clarify the fundamental question, based on the experience gained from hindsight, of how we can proceed in the future in order to increase the prospect of achieving economically effective protection, especially in the early stages of development of what is initially a self-contained research area like 'house building.'"[74]

With its ascension to power in 1933, the National Socialist administration asked Junkers to assist them with rearmament efforts. When the company's founder Hugo Junkers declined to assist the new regime in this regard, the National Socialists went after him where they knew he was most vulnerable: his patents. Because he was unwilling to assist Hitler's regime, the authorities demanded ownership of all of the company's patents as well as its market shares. Even under the threat of imprisonment on the grounds of high treason, Junkers refused to comply. His punishment was house arrest in 1934, from which he continued to fight for as much of his company's IP as he could retain. The stress and grief this caused him likely contributed to his death in 1935, which prompted his wife, Therese Junkers, to ultimately cede full control of the company to the Third Reich. The Reich utilized Junkers's patents and facilities to create some of the most impressive, as well as destructive, aircraft deployed in World War II. The full industrial creativity that could have flourished in Dessau, particularly in Junkers's alliance with the Bauhaus, is something we will never know.

Hugo Junkers was in many ways the verso of Willy Messerschmitt, the founder and director of the aircraft company bearing his name. Whereas Junkers had resisted Nazi collaboration, Messerschmitt sanctioned his firm's complicity with the Reich both before and during the war. This included the use of slave labor, which brought Messerschmitt two years in prison at the denazification proceedings of 1948.[75] By 1950, however, he was already back at the helm of his company, directing the recalibration of the assembly lines in Augsburg for his newfangled housing system Messerschmitt Bauart (as well as the production of sewing machines and small cars known as Messerschmitt Kabinenrollers).[76] *Time*

View of a house under construction using the concrete housing system by Messerschmitt, c. 1949. Property of the author, acquired on eBay.

magazine published a pithy and unduly innocent summary of Messerschmitt's new pursuit in 1949:

> He read dozens of books on housing, hired 20 architects, put them to work on the same drawing boards that once held his aircraft blueprints. Three months ago, the first mass-produced parts of his houses began rolling off the assembly lines. When the buildings were put up in Munich, Germans gasped with joy and wonder. The Messerschmitt houses were almost as ingenious as the Messerschmitt planes. The houses made of steel and "foam concrete" went up singly or in six-family units. His one-family houses sold for $14,000 Deutsche Marks ($4,200 at the official exchange rate). Within a few days Messerschmitt was swamped with orders. In his old Augsburg aircraft plant, Messerschmitt last week was busy buying new machinery and adding to his original staff. The bustle reminded him of the wartime plane-building days. He is currently producing 40 prefabricated houses a month. "This is only the beginning," he said. By September, he hopes to have stepped up his production to 300. Says Messerschmitt: "You must conform to the needs of the times. I would rather be building planes. When the opportunity presents itself, I will."[77]

It seems highly likely that Messerschmitt was inspired by the legal strategy adopted by Junkers, as he, too, took out patents for discrete elements of his system that could also be used independently of it. These included patents for prestressed concrete beams, windows, insulating structural panels, and roofing systems, among other things.[78] Messerschmitt's new commitment to housing, born as it was out of necessity, was something he drummed up as part of a new attention the firm would pay to social issues and perhaps the slightest gesture of regret for his sins during the war. We know this from an unpublished typescript by Messerschmitt about the problem of social housing in the immediate postwar period, entitled "A Way toward a Solution for Social Housing [Ein Weg zur Lösung des sozialen Wohnungsbaues]."[79] Messerschmitt's approach to the postwar housing problem in this text is essentially econometric, noting that before the war a cubic meter of enclosed space typically cost 28 marks, compared to 45 marks after the war.[80] Messerschmitt's assessment of the housing crisis extends not only to the volatility of the economy but also to Germany's new reality as an occupied country:

> All offers that are lower [than 45 marks per cubic foot] today in times of economic hardship are competitive prices and cannot be maintained in the long run. They come at the expense of healthy profits and at the expense of depreciation. In a healthy economy, this increase in construction costs would of course lead to an increase in rents. Since a bad economy naturally leads to an increase in rents. However, since the rent increase would be a massive intervention in the social fabric of our people, this step was not chosen, rather the rents were kept low through public subsidies. But this procedure is very dangerous. The occupying powers have already declared that they prohibit subsidies, with the aim of being able to compete on the world market. It is to be feared that one day the occupying powers will forbid subsidies for lower rents and will instead demand that rents be adjusted to reflect the true cost of construction. This means a great burden for the workers and leads to wage wars and thus to an increase in the price of our products, which we have to sell on the world market and which are becoming increasingly difficult to sell. The path that the new technology has taken on the basis of the investigations is not new in itself. Prefabricated houses are being built everywhere. We know, however, that for lack of space and for cost reasons we cannot limit ending the housing shortage to single houses. Rather, we have to build multistory buildings to make housing cheaper. The new technology has solved this problem of how to build multistory houses with prefabricated parts.[81]

Messerschmitt, ever aware that his recent past could potentially blemish his corporate success, frequently recruited character witnesses and employed rosy social rhetoric to bolster his product. For example, he recruited Sep Ruf, a well-known architect, to write favorably about the building system for publicity materials and in those same materials commonly urged that the buildings could easily be erected "with the support of our foreign friends" (referring to the "guest workers"—mainly Turks and Italians—who were pouring into the county to gain work in the vast postwar reconstruction efforts).[82]

With the new practices of the postwar housing industry came new materials, which brings us back to the other side of the Atlantic. DuPont's single greatest contribution to the construction sector was, without a doubt, Tyvek. DuPont researchers first developed Tyvek's progenitors under the auspices of a larger research effort around nonwoven fabrics. Tyvek's origin, as such, is attributed to a DuPont research chemist by the name of Herbert Blades. In the fall of 1956, Blades was conducting experiments with linear polyethylene for research relating to his specialty—spinneret geometry—when he made two key technical advances that, when combined, allowed him to prepare a continuously strong and uniform flash-spun yarn.[83] Blades was helped along the way by J. R. White, R. Dean Anderson, and Louise Jones, all DuPont scientists.[84] When spun randomly from a surface with several small holes, the yarn facilitated the creation of the sheet product that would come to be known as Tyvek. Blades and DuPont, as per protocol, immediately pursued patent protection for the discovery (US Patent 3,081,591) and for all of the refinements that would emerge using the flash spinning process over the ensuing years, and were successful in doing so each and every time. Refinements in Tyvek included using photomicrography to study composition and spinning behavior, refining appropriate coating materials, and perfecting Tyvek's key metric: tear strength. Blades and other male scientists could count on annual bonuses in the ballpark of 30 percent when developments in Tyvek merited further patents.[85] The gender caveat here is because Louise Jones, a mid-ranking research scientist working with Blades and one of the company's few female research scientists at the time, was conspicuously passed over for these annual bonuses. "In my opinion," management wrote, "[we] do not have the evidence to support an 'A' bonus case for . . . Louise Jones . . . in connection with the flash spinning patent."[86] Her contributions, they argued, did not go "above and beyond" what would be expected for work to which she was assigned. However, it was actually Jones who discovered that linear polyethylene (LPE) could be made into high-strength fibers, a key foundation for the entire Tyvek endeavor.[87] What is more, Jones's name is conspicuously absent from the original 1963 patent.[88]

Well into the 1970s, Tyvek remained an invention without an audience. Applications for Tyvek did ultimately emerge: in hazard

Unknown, photomicrographic analysis of the structure of DuPont's flash-spun Tyvek, c. 1970.

attire, for example, where people sought to keep blood, bodily fluids, and liquid chemicals off of their clothing. Tyvek quickly usurped all manner of competition for lab coats, coveralls, gowns, and trauma suits.[89] Made in rolls and easily printed upon and customized, Tyvek was dynamic, and the sales numbers were sky-high.[90] By 1985, spurred on by an executive, Russell W. Peterson, who was eager to get deeper into the hot American construction sector, DuPont had assimilated Tyvek fabric into a sheathing product for construction called Tyvek HouseWrap (later known as Tyvek HomeWrap).[91] DuPont recognized a widespread need in the American construction sector for a product that could provide an air filtration barrier that was capable of keeping out wind while also letting water escape so as to prevent in-wall condensation, which could lead to mold and rotting in timber frame structures.[92] HouseWrap was ten times as effective as any other sheathing product on the market.[93]

For every DuPont product that was as successful as Tyvek, there were dozens more that were scrapped or failed outright. Successful and unsuccessful patent efforts alike almost always originated with the efforts of DuPont's research personnel. One of the few examples where this was not the case is also one of the most instructive: a scheme for "growing" houses with foam, a project conceived by Domenico Mortellito in 1964. Mortellito was not a research scientist, but rather the head of DuPont's packaging design department, a department that rarely came onto the radar of either upper management or the research divisions.[94] Before joining DuPont, Mortellito had been a moderately successful artist who was perhaps best known for his experimental use of materials like carved linoleum and etched plastic.[95] Mortellito had attended the Newark School of Fine and Industrial Arts and later the Pratt Institute, where he was a decorated student.[96] While running his own studio between 1927 and 1942, Mortellito also worked as an artist for the Treasury Relief Art Project (TRAP), a New Deal arts program that commissioned visual artists to execute new artistic work and decorative schemes for extant federal buildings and was a highly public component of the Works Progress Administration.[97] His work for TRAP included murals for world's fair pavilions, churches, and luxury liners, and architectural decorations for public buildings, sculpture, and furniture.[98] Then, through 1945, Mortellito worked as both a civilian and an air force lieutenant at the Pentagon, designing exhibits, booklets, and brochures and

View of Tyvek HomeWrap in contemporary house construction.

supervising the Army Air Forces' graphic presentations.[99] After the war, Mortellito relocated to Wilmington to take up a position at DuPont, where he designed product packaging, graphics, and logos. Mortellito, who had been known to incorporate some of DuPont's own patented materials into his work—lucite, nylon, Teflon, and Corian, among others—was, by all accounts, the perfect person to assume the role of chief in-house artist at DuPont.[100] His verve for both synthetics and art emblematize the burgeoning profile of the "corporate creative."

The Growing Shelters Synthetically scheme, comprising a series of "integral grown-in-place units" of expanding foam, was by all accounts a fantastical project that may or may not have been visionary in nature. What it certainly did exhibit, as we know from Mortellito's extensive writing and research files on the project, was a vivid aesthetic philosophy. In his solicitations for internal corporate support, Mortellito politely but forcefully made the argument that the race for patents and products had dulled the company's ability to deploy its work with any kind of aesthetic rigor. His opening salvo for his proposal, which he sent unsolicited to upper management in the summer of 1964, speaks to this absence and bears repeating at some length:

> One night, one of our chemical engineers and myself engaged in a discussion concerning the lack of aesthetic values in some of our new products. We were discussing some of the materials which are being considered for structural and decorative use in housing. The question I ask is: Why can't we get more aesthetic qualities into many of our new products? This tickled my imagination and started me thinking about the questions of "missing aesthetics." My first reaction was that aesthetics are not so much missing as they are elusive, but more importantly, too many times they happen as an afterthought.... I began to think that in nature, "aesthetics" is always an integral and inseparable part of everything that "grows" or comes into being. I was thinking of such things as: Trees, flowers, snow crystals, pearls, diamonds, etc. I also know that "aesthetics" is not only an inherent part of these things, but it is the measure of their value and significance. I also thought, in nature form follows function and function follows form, and that they are always directly related. When I compared some of

Domenico Mortellito, *Man with Horse*, relief carving in lacquered linoleum for the WPA arts program, c. 1935.

Scientists testing DuPont's Ludox colloidal silica foam insulation in its Chestnut Run laboratory, 1968.

nature's aesthetic things with some of our man-made things, it became obvious that many of our synthetic materials are lacking in aesthetics.[101]

Mortellito proceeded to furnish a broad-brush image of his house design:

My thinking went something like this: in order to create a synthetic shelter with built-in aesthetic values, we must first think of the joys, comforts and feelings of warmth and security which must be inherent in this shelter to make it successful. This led me to thinking about rooms which permit you to look out without the feeling of framed windows, ceilings which went up and overhead without limiting trimmings, walls and floors shaped to provide service units without additional furniture pieces, floors of different levels for different functions, etc. It is then necessary to assume that we can create products which will flow, grow, foam up, or simply expand into shapes and will rigidize at the right moment. This might be one product which changes into various characteristics, or it might be a number of projects. . . . This approach to creating new products has in its conceptual approach the same potentials for beauty and aesthetics which we find in a growing plant. If our designs meet the aesthetic needs for living, there is no question that the products and production will fall in line. I believe it is possible to achieve this idea, and that we can "grow" houses which will have good structural values. These can be houses which can be designed and structured to meet the conditions of climatic, geographic and socio-economic conditions.[102]

Mortellito concocted Growing Houses Synthetically in his own spare time. Along with his lengthy proposal, he provided management with sketches and newspaper clippings of what he designated as precedents and corporate competition on the topic of biomimetically inspired housing design.[103] Management was slow to answer Mortellito and, when they did, they did so in a manner that was friendly but noncommittal. In pushing harder for a concrete response, Mortellito noted: "I believe if our Research and Development people were exposed to such an idea, as impractical

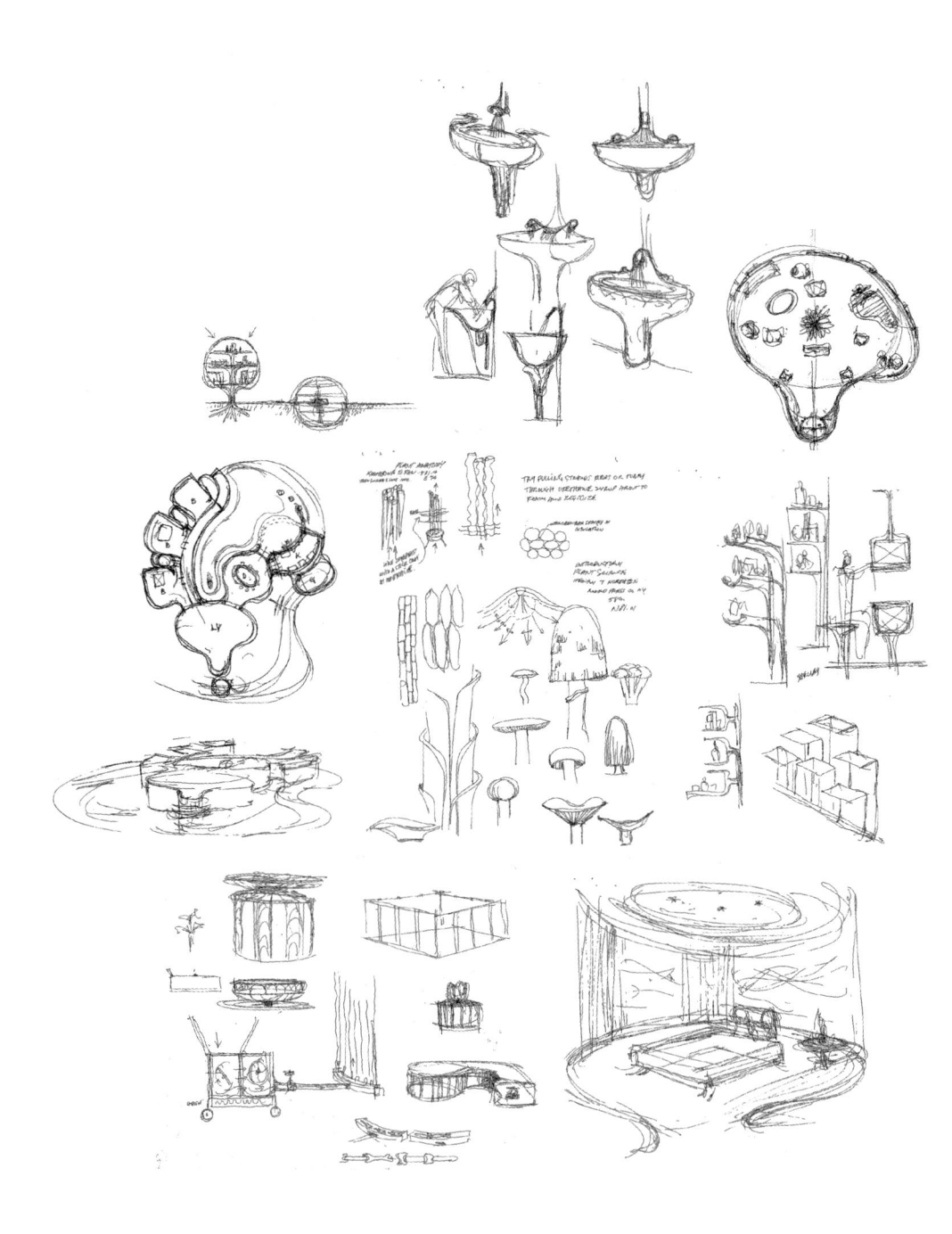

Domenico Mortellito, composite of sketches for the Growing Shelters Synthetically housing system, 1964.

as it may seem, it would widen their perspective and pique their imagination."[104] Ultimately, Mortellito was politely rebuffed, with management citing the inability of the company structure to undertake research ventures from his department, despite his relatively senior role within the company. In reality, Mortellito's architectural invention, and the patents he sought to produce from it, were probably simply too farfetched for DuPont's management.

With his creative dream not taken seriously, there is a palpable frustration in Mortellito's writings both throughout and after the process of advocating for the project, suggesting a well-founded and deep-seated frustration with DuPont's resistance to aesthetic—indeed humanistic—endeavors. He believed the company was unduly setting these endeavors aside out of the misguided belief that aesthetics were irrelevant to industrial progress. Mortellito, who held several patents himself, also hinted at the sentiment that his status as an "artist" relegated him to a lower status than others in DuPont's corporate structure. Mortellito's frustrations were not unique: manufacturing culture seemed quick to pay lip service to the value of aesthetic creative activity but was often loath to support it financially because its effectiveness was so difficult to measure in a way corporations would understand.

4 *The Repository*

The positivist pursuit of encyclopedic knowledge emerging from the Enlightenment produced a need for databases, repositories where new and vital knowledge could be recorded and consulted. These databases, specifically the predigital kind, took on many forms, and their architectural settings were often essential to their presentation. For example, Zeynep Çelik Alexander has shown how the herbarium at Kew Gardens in London, a facility established in the 1850s that today holds upward of seven million preserved vascular plant specimens from around the world, was a classic Victorian manifestation of sedimentary forms of knowledge that dictated architecture's manifestation as a kind of spatial database.[1] The library and the museum are, of course, the touchstone examples of this genre. The patent office, although more rarified and with a more circumscribed public, is another. As patents proliferated exponentially throughout the nineteenth and twentieth centuries, the spaces where the records of those inventions resided became ever more critical to the inventors who needed to consult them. So too were they important for the patent agents, lawyers, and inventor-researchers seeking out prior art to determine just how new an invention really was, if it was new at all. On both sides of the Atlantic, these were spaces and databases critical to the culture of invention and to the technological dynamism that had become the economic spur behind nations that touted themselves as modern.

These predigital database environments all centered on some kind of object, be it a plant specimen, a book, an artwork, or a patent dossier. Just as the patent repository was not a monolithic typology, neither was the patent dossier, whose constituent parts changed over time as well as from place to place. Those parts could include models, drawings, and textual explications, each with its own set of rules and historical vicissitudes. The repositories in which these dossiers were stored and made available have also tended to have programmatic functions that supplement the core purpose of the database, ranging from museum functions where patent materials are displayed to spaces for prosecuting patents and facilitating meetings with patent examiners. Exploring the patent repository through its constituent media—models and drawings—is a productive way to decipher its contours. So too is the historical condition of seized patent repositories in times of war or conflict, when IP abruptly changes hands to the great benefit of the victor and the irreversible detriment of the loser.

The Patent Model

Patent models are the least common of the patent media that one might find in a patent repository. This is because they were only required for a certain period (1836–1880) in a certain place (the United States).[2] Some inventors built what they describe as patent models at other places and other times, although many of these examples would seem to be more correctly described as scaled prototypes, not patent models.[3] The models in the wondrous collection on display in the construction history division of the Musée des Arts et Métiers in Paris, for example, are yet something else: models produced by architects and engineers to represent a specific project or site. Although a number of these projects were related to patents in some manner, these models were not part of the evaluative process that led to a patent's approval or rejection.

Patent models that relate to the world of architecture and construction perform many of the same functions as normal architectural models do.[4] As Paul Emmons notes, "Models are sometimes tools of thought and other times toys of play. Wittgenstein observes that when a child pretends a chest is a house, 'he quite

A miniature model of a Bessemer factory, 1860, on display at the Musée des Arts et Métiers, Paris, 2019. Photograph by Roland Nagy.

forgets that it is a chest; for him it actually is a house.' Models can magically transport us to other times and places. Ultimately, models know more than even their creators intend."[5] Patent models also reinforced broader cultural associations between patents and machines.[6] One historian argues that patent models "helped to shape broader conceptions of invention as material as opposed to an intangible idea . . . patent models could be interpreted as both idea and embodiment, designed to represent an intangible concept in material form."[7] Patent models represent the intellectual component of IP perfectly. They pull at our desire to conjure a theoretical embodiment of something that does not yet exist and they give their examiners special license to decide whether their embodiment in miniature—the closest thing an abstract thing can be before passing into the threshold of the real—is enough to imagine the possibility of that object as one that can both function as promised and make some object or process of our world more perfect. The act of making models enables us to assess our own ideas from a new position.

Given the rich culture of modelmaking and miniatures not only in Europe but also in Islamic, East Asian, and Australasian cultures, it is something of a surprise that the patent model first became a kind of formal bureaucratic entity in the upstart United States, in the early nineteenth century. Some context regarding the early American patent system can offer insights. The War of 1812 had brought tumult to the US capital. The first superintendent of the US Patent Office, beginning in 1802, was William Thornton, the son of the governor of Tortola and a man with an avid interest in architecture (he had served as the first Architect of the US Capitol). Thornton ran US patent operations from temporary premises in Blodgett's Hotel, which was located near other federal offices.[8] By his own account, Thornton had anticipated a British attack on the US capital days before the British set fire to most government buildings on August 25, 1814.[9] Thornton secured as many public papers relating to patents as he could and relocated them to a "place of perfect safety."[10] When he learned of the imminent British plan to destroy Blodgett's Hotel, which still contained a smattering of models that had voluntarily been submitted by inventors, he returned to the hotel, where he says he pleaded with the British major on duty: "I told him that there was nothing but private property of any consequence, and that any public property to which he objected might be burnt in the street, provided the building might be preserved, which contained hundreds of models of the arts, and that it would be impossible to remove them, and to burn what would be useful to all mankind, would be as barbarous as formerly to burn the Alexandrian Library, for which the Turks have been ever since condemned by all enlightened nations."[11] Even if Thornton exaggerated his own role in sparing what remained of the records of the Patent Office, including its first models, he can nevertheless be credited with sparing the records from disaster.

That Thornton spared the sundry models from the torch of the British imbued them with a certain patriotic value that they carried over into policy decisions in the decades to come, especially those made after another disaster: on December 15, 1836, with the Patent Office still housed in its makeshift quarters in Blodgett's Hotel, another fire (ultimately ruled accidental) wiped out all existing records of patent files, drawings, and models, save one volume.[12] Members of Congress lamented that "[o]f all the amount of loss of papers and property sustained by this disaster, that which is most to be regretted (because irreparable) is that of the whole of the great repository of models of machines in the Patent Office. The smouldering ashes now only remain of that collected evidence of the penetration, ingenuity, and enterprise which peculiarly distinguish the descendants of Europe in the Western World."[13] The lost property was estimated as 168 volumes of records, 26 portfolios comprising approximately 9,000 drawings, the entire library of 230 volumes, and approximately 7,000 of the patent models voluntarily submitted by American inventors prior to 1836. Officials proposed that 3,000 or so of the most interesting models should be replaced, and they proposed seeking the inventors and information needed to restore them.[14] The robust federal reaction, similar to the rebuilding that is promised in the wake of any major natural disaster, fixed the patriotic status of the American patent model in the public imagination as well as in the patent process.

The official patent model originated in a bill in 1836, signed into law by President Andrew Jackson in the wake of the disaster at Blodgett's, that comprehensively overhauled the American patent system.[15] Whenever an invention could be so represented, the law stated, a patent model should be deposited along with drawings and textual materials. This requirement remained in effect as a statutory requirement for 34 years, until 1870, and as a requirement of the US Patent Office for an additional 10 (until 1880). "The model," the law noted, "not more than twelve inches square, should be neatly made, the name of the inventor should be printed or engraved upon or affixed to it, in a durable manner. Models forwarded without a name are disregarded by the commissioner and are not entered in record."[16] The law made no further mention of any kind of regulation of the models apart from their size: they were to fit within a one-cubic-foot envelope, which meant that the longest dimension could not exceed twelve inches, regardless of the actual size of the invention being represented.[17] The model functioned as a key element of the disclosure of an invention, not only to the patent examiner, but to anyone who might examine the patent at a later date for research on prior art.

In 1856 the Patent Office, now in a stately purpose-built space designed by the architect Robert Mills, included rooms for exhibiting the patent models in its east and west wings, and the galleries (known as the Patent Museum) quickly became a popular attraction in the capital.[18] The press heaped praise on both Mills's architecture and the model galleries. The exhibits ostensibly made the visitor feel "prouder of history, and feel that while we are free, we are also independent."[19]

As the loss of 7,000 models in the 1836 fire *prior* to the 1836 model requirement demonstrates, the model requirement was not an ex nihilo convention imposed by Congress. Inventors sent models to compensate for poor or incomplete drawings or textual explications that failed to convince. "It is believed," a representative of the Patent Office said, "that the nature of the machine will be more clearly shown by [a model] than by drawings alone."[20] Nevertheless, when the overall patent system was tightened under Andrew Jackson, the model was no longer allowed to fill in gaps in other aspects of a patent application. At best, the model could allow a patent examiner to suggest ways in which a drawing or a text could be revised. Thus, the great irony of the patent model is that its institutionalization in 1836 foreshadowed its ultimate irrelevance later in the century, when the paper documentation (drawings and texts) in patent applications would reign supreme.

Inventors who could afford to have their patent models fabricated by professional modelmakers typically did so.[21] A small but not insignificant cottage industry of patent modelmakers had materialized around the country, especially in Washington, DC, with advertisements abounding in the classified sections of journals and trade papers for artisans, mechanics, and inventors, such as an advertisement by John B. Waring of Jersey City, New Jersey:

View of storage of the patent model collection at the Hagley Museum, January 2022. Photograph by the author. Also color plate 13.

"To the Inventors and Manufacturers[:] The subscriber would respectfully notify inventors, that he is prepared to make models of new inventions of any description, in the most perfect manner, and at reasonable rates. Being possessed of a fine stock of tools, especially adapted to the business, and extensive practical experience in it, he flatters himself that he will be enabled to do full justice to any orders that may be committed to him."[22] A high-quality model was one of the most significant expenses an inventor could expect in the patent process; in 1870, the cost of a typical model was calculated by adding the modelmaker's labor—50 cents per hour—to the market price of the materials used.[23]

Even after the statutory dissolution of the patent model requirement in 1870, a handful of modelmakers survived, making models for voluntary submissions to the Patent Office or simply as evidence of an inventive idea to be shown to prospective financial backers. Most prominent among the modelmaking firms was the D. Ballauf Manufacturing company. Founded in Washington in 1855 by German émigré Daniel Ballauf, the firm quickly established a reputation for versatility and craftsmanship, fashioning models for virtually every category of invention that existed in the Patent Office's register: spyglasses, nail machines, clothes wringers, hayforks, guns, bedsteads, churns, locomotive brakes, ice machines, stereoscopes, and so on.[24] Ballauf's clients included prominent inventors such as Alexander Graham Bell, Charles van Depoele, Thomas Armat, Ottmar Morgenthaler, and Emile Berliner.[25] In Ballauf's shop as well as elsewhere in the industry, German émigré craftsmen dominated the modelmaking business. Rudolf Schneider, who inherited the Ballauf business, was a trusted personal friend of Thomas Edison and advised him on numerous mechanical and modelmaking issues during his career.[26]

Congress was also cognizant of the didactic value that patent models held in the promotion of knowledge amongst the general public. In an 1836 report, Senator John Ruggles of Maine, Chairman of the Select Committee on the Patent Office, boasted that the records of the young US Patent Office were the ultimate testimony to America's rise as a manufacturing, agricultural, and commercial world power: "The accomplishments registered [in the Patent Museum's display vitrines of patent models] in just a few decades would have taken Europe a century."[27] Displaying patent models en masse was the clearest way to articulate American superiority, whether or not it was actually true. But already by the

View of the interior of the Ballauf patent model workshop in Washington, DC, 1894.

1860s, before the Patent Office was completed in its entirety, issues surrounding the exponentially increasing collection of models began to vex Patent Office staffers and administrators. Staff shortages meant that models often went untagged and uncatalogued and that the vitrines were crowded and poorly labeled.[28] These problems prompted the statutory abolition of the model requirement in 1870. Nevertheless, enmeshed as they had become in the practice of patenting, models kept arriving on the doorstep of the Patent Museum.

US Patent 175,765 for Improvement in Elevator-Towers, invented by Lemuel Sawyer in 1875 is one of the many examples of what one could have seen on display in the ground-level vitrines of the Patent Office's museum in the heyday of the museum around 1876.[29] Improvement in Elevator-Towers is "a structure for the convenience of persons who wish to view a town and the surrounding country from an elevated stand-point," providing "a place where large numbers of people can be safely and expeditiously raised to a sufficiently elevated position to obtain a good view of the city, without the severe toil of climbing."[30] The model, made of brass, steel, wood, and rope, reached just under one foot in its longest dimension, its height, so as to fit the Patent Office's one-cubic-foot requirement. Like many patent models, the model is operable, with a hand-operated crank on one side of its base that lifts two cylindrical brass cars up the tubular elevator shafts by means of ropes and pulleys. Sawyer's invention was not without precedent, as towers whose primary function was to survey the surrounding vista had existed in the form of lighthouses since antiquity.[31] Yet the model hints at the ways in which technology for technology's sake (in this case the steel frame construction and the elevators) was a leitmotif of patents in the arena of construction and architecture. That this patent precedes Gustave Eiffel's patenting of the underlying design for the Eiffel Tower, essentially the same invention with a much grander and more eloquent form, testifies to the broad influence of certain patents that celebrate technology for its own sake.[32] One can see this in a number of patent models at the zenith of patent model construction: William C. Phillips's Fire Escape (1878), Philip Jarvis's Bridges (1879), Charles and Norton Otis's Hoisting Apparatus (1880), and Mark J. Sullivan, J. Kessler, and J. R. Foster's Wooden Truss Bridge (1880).[33]

However, the Patent Museum suffered yet another calamity in September 1877, when the largest fire in the history of Washington,

Lemuel Sawyer, patent model for Improvement in Elevator-Towers, US Patent 175,765, filed December 20, 1875. Also color plate 2.

William Phillips, patent model for Improvement in Fire-Escapes, US Patent 202,460, filed October 6, 1877.

Philip Jarvis, patent model for Bridge, US Patent 212,941, filed October 29, 1878.

Charles R. Otis and Norton P. Otis, patent model for Elevator, US Patent 226,673, filed February 5, 1880. Also color plate 3.

Mark Sullivan, Jacob Kessler, and Josiah R. Foster, Wooden Truss Bridge, US Patent 224,491, filed August 13, 1879.

DC broke out at the Patent Office Building. Apparently, a worker had started a fire in a stove on the first floor after patent copyists complained of being cold. Sparks that traveled up the chimney set fire to a series of wooden grates that were resting on the roof, which fanned a conflagration from the roof downward into the building. Neither the fire trucks nor the hydrants of the day were adequate to douse a fire at such a height. A total of about 100,000 patent models residing in the building's Model Room (on the floor directly beneath the roof) were lost, including several models of prominent inventions, such as Eli Whitney's Cotton Gin. The Patent Commissioner was out of town and unable to supervise the rescue efforts. What was saved, including an original copy of the Declaration of Independence and all of the Patent Office's textual records, was saved in large part thanks to the leadership of the former territorial governor Alexander Shepard, who arrived at the height of confusion and stepped in to guide the rescue efforts.[34] Shepard formed Patent Office clerks into assembly lines along the central staircase leading to the model room, having them pass what models could be saved down to the street one at a time. The destruction included an estimated 76,000 patent models, most of them agricultural and mechanical inventions.[35] Facing yet another expensive rebuilding scheme, Congress appropriated $45,000 for the restoration of the lost patent models, or about 59 cents for each lost model.[36]

The calamity marked a decisive moment in the history of patent models. In addition to the numerous lamentations regarding the lost patent models published in a variety of outlets, the fire also prompted many to question the necessity of the patent model in the first place. *Scientific American*, long the go-to resource for inventors seeking patent news, led the charge, noting that patent models had become an expensive, useless formality: "There is more sentiment than sense in the oft repeated claim that they constitute a great national museum," an editorialist wrote. "While some are intrinsically interesting as historical relics, the majority constitute a monument showing only in the aggregate how prolific is the genius of the American inventor."[37] This argument—that the models testified merely to the quantitative scale of American ingenuity and not its actual quality—was by extension a condemnation of the quality of the models themselves and of the entire idea that patent models should automatically be objects of museological scrutiny. Fatigue from managing the patent models, the

Lewis E. Walker, *Interior View of the West Hall, United States Patent Office Model Room, after the Fire of September 24th, 1877.*

prospect of needing to rebuild several thousand of them, and the public critique of their qualitative shortcomings together led to the outright prohibition of patent models in 1880, with exceptions made for flying machines, perpetual motion machines, and other specific instances.[38] All extant patent models were removed and placed in storage.

Moving the models offsite was not the solution that it at first appeared to be. In relinquishing patent models to rented storage, the Patent Office was forcing inventors and its own examiners to regularly travel to the storage facility when doing due diligence in prior art dated 1836–1880 for new patent applications. The abolition of patent models may have gained the Patent Office some much-needed space, but it was making the work of many examiners more difficult. Many patent experts also still held a fondness for the patent models. During his short tenure as Commissioner of Patents, Charles E. Mitchell lamented the demise of the patent model in no uncertain terms:

> I regard it as nothing less than a public calamity that the office was several years ago compelled to suspend the reception of models, excepting in special instances, for want of space in which to store and exhibit them. . . . I venture to express the hope that the time will come when models will again be required in connection with all applications, and that when that time arrives an effort will also be put forth to obtain specimens of the more important inventions which have been patented during the intervening period. I recommend Congress to make some provision which will enable the department to return these models to the galleries in this building, designed and constructed for their exhibition.[39]

Mitchell's pleas fell on deaf ears. In 1907, Congress announced that it would gradually deaccession patent models in federal custody and gave the Smithsonian Museum a mere six months to select the models it might like to save for its own collection. Worse, the Smithsonian apparently appointed a committee with very niche interests to choose a mere 1,061 models: many models related to the development of sewing technology were spared, for example, while many more important inventions were not.[40] A few years later, inventors and their descendants were invited to collect

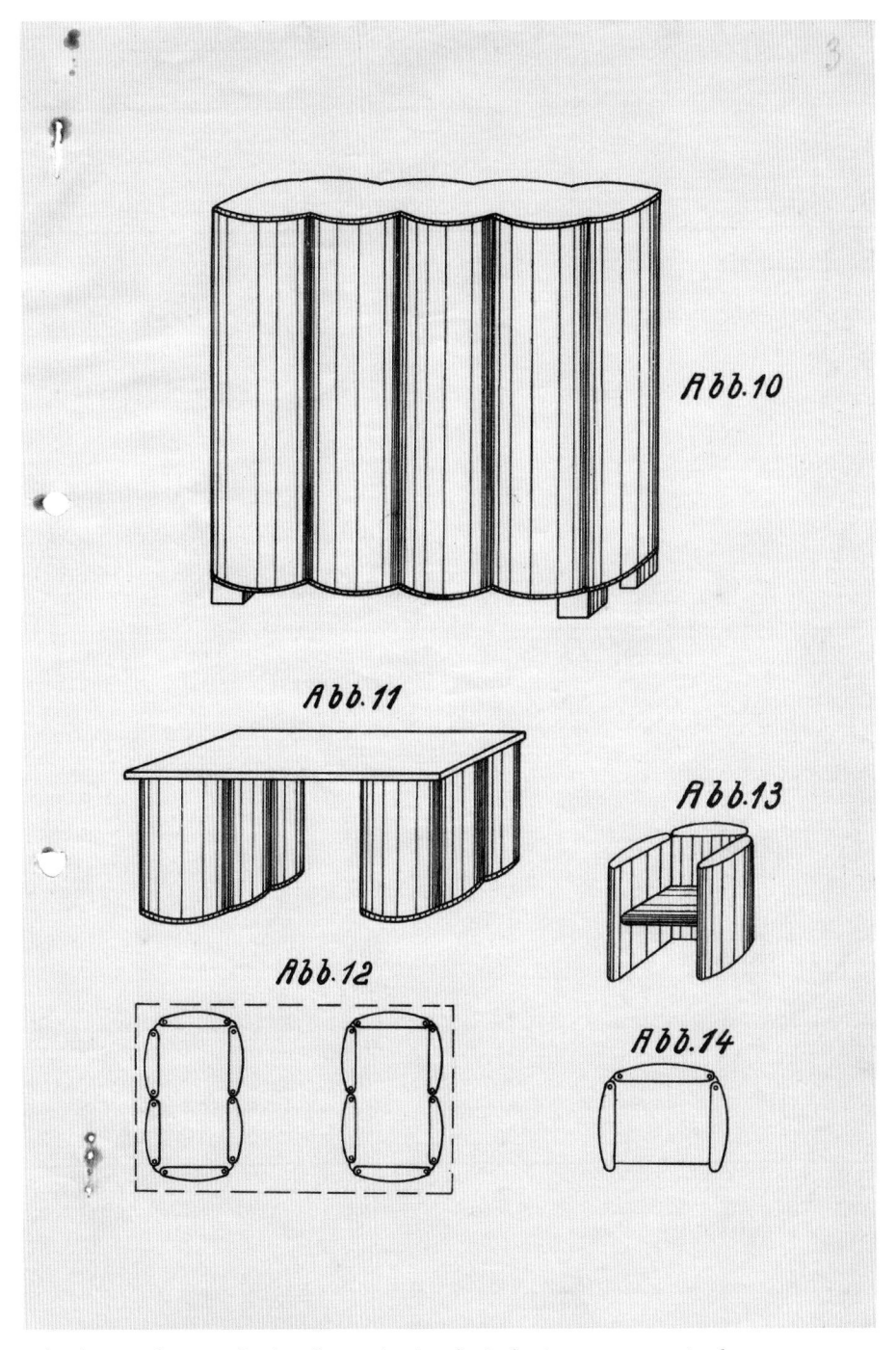

Color Plate 1   Unknown, collection of patent drawings for the furniture system associated with the Junkers *Metallhaus* system, c. 1931.

Color Plate 2 Lemuel Sawyer, patent model for Improvement in Elevator-Towers, US Patent 175,765, filed December 20, 1875.

Color Plate 3 Charles R. Otis and Norton P. Otis, patent model for Elevator, US Patent 226,673, filed February 5, 1880.

Color Plate 4   James Johnson, original color patent drawing for Fire Ladder, US patent issued April 1831. NAID 178329524.

Color Plate 5 Mariano Taccola, illustration of Brunelleschi's "Il Badalone" from *De ingeneis*, c. 1449, Codex Palatinus 766, p. 40 verso / p. 41 recto.

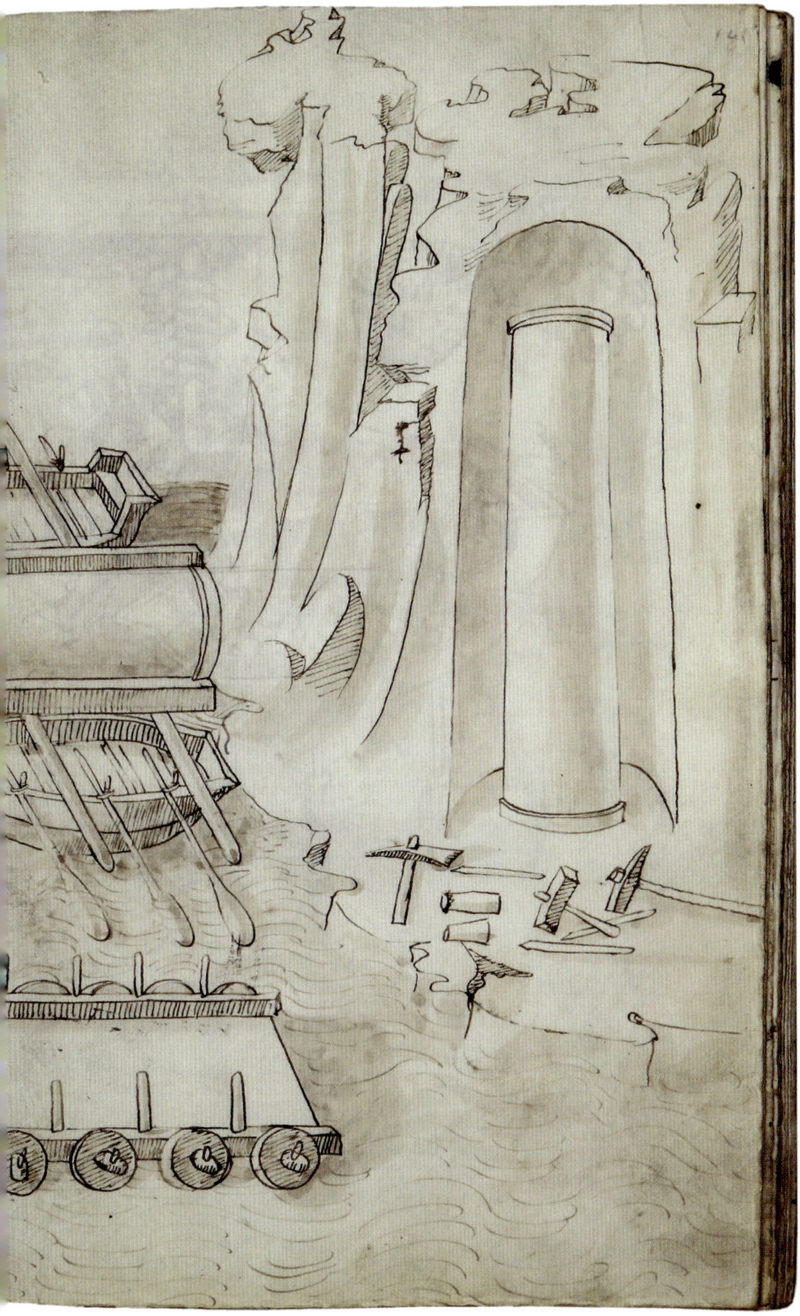

Color Plate 6 Jorge Ferrari Hardoy (in conjunction with Antonio Bonet and Juan Kurchan), iteration of an axonometric drawing of the BKF chair, design 1938, drawing c. 1941. Harvard University, Loeb Library Special Collections D083a.

Color Plate 7   C. L. Fleischmann, original color patent drawing for Francis Follet's Self-Balancing Sashes, US Patent 3596X, issued October 15, 1822.

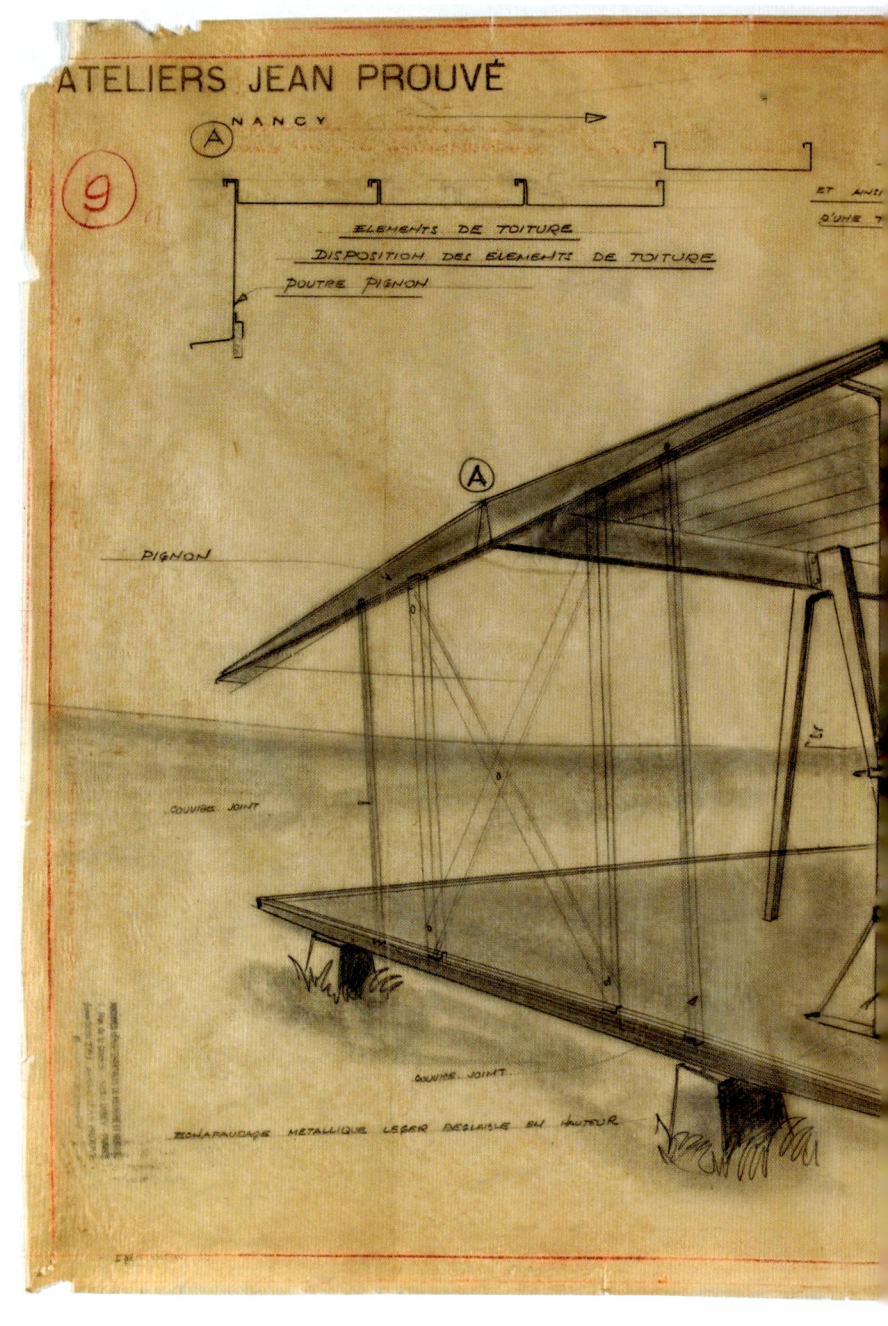

Color Plate 8 Jean Prouvé, Demountable House, perspective drawing showing the installation of the first roof span, 1947. Pencil on tracing paper, 48.5 × 71 cm. Musée National d'Art Moderne / Centre Georges Pompidou, Paris, France. Inv. AM 2007-2-342. Photo: Georges Megeurditchian.

L'AUTRE PIGNON. MAIS PAR MONTAGE SIMULTANÉ
PUIS D'UNE TRAVÉE PANNEAUX. VOIR DESSIN N° 11.

SABLIÈRE D'EXTRÉMITÉ A
INTRODUIRE DANS LES AGRAFES
DES TOITURES ET A ASSURER EN BOUT
PAR UN BOULON SUR LES PIGNONS.
PROCÉDER DE MÊME POUR LES RIVES
COURANTES SE JUSTAPOSANT BOUT A BOUT SANS
FIXATION.

ATELIERS JEAN PROUVÉ
NANCY

PAVILLON DÉMONTABLE

BREVET N° 849.762

NOTICE DE MONTAGE

POSE D'UNE PREMIÈRE TRAVÉE
DE TOITURE. 8 ÉLÉMENTS DE 0°50 POUR TRAVÉE 4°00
12 ÉLÉMENTS DE 0°50 POUR TRAVÉE 6°00

⑨

Color Plate 9   Elizabeth Burbank for Luther Burbank, Peach, US Plant Patent 15, granted April 5, 1932. The Alien Property Caché.

April 5, 1932. L. BURBANK Plant Pat. 15

PEACH

Filed Dec. 23, 1930

Color Plate 10 Postcards depicting cannetis, a building material made from split plaited reeds, 1946. National Archives (UK) DSIR 4/2495.

Une expédition à la sortie de l'USINE
(Construite en Cannetis)

Une Plantation de Roseaux

Color Plate 11   J. C. Lanchenick, *View of the South Kensington Museum (then the Patent Museum), South End of Iron Building*, 1863. Watercolor on paper. Victoria and Albert Museum, Prints, Drawings & Paintings Collection, accession no. 2816.

Color Plate 12 View of a patent litigation meeting at the European Patent Office, Munich, with synchronous translation booths.

Color Plate 13 View of storage of the patent model collection at the Hagley Museum, January 2022. Photograph by the author.

the patent models that remained. Some private collectors, most prominently Henry Ford, were also interested in collecting patent models for their own upstart collections; Ford selected 25 models, which would help build the collection of his namesake museum in Dearborn, Michigan.[41] By 1950, the US Patent Office's storage facilities were so pressed for space and resources that the office was compelled to bring a lot of 5,000 patent models to auction, with asking prices ranging anywhere between $1 and $1,000 (itself a testament to the rather wide range of their quality).[42] This last lot of patent models forms the basis of the second-best patent model collection after the Smithsonian's, in the Hagley Museum in Delaware, which has a particularly strong collection in patent models related to architecture and construction. Occasionally, extant patent models also found their way into museum collections, such as Isaac Cole's 1873 patent for a One-Piece Plywood Chair, which was acquired by the Museum of Modern Art in 1956 under the Department of Architecture and Design's chief curator, Arthur Drexler.[43]

The only museological treatment of patents comparable to the US Patent Office exhibitions took place in England. The Great Exhibition of 1851 generated significant surplus revenue as a result of its international popularity, and Prince Albert, the exhibition's patron, believed that the surplus would most fittingly be used for a number of educational establishments on land that was available nearby, in the district of South Kensington.[44] By 1857, the South Kensington Museum, housed in a hastily constructed industrial structure of iron framing and sheathing, fulfilled that mission through its display of the industrial and decorative arts as well as miscellaneous scientific collections of animal products, food, building materials, and educational tools, adopting the same encyclopedic aims of the original exhibition. In the same building, there was also a separate exhibition organized by the Superintendent of the Patent Office, Bennet Woodcroft, that came to be known as the Patent Office Museum; here he displayed contemporary and historical patented inventions, including locomotives, steam engines, and model ships. The British Patent Office Museum opened nearly simultaneously with its American counterpart, but the clarity of its mission and collection was hindered by its disorganized approach and the lack of distinction between its content and that of the South Kensington Museum. This perhaps explains why, by 1883, the holdings of the Patent Office Museum were formally transferred to the South Kensington Museum and the Patent

J. C. Lanchenick, *View of the South Kensington Museum (then the Patent Museum), South End of Iron Building*, 1863. Watercolor on paper. Victoria and Albert Museum, Prints, Drawings & Paintings Collection, accession no. 2816. Also color plate 11.

Office Museum ceased to exist as a separate entity.[45] Today, the collections of the South Kensington Museum form the basis of the Science Museum on the campus of the prestigious Imperial College, an important center for science and technology.

Unlike the US Patent Office's exhibitions, the Patent Office Museum at South Kensington comprised all manner of didactic materials, not just models. In fact, none of the material exhibited to tout English inventive activity was official material that had been made for the patent application process, although some patent drawings were reproduced. Rather, the material that the museum exhibited was auxiliary material or items at full scale (as with the locomotives, for example) and thus privileged the goal of entertaining viewers in much the same way as the Crystal Palace had. A rationalist might have much preferred the American exhibition, in which models, with their neat and similar dimensions, could be compared serially and practically ad infinitum. Woodcroft had a different approach, which was outlined in the mission of the Patent Office Museum in 1859 as follows:

> It is intended to make the Patent Office Museum an historical and educational institution for the benefit and instruction of the skilled workmen employed in the various factories of the kingdom, a class which largely contributes to the surplus fund of the Patent Office in fees paid upon patents granted for their inventions. Exact models of machinery in subjects and series of subjects, showing the progressive steps of improvement in the machines for each branch of manufacture, are to be exhibited; for example, it is intended to show in series of exact models each important invention and improvement in steam propellers . . . from the first engine that drove a boat of two tons burden to the gigantic machinery of the present day, propelling the first-rate ship of war or of commerce. The original small experimental engine that drove the boat of two tons burden, above referred to, is now in the museum, and is numbered one in the series of models of propellers.[46]

Woodcroft's intent, whether or not it was in direct response to his American counterpart, was the inverse of the curatorial approach adopted in Washington. By emphasizing the steps of a single type of invention over time, Woodcroft was placing a premium on narrative, and this meant that even if the objects on display were tailor-made for the museum and not artifacts of the patent process, they in fact emphasized the qualitative experience of patenting as a process reliant on prior art. In other words, this was not a quantitative showpiece.

H. T. Wood, a secretary of the Royal Society of Arts, outlined in a letter to the editor of *Nature* why, in his opinion, the Patent Office Museum's curatorial approach nevertheless failed miserably:

> Unhappily, this brilliant project rested unfulfilled. "No 1" of the series of models of steamboat propellers had but few followers, while other branches of mechanical science did not get so far as to have even a "No 1." The conception was excellent, the execution lamentably deficient.... Luckily there is a chance of something better now, and it is to be hoped that we shall soon have the collection belonging to the Patent Office divided into two parts—one part to be sent to the Science Museum, and the other to the nearest dust-heap. So long as it belongs to the Patent Office, the aggregation of rubbish will be sure to continue. The Commissioners have never exercised a power of selection, and any foolish invention, so long as it is only the subject of a patent, has the right of *entrée*.... It is hardly possible to imagine an invention which—at least to an expert—cannot be as clearly explained by descriptions and drawings as by a model.[47]

Whereas Joseph Paxton's iron architecture proved to be part of the overall spectacle (intended to be temporary) of the Great Exhibition, Prince Albert and his associates at the top of England's culture sectors knew that the South Kensington and Patent Office Museum's hastily erected edifice needed a better, more permanent form. In 1864 the architect Francis Fowke won a competition for a new building that would house both a Natural History Museum and the Patent Office Museum, which at this point was still considered an institution that should remain independent. Fowke's design for the Patent Office Museum comprised a central quadrangle covered with glass and surrounded by galleries lit from the outside, providing a total of no less than 23,000 square feet across two floors.[48] The ground-floor quadrangle was sunken five feet lower than the surrounding space so that, in Fowke's words, "the larger class of objects belonging to this museum could be exhibited without interfering with the light of those parts under the galleries."[49] But Fowke's proposal never came to fruition; Woodcroft's inaugural installation, for better or worse, had failed to curry support for a more permanent institution, and, as happened with its American counterpart, the display of patent activity would fail to take off as a museological venture.

Patent Drawings

The seventeenth-century British and German patent systems were both characterized by a rise in the importance of textual description and written technical specifications.[50] This can be seen as a recognition of the importance of embodied knowledge and a recognition that lilliputian models, vague descriptions, and abstract drawings alone could not convey the full knowledge embedded in an invention (which thus rendered them rather useless as evidence of prior art). Yet, by the early nineteenth century, attention had again returned to drawings, hinting that, in tandem with text, these were optimal for conveying both the explicit and the implicit knowledge embedded in any inventive act. To be sure, patent drawings constitute a singular representational mechanism unlike anything else that exists in the world of images. They are absolutely critical to the patent system, exemplifying the well-worn cliché that images do things that words cannot, even in a domain like law that is so firmly entrenched in the word. Like models, drawings too form a repository of sedimentary inventive steps and in their very making capture the creative decision-making and communicative efforts of their inventors, helping to define the nature of invention itself. They help to fix legal tenets not easily expressed in words. Indeed, information articulated through drawings but not through text remains a form of legal disclosure today.[51]

Prior to standardization, patent drawings in all of the major international patent centers were characterized most of all by their heterogeneity. Inventors often rendered their drawings in watercolor, depicting people and landscapes with all sorts of flourishes that approximated closely the tradition of illustration as a medium of publicly oriented art. Professional draftspeople executed drawings as well, commonly placing inventions in their domestic, commercial, or industrial settings. The earliest extant patent drawings

John Bouis, original color patent drawing for Tin Copper and Zinc Roofs, US patent issued June 26, 1835. NAID 178329447.

in the US date from 1836; these were specifically requested by the Patent Office to replace the drawings that had been lost in the fire of that same year.[52]

John Bouis's redrawn drawing for his 1835 patent for Tin Copper and Zinc Roofs is an example of a patent drawing by the inventor; there is no third-party signature indicating a draftsperson's hand, and the drawing, while adequately descriptive, lacks the precision, particularly in the drawing's perspectival foreshortening, that a professional draftsperson would bring to it. Here it is the tableau of elements that is crucial, not any kind of geometric precision. Five annotated elements indicate the sequence of the procedure: a standard tin plate (a), attached to other sheets through a press (b), laminated with a layer of copper spun from a wheel (c, on the drawing's verso), slotted into a zinc mullion (d), and placed on a building's roof (e). Of course, a normal viewer would be much more consumed by the pictorial elements of the images, the lovely setting of the river embankment, the rolling hills and craggy mountains in the background, the stone bridge in the distance, the bird overhead. It is almost as if Bouis was trying to distract us from the actual technical content of the drawing, which of course raises the question: Why would a technical drawing be so pictorial? Arguably, since the item in question was roofing, Bouis's larger pictorial strategy was an attempt to show us how the roof and its shiny, modern materiality could in fact be naturalized by its surroundings, as if to allay any concerns that it would wind up destroying the character of the pastoral environment around it, which in the drawing looks much more like the old world than the new.

In contrast, a 1841 redrawn drawing for Francis Follet's 1822 patent for Self-Balancing Sashes is an example of a patent drawing by a professional draftsperson; the signature of the draftsman, a certain C. L. Fleischmann, can be seen in the lower left-hand corner of the window. Fleischmann's rendering of the window incorporates elements of context and abstraction: The window, the top and bottom segments of which are open to demonstrate the balanced sashes, looks out onto a vista of a placid lake with rolling hills beyond. The spectral quality of the light, indicating that it is either dusk or dawn, is seen not only in the window itself but also in the shadows the light casts through the sashes and onto the window frame itself. Perhaps this is why the draftsman chose to render the "frontal" view in a subtle two-point perspective,

C. L. Fleischmann, original color patent drawing for Francis Follet's Self-Balancing Sashes, US Patent 3596X, issued October 15, 1822. Also color plate 7.

allowing the shadows themselves to frame the invention. That perspective is abandoned in the transverse section on the right side of the drawing, which cuts through one of the window's vertical mullions to reveal the pulley system behind it. It also shows us the brick of the building envelope, an element omitted in the frontal view in order to retain a focus on the window itself. Bouis's and Fleischmann's are two of several excellent examples of color drawings pertaining to architectural patents and how they could be indistinguishable from the genre of building renderings.

As Bill Rankin has shown in the case of the United States, patent drawings underwent a revolution of standardization in the 1870s, right around the time when the United States was eliminating its mandate for models.[53] This change was also mirrored by patent reforms in London, Berlin, and Paris. For a number of reasons, the idiosyncrasy and fragility of the color drawings became a liability for prior art research. For one, the drawings were not easily duplicable when they had lavish pictorial embellishments and their own methods for representing depth through varying (and mostly inaccurate) kinds of perspective. In addition, repeated handling made them wear out quickly. Most importantly, while they were certainly enjoyable to look at, they often lacked the technical details that would make them truly instructive, which was of course their very purpose. Commissioners of patents on both sides of the Atlantic sought to eliminate these problems by mandating black and white ink drawings that would be easily reproducible using photolithography.[54] The verisimilitude and pictorialism in drawings like those for Bouis's and Follet's inventions were framed in a new light under the new standards: although perhaps making the drawing appealing to the imagination of the patent examiner, these strategies hindered the inventor's ability to make the broadest possible claims, by virtue of their intense pictorial specificity. The new emphasis would be on generality, or what Rankin describes as "a balance of prolixity and ambiguity."[55] This dialectic of prolixity and ambiguity, which has changed very little during the last 150 years, would appear to show objects that exist in the world even when they do not, which leaves all sorts of unanswered questions about materiality, weight, manufacture, and the business aspects related to bringing a product to market. The mysteriousness that patent drawings thus evince is anything but an accident; it is in fact the very best summation of the theoretical

The Repository 195

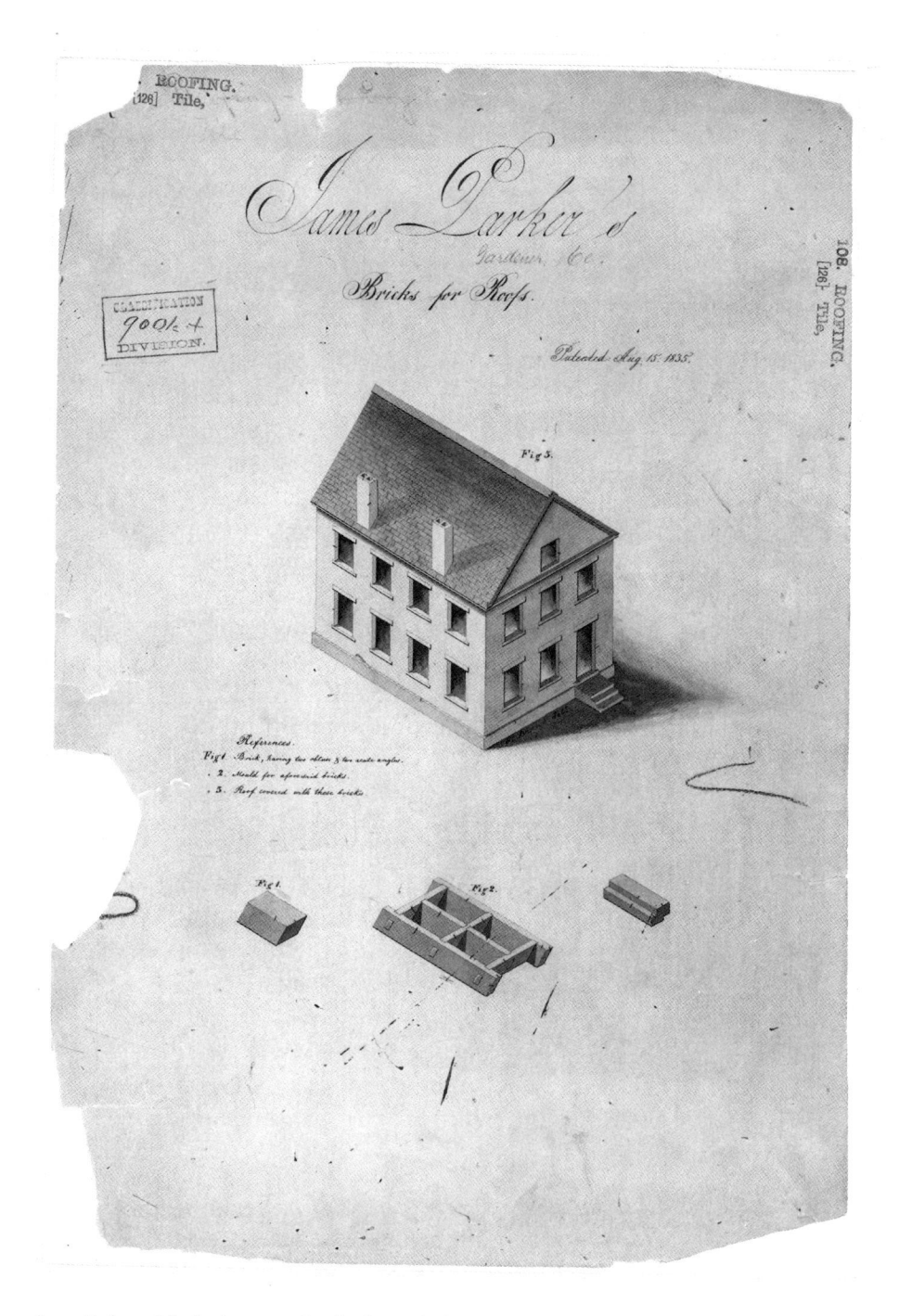

James Parker, original color patent drawing for Bricks for Roofs, US patent issued August 15, 1835. NAID 178329524.

James Johnson, original color patent drawing for Fire Ladder, US patent issued April 1831. NAID 178329524. Also color plate 4.

William Bryant, original color patent drawing for Supplying Houses with Water, US patent issued October 25, 1832. NAID 169378725.

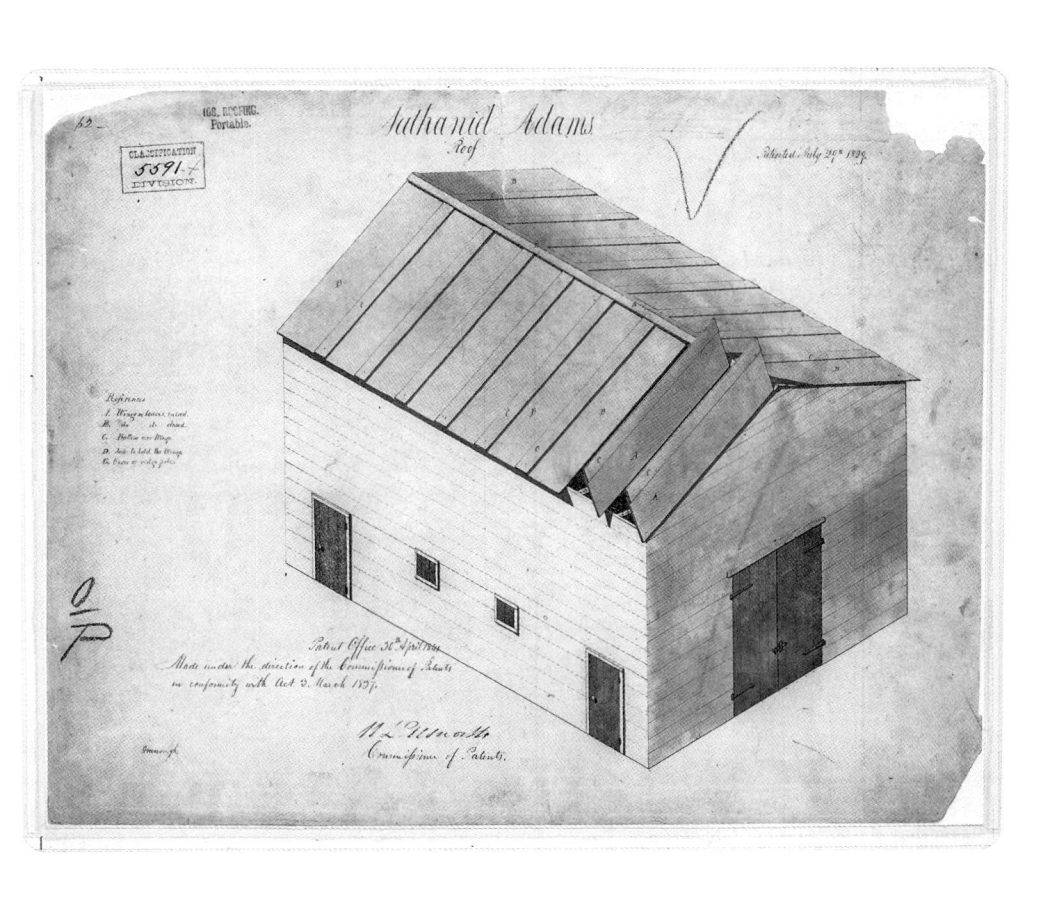

Nathaniel Adams, original color patent drawing for Roof, US patent issued July 29, 1829. NAID 158587566.

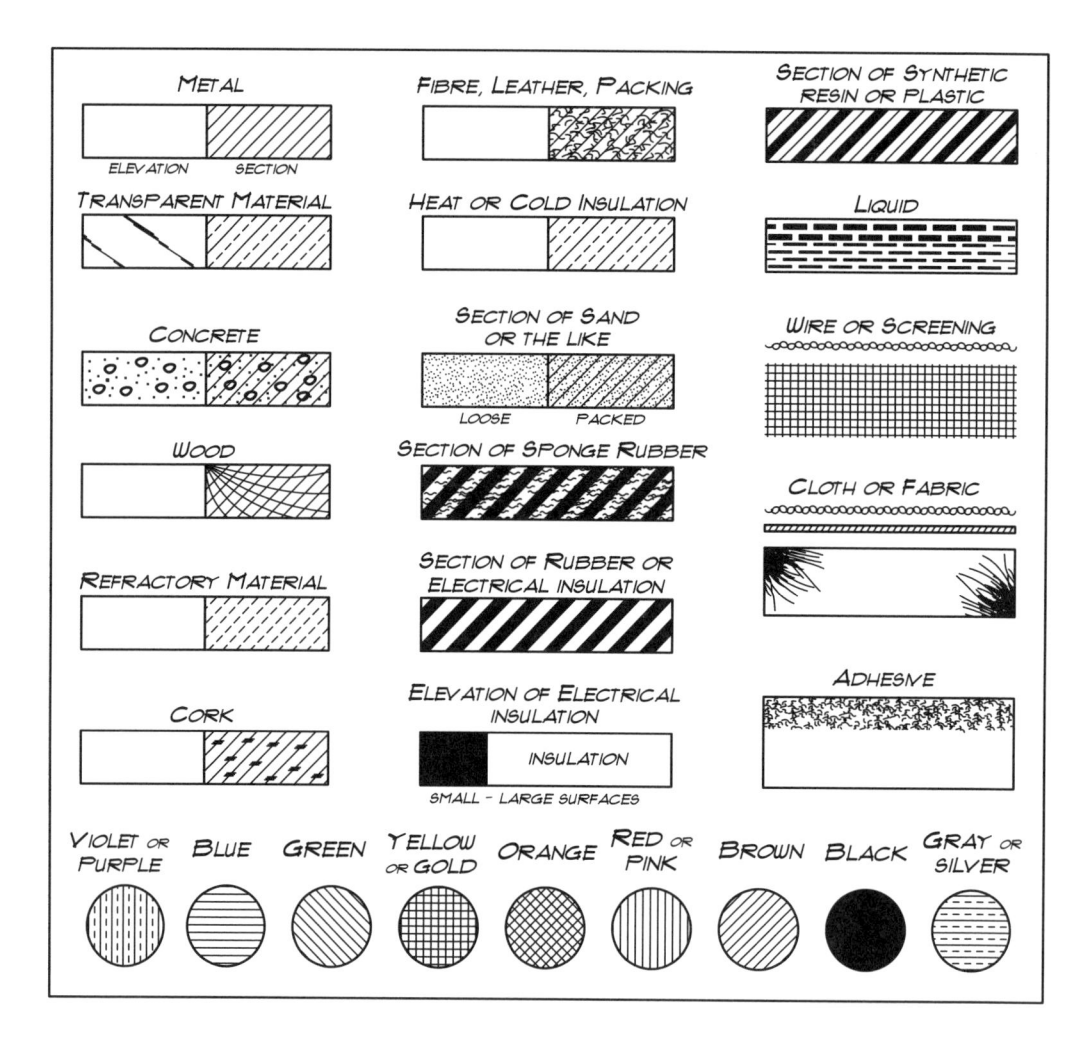

Diagram showing standard hatch patterns from the US Patent Office. These have been in use since at least the 1940s. This image is adapted from the *Manual of Patent Examining Procedure* (Washington, DC: US Patent and Trademark Office, May 2004), fig. 600-99. See also Rankin, "'The Person Skilled in the Art,'" 63.

nature of patents: items with a potential applicability that is neither certain nor obvious.

There exists a distinct visual corollary to the dialectic of ambiguity and prolixity. This is the interplay of abstraction (axonometry and isometry, a lack of context) and a highly specific set of requirements for depicting light, materiality, and depth so as not to lead to any kind of technical confusion. As Rankin explains,

> Monolithic parts are idealized for the sake of generality, and strict adherence to scale is not always important, especially for drawings of processes, assembly lines, or clothing. . . . Likewise, individual parts are never shown on their own but as part of a working whole: the goal is to patent a set of relationships, not a particular object. . . . Materials are labeled as generically as possible, often identified as simply "metal" or "wood." Dimensions, centerlines, and milling tolerances are omitted. When drafting a claim, patent lawyers begin by describing the drawing itself—what they call the "picture claim"—and then incrementally broaden subsequent claims to include as many similar ideas as possible. The original drawing must be ambiguous enough to allow these broader claims.[56]

Noting that the United States Patent Office instructs draftspeople that "light should come from the upper left corner at an angle of forty-five degrees," Rankin elaborates:

> edges to the bottom and right should thus be made graphically thicker, to indicate a shadow. In combination with surface shading of curved parts, these shade lines can aid greatly in understanding an object three-dimensionally. Whereas modern engineering drawings will show an object in several standard views (front, side, top, etc.), patents will usually only show one such view, and shade lines might be the only way to differentiate between a hole and a protrusion, or a surface seam and a hard edge. . . . In a similar way, line weights on axonometric and perspective drawings were often used to add depth and eliminate optical gestalt shifts. Edges that point toward the viewer were made heavy, while all other lines were left light. . . . Together with hatch

symbols and standardized reference labels, these visual effects gave patent drawings a surprisingly uniform flavor.[57]

Leroy Buffington, an architect who claimed to be the first person to patent a skyscraper, held a 1888 Canadian patent for an Iron Building Construction that demonstrates how many of the graphic sensibilities described by Rankin were cemented internationally by the 1880s.[58] Evidence of the human hand remains in some of the imperfections of the ink lines, and line weights, poché, and materials are clearly articulated in a set of intrinsically abstract views: a detail of a typical floor plan, a section detail, a frontal elevation, an array of floor plans, and a transverse section. These elemental views are coupled with construction details and an orthographic projection of the iron framing system, detailed down to the rivets, welding lines, and revetment bracing for brick. A 1978 German patent for a building by the Italian builder Giuliano Viviani demonstrates in many ways, despite the radical differences, how little in the drawing conventions changed over the course of much of the twentieth century.[59] Although this array of six axonometric views of the deployment of a prefabricated construction system for a house is also drawn by the inventor's hand, there are fewer imperfections due to the advances in drafting technology over the course of 90 years.[60] Beyond this, little is changed: callouts, poché, and the primacy of the axonometric view prevail, despite the fact that these two examples, two of millions that could be mentioned, are separated by both an ocean and nearly a century.

There are some notable exceptions to the rigid drawing regulations that took hold in the late nineteenth century. Patents for new varieties of plant species specifically bred by horticulturalists became possible with the US Plant Patent Act of 1930 and were exempt from the drawing guidelines, due to their status as organic entities and as entities whose shape and color were prone to natural variation.[61] Luther Burbank, a prominent California horticulturalist, had lobbied for the law for most of his later career, and his efforts were championed by the likes of Thomas Edison.[62] The Act passed four years after Burbank's death, and several patents were posthumously granted to his wife, Elizabeth, for his innovations in plant breeding. One drawing depicts an improved (plumper) peach, the fifteenth such patent to be granted, and returns the patent drawing to some of its early nineteenth-century character, with clear evidence of the human hand and a romantic use of color.[63]

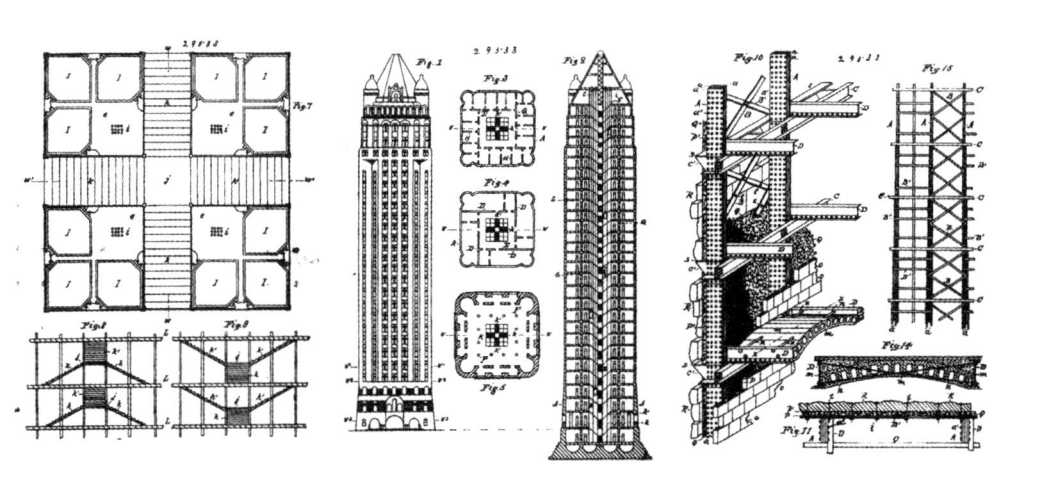

Leroy S. Buffington, Iron Building Construction, Canada patent 29,533A, granted July 23, 1888.

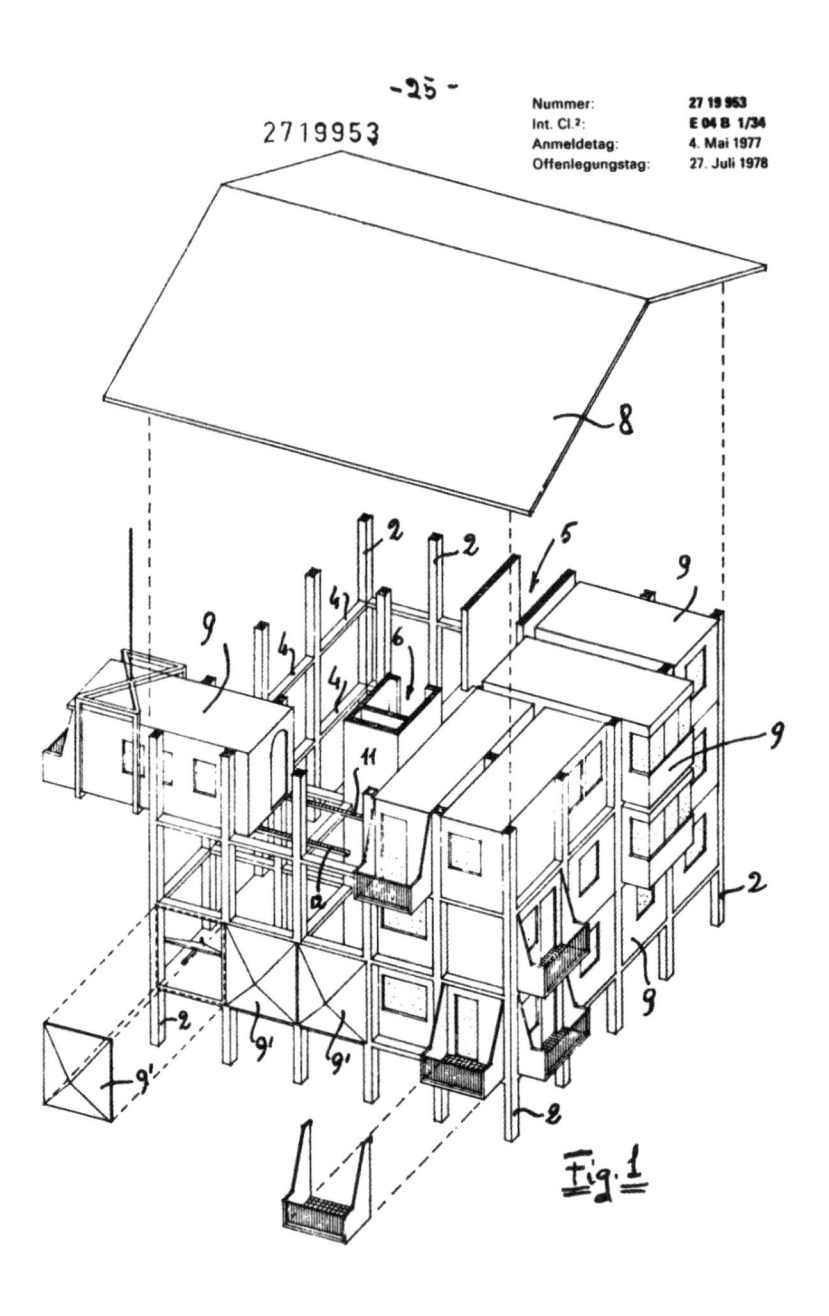

Giuliano Viviani, Building Formed of Modular Components—Has Roof on Frame with Access Ways to Different Storeys and Mountings for Prefabricated Modules, German patent 2719953A1, issued July 27, 1978.

Building safe and effective repositories for patent models, drawings, and textual records was a necessity for all countries that sought to build a robust national IP apparatus. These repositories were sacred ground when it came to advancing technology, assuring economic dynamism, and incentivizing innovation. The experience of fires and world wars, however, proved that these repositories were neither untouchable nor eternal. In the case of war, this was due not merely to the specter of destruction that always comes with it but also the possibility of what a loss would mean for an entire nation's corpus of intangible IP: a patent seized was a patent rendered void. Such was the case with the vast majority of German industrial patents after the country's defeat in World Wars I and II.

The administration of repositories of alien property such as foreign patent offices first fell under the wing of legislation in the United States with the Trading With the Enemy Act (TWEA), enacted on October 6, 1917. The TWEA gave the US President (then Woodrow Wilson) the broad right to terminate any and all trade with enemies during wartime. The TWEA also contained a proviso that made it possible for foreign patent holders whose property was seized to seek remuneration for the lost royalties associated with them. Yet after protracted legal proceedings, companies like BASF discovered that the proviso was ultimately superseded by international law.[64] As a result, like numerous other German companies, BASF was unable to retrieve either its patents or its losses from patent royalties. It was not until another court decision in 1931 that BASF was awarded direct compensation for German patents seized in World War I. BASF's patent for its Haber-Bosch process, for example, a nitrogen fixation process that was the main industrial procedure for the production of ammonia, was widely used by the US army during World War I.[65]

World War II was a much different beast, and the fate of the German patent repository this second time, with its advanced array of military and industrial patents, was more critical than ever. On July 27, 1946, the so-called London Agreement reached by the Allied powers determined that all German and Japanese patent holders would have their property expropriated without compensation and that this property would be exploited jointly by the victorious powers.[66] This included patent applications with a pending status.[67] The United States, France, and Britain shared the view that a Germany weakened in the extreme was in fact the

The Repository

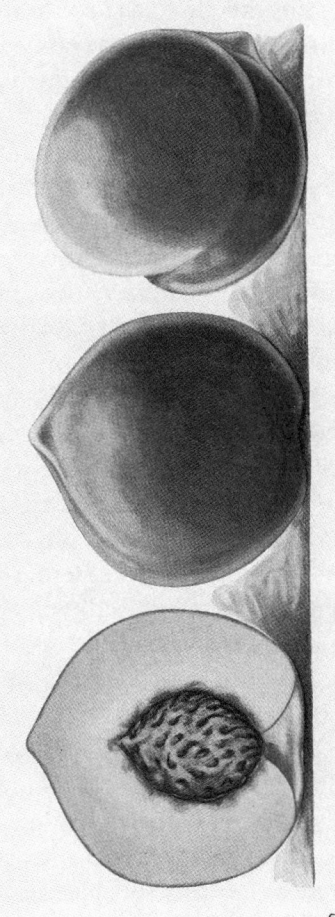

Elizabeth Burbank for Luther Burbank, Peach, US Plant Patent 15, granted April 5, 1932. The Alien Property Caché. Also color plate 9.

most dangerous prospect, a lesson perhaps of the aftermath of World War I. Instead, they believed that the expropriation of IP and the entire corpus of German patents should be leveraged in a manner that would curb nationalistic and militaristic activities without economically crippling the country. Foreign management of the German patent office was seen as the hinge on which this tricky balance of submission and sustenance could be achieved.

Although the Allies assisted in the custodianship of "enemy assets" in both wars, the United States was the most active administrative center for the management of so-called alien property. Most famously to this end, the United States masterminded the semi-covert Operation Paperclip, a program that included brainpower in its broad definition of assets, leading to the detainment of hundreds of top German scientists, engineers, and other highly skilled professionals. These men were settled in the United States to assist with the military and space research necessary to counter the postwar ambitions of the Soviet Union.[68] A good number of these men had been active contributors to the Nazi war machine, and their participation in the program thus represented one of the boldest extrajudicial efforts of the United States in the immediate postwar period. Collectively, the men in the program held around 10,000 German patents. While their patents would still be null and void, they would not be deprived of fairly remunerated work in their areas of expertise. Among the participants in the program were a handful of professionals with ties to the architecture and construction sectors, including the architects Heinz Hilten and Hannes Luehrsen. Hilten designed several buildings for NASA, including the Propulsion and Structural Test Facility in Huntsville, Alabama.[69] Luehrsen designed a number of the other buildings comprising the Marshall Space Flight Center, where Hilten's structure stands.[70] The Paperclip men were at the core of the entire United States space program during the second half of the twentieth century. There were similar, although markedly smaller, operations in the UK (Operation Backfire, Fedden Mission) and the Soviet Union (Operation Ossavakim).[71] Although the British and American "bi-zone" considered creating a joint patent office for new German patents, a plan for an all-German solution under foreign supervision ultimately won out.[72] Britain, for its part, truly globalized the German patent technology it acquired, extending the rights to seized German IP to the entire British Empire, which

Canada, Australia, India, and South Africa were more than happy to have.[73]

In addition to claiming German patents as war booty, the United States and Britain seized the German patent office's actual facilities in Berlin, which had been severely damaged over the course of the war (many patent files were brought to a potash mine in Heringen in the state of Hessen).[74] As Douglas O'Reagan relates,

> American teams microfilmed its contents and shipped copies to London. This created considerable ire from France and the Soviet Union, who were denied full access. Only Britain's "excellent relations" with the United States "accorded, unofficially, certain privileges," among them this access to the full body of German patents. In occupied Germany, not only were all patents up for grabs, but there was no system for filing new patents. Until a uniform policy could be decided upon with quadripartite consent, each zone was free to create and enforce (or not) its own regulations for allowing trademarks, patents, and copyright. From 1945 to 1949 it was impossible for Germans (or anyone else) to register new patents or trademarks within Germany, or for Germans to patent abroad (other than refugees or others who might be declared "not Germans").[75]

Stefan Szegö was one such example of a refugee German citizen authorized to patent in the United States. Industrial, military, and intelligence "investigators" also scoured (and occasionally ransacked) all manner of German research institutions and laboratories for information seen as auxiliary to the IP that formed the basis of the Allies' "intellectual reparations," with lab notes, journals, and patent application drafts among the materials they seized.[76]

Reconstituting a German patent system, although certainly necessary, was not a priority for any of the occupying powers. The Soviets and the French saw this as an effort that would explicitly undermine their own efforts to gain access to German technology.[77] The Soviets also tended to see patent systems as handmaidens of excessive capitalism. The French, long known to have one of the weakest patent systems of the advanced nations, rejected the idea of any kind of strongly centralized body in a new Germany.[78] The Anglo-US alliance was truly at the helm in redesigning the German patent system.

Windfalls of IP cannot be addressed without also discussing brainpower. Most prominent is the story of German Jewish émigrés joining American industry.[79] Measuring the direct impact of immigration on patent activity is difficult, as patenting is, generally speaking, perpetually on the rise. The economists Petra Moser, Alessandra Voena, and Fabian Waldinger have nevertheless estimated that US patenting in certain subfields (particularly in chemistry) grew by as much as 31 percent as a direct result of German Jewish inventors emigrating to the United States.[80]

The management of property (instead of people, as with Operation Paperclip) took place within the US Department of Justice under an evolving office that more or less did the same work despite frequent name changes: the Office of Alien Property Custodian (1917–1934), the Alien Property Bureau (1934–1941), the Alien Property Division (1941–1942), the second incarnation of the Office of Alien Property Custodian (1942–1946), and the Office of Alien Property (1946–1966).[81] The office, here identified as OAP for convenience, was established by Woodrow Wilson in the throes of the First World War and performed critical tasks relating to the management of enemy property: dissolving and selling property, liquidating assets, and voiding IP, mainly as a strategy to prevent enemy nations from retaining industrial power in key military-industrial sectors. In just one year, the OAP had turned into "the biggest trust institution in the world, a director of vast business enterprises of varied nature, a detective agency, and a court of equity."[82] The OAP's seizure of the assets of the German chemical company Bayer is a case in point. In 1919 OAP auctioned off Bayer's property, including its extremely lucrative patent for the pain reliever known as Aspirin.[83] That same year, over 4,500 German chemical patents valued at over 8 million US dollars were sold to the American company Chemical Foundation, which paid about 3 percent of that estimated value and then licensed the patents directly to American companies.[84]

In World War II and its aftermath, the Roosevelt, Truman, and Eisenhower administrations approached the use of OAP in a manner that can best be described as preventative, certainly more so than punitive. All three men understood that Hitler's success in terrorizing Europe depended in large part on the support of major industrialists and their firms. As the roots of this support lay in ideology more than anything else, economic penalties alone, such as selling off patents, would not be sufficient. The OAP used its proprietary leverage to tell companies, particularly those critical to the military-industrial complex, such as Junkers, what they could and could not manufacture and bring to market.

In England, the nerve center for collecting and exploiting German patents was the British Intelligence Objectives Subcommittee (BIOS).[85] BIOS commissioned a number of intelligence reports on German industry in the wake of World War II. One, dubbed the Report on German Patent Records, had the stated objective of surveying "the potentialities of patent files of official bodies and industrial concerns as a source of German wartime patent specifications, and especially of secret and other unpublished ones, and [obtaining] any other useful information available in relation to patent records in Germany."[86] The major "targets" that were visited included titans of German industry: Krupp, Henkel & Cie., Vereinigte Stahlwerke, Deutsche Gold- und Silberscheideanstalt, I. G. Farben, and Maschinenfabrik Augsburg-Nürnberg (which would ultimately be dissolved in the Nuremberg trials).[87] The BIOS team found that a crypto-patent system (or "secret patent system," as it was sometimes known) covering about 12,000 patents and about 25,000 patent applications had been in effect during the war, a system that sought to guarantee the continuity of advanced German industrial patent activity throughout the course of the war and thereafter. The BIOS team destroyed all but the most important of the secret system's files and microfilmed the remainder. What the investigators found only confirmed the scope of German industries' involvement in Nazi operations. Krupp, the steel manufacturer, was foremost in this, with almost 1,190 patent applications pending by the end of the war, a number of which were secret.[88] Prior to the Allied capture of the Ruhrgebiet, where Krupp was based, Nazi leadership had issued the secret patents that they planned to ultimately destroy to prevent their falling into the hands of the enemy.[89]

Pondering the difficult future of the German patent office and the tremendous backlog it had to contend with, the BIOS report noted that "the Patent Office, when re-established, should be given power to waive the 'novelty' examination until such time as it becomes practicable to resume more normal Patent Office procedure."[90] The British ardently desired a future German patent system that would resist another rampant build-up of the military-industrial alliance, and also one that could be a "permanent harness" for German minds, which the British could exploit through good trade relations.⁹¹ Douglas O'Reagan notes that the historical echoes of the paternalistic approach taken by British administrators show how "the British patent system was built on concerns of acquiring foreign technology and keeping British innovation in-house. 'Letters of Protection' under King Edward II in the fourteenth century encouraged foreign craftsmen to settle in England and teach apprentices, a practice renewed in the mid-sixteenth century."⁹²

Nikolaus Pevsner, Russian-German émigré and well-known architectural historian, was about to assume a critical role in Britain's management of its own industrial assets, including patents, both at home and in British-occupied territory in Germany and Austria. The British Council of Industrial Design (COID) summonsed Pevsner to lead a study expedition to Germany in July and August of 1946 to report on the state of German industry, using his familiarity with the country's manufacturing scene, his fluency in the language, and his excellent connections to great effect.⁹³ Britain feared that the quality of its industrial design had fallen behind that of Germany and the United States, among others, during the interwar period. Studying the German industrial landscape would give the British insight into what changes had been under way and how their country could compete accordingly, while also giving it crucial information on how (or whether) German industry was obeying international restrictions placed on it after World War II. What Pevsner may have lacked in everyday charm he certainly made up for with his verve for this kind of research, which by all accounts suited him perfectly. The final aim of this expedition, part overt industrial espionage and part IP management, was a thorough report to be made to BIOS.⁹⁴

Nine years prior, Pevsner had written a book for the Cambridge University Press titled *An Enquiry into Industrial Art in England*.⁹⁵ The book was widely read and somewhat controversial in its assertion that 90 percent of British products were "devoid of any aesthetic merit" and that the state of design (and patents thereof) in England was, generally, "artistically objectionable."⁹⁶ The text was nevertheless influential, particularly among men of industry who recognized it as spurring the public's heightened sensibility for industrial design in the wake of the International Style's broad changes to the design world. One would have thought that Pevsner, as a German examining English design (not unlike Hermann Muthesius before him), would have to wade into the

English industrial landscape with caution so as to neither offend his host country nor push the same clichéd criticisms of English design and craftsmanship that had been bandied about in Germany for decades. Pevsner was successful in this tightrope walk despite his sweeping criticism, taking to task manufacturers that seem to have deserved it while bringing forward those that were, in fact, doing good work. Pevsner advocated for a kind of rational, post-Victorian design sensibility that was neither nationalist nor uncritical. His more pointed critiques of industry came when industrialists or designers demonstrated shoddy logic, as when the proprietor of an old-fashioned firm manufacturing electric and gas fittings declared outright and unapologetically that he was "not interested in the question of public taste in design."[97] For BIOS, Pevsner and his team utilized his long-running practice of memorizing questions to ask interviewees and asking them without taking notes on-site so as to make the interviewees feel more at ease. Then, in the car outside the factory, they would frantically fill their notebooks with their observations before they had a chance to forget anything.[98] Indeed, Pevsner's tone as an industrial spy for the BIOS report on Germany differs so little in character from his mass market book on England that it prompts the question of what, if anything, "intellectual reparations" actually meant to him.[99] The undated final report was probably completed in May 1947 and circulated as a small typescript in August 1947 among an unknown number of British industrialists and government officials.[100]

Pevsner traveled in his native Germany with the rank and uniform of a British officer, which imbued his visits with the gravitas that came with these.[101] One of the first things that strikes the reader about Pevsner's portion of the report was his interest in a cross-section of industries. For each city, there are studies of small, medium, and large firms, as well as firms that are known for their cheap wares along with others known for being refined and more expensive. Also, Pevsner (and his colleagues) interviewed and assessed the work of the lowly factory assistant, the factory manager, and everyone in between. The report displays an ethnographic rigor that is the hallmark of Pevsner's work as an architectural historian.[102] Another striking thing is the extent to which German companies, and their design and manufacturing practices, were inflected by Nazi aesthetics, which Pevsner found ham-fisted and retrograde even when disabused of its political content (to the extent that that was possible). Huge companies holding hundreds of patents between them—Siemens, Telefunken, SABA, WMF, Pelikan, Vorwerk, Junker & Ruh—were guilty of letting politics surpass design aesthetics: "The German idealization of craftsmanship in many cases led to 'heavy and pompous' ornamentations, especially in cases of commissioned work for the bourgeois elites or representatives of the Nazi regime."[103]

Pevsner was particularly sensitive to the labor practices and the IP (when it existed) of the construction industry. In his discussion of furniture design and interior architecture (*Innenarchitektur*), he outlined the conditions and limits of attribution and IP among the designers: "Designs purchased from an *Innenarchitekt* could only be modified or altered with his approval, and he was of course obliged to give working-drawings of any complicated details. Generally speaking, there was no obligation to mention the name of the designer but leading *Innenarchitekten* could naturally dictate their own terms and conditions. It was understood that a design, once accepted by the firm, was not submitted to another, but the *Innenarchitekt* could work for different types of firms without breach of etiquette."[104] Pevsner noted that, as a general rule, an *Innenarchitekt* made approximately 100 marks for the design of a single piece of furniture, or, less commonly, a 1 percent royalty on the selling price.[105] Pevsner was quite adamant about the importance and agency of the *Innenarchitekt* in German industry, noting that "the fact which impressed us most which seemed to hold the best lesson was the universal acceptance of the *Innenarchitekt*, a professional man who does not exist in [this] country where the furnishing and equipping of houses is more often than not undertaken by interior decorators or stores who have only a sales interest to spur them on. The *Innenarchitekt*, with his catholic training at Fach-, Kunstgewerbe- and/or Technische Hochschulen, is able to bring to the problems of design all the advantages of an unbiased, professional mind, and a keenly developed intelligence and imagination."[106]

In the town of Hattdorf, northeast of Göttingen, Pevsner met a carpenter by the name of May who eagerly showed him a catalogue of kitchen units that were unapologetically based on the work of the designer Otto Escher. May proudly told Pevsner that Escher's designs were unpatented, as if he had found an ingenious loophole for marketing his products. Pevsner, disapproving of May's pride in design plagiarism, indicated to the reader that, despite May's cheaper versions, "Escher achieved his place in the market through

The Repository

ingenious and well thought-out technical construction which was the result of exclusive and close co-operation between the designer (himself) and the manufacturer. [Escher] maintained his place in the market by his impeccable methods of production."[107] Buyers, he noted, were generally willing to pay 100 marks more for Escher's products, even if unpatented and copied elsewhere at cheaper prices. Pevsner nevertheless spoke favorably of the practice of patenting when visiting the plant of Wilhelm Knoll in Stuttgart, the well-known furniture manufacturer, whom he praised as the patent holder for the famous springs used in sofas and armchairs and known as Knoll Antimot in Germany and Parker-Knoll in England.[108] Pevsner admired Knoll's independence and singular position in the field of furniture production: "All his trade went, of course, to retailers whom, in contrast to the usual practice, he did not consult on matters of design: rather were they induced to take what he produced. All his chairs bore a little nameplate and a cardboard label, so that it can be presumed the public knew his designs in spite of the fact that he did not [do any] national advertising in Germany, where such advertising is usually undertaken only for office furniture and bedding."[109]

While exploring cutlery production in Solingen, southeast of Düsseldorf, Pevsner commented on the recognition of official registrations of design (*Musterschutzen*). In speaking with a manufacturer who found its registrations to be regularly infringed upon, Pevsner was informed that a strongly-worded letter from the firm's attorney was typically enough to get an infringer to cease and desist or otherwise prove that they had in fact invented the item in question.[110] "Only small firms of no standing would disregard such a warning," the firm's director told him.[111] Finally, in reviewing kitchen and bathroom equipment, Pevsner was impressed with a manufacturer that had its own drawing office where engineers and draftsmen were supervised by a professional architect (*Diplom-Ingenieur*).[112] The overseeing architect followed the state's so-called industrial norms (*Normen-Ausschuss*), and the company had a demonstration room and model kitchen where a trained cook demonstrated the tools and answered the questions of onlooking housewives.[113]

Pevsner's survey of the postwar German industrial landscape is so useful because of its ethnographic rigor. That rigor gives us candid snapshots of how and when knowledge (or lack thereof), appreciation (or lack thereof), and creativity with patents were approached at the level of the individual inventor, the small firm, and the large firm. Precisely this kind of survey is what gives us a tactical understanding of how patents did and did not factor in the everyday lives of the design world's denizens: designers, manufacturers, laborers, and consumers. It is a much-needed counterpoint to the world of patent strategy, formulated by governments and represented by museums through models and drawings. Where these two parties—the denizens of the design world and the legal-bureaucratic apparatus—meet face-to-face is in the patent office. There the dynamics will shift once again.

5 *The Patent Office*

In 1850 Charles Dickens published a short story in the periodical *Household Words* entitled "Poor Man's Tale of a Patent." It is the story of a poor and beleaguered provincial inventor named John who travels to London to patent an unspecified invention that he has toiled over for some time and in which he has invested the little spare money he has. In London, John is faced with the bureaucratic realities of the Patent Office: The process is much slower and more expensive than John had imagined. John's journey through the Patent Office is as arduous and expensive as it is comical for Dickens's reader:

> At the Patent Office in Lincoln's Inn, they made "a draft of the Queen's bill," of my invention, and a "docket of the bill." I paid five pound, ten, and six, for this. They "engrossed two copies of the bill; one for the Signet Office, and one for the Privy-Seal Office." I paid one pound, seven, and six, for this. Stamp duty over and above, three pound. The Engrossing Clerk of the same office engrossed the Queen's bill for signature. I paid him one pound, one. Stamp-duty, again, one pound, ten. I was next to take the Queen's bill to the Attorney-General again, and get it signed again. I took it, and paid five pound more. I fetched it away, and took it to

the Home Secretary again. He sent it to the Queen again. She signed it again. I paid seven pound, thirteen, and six, more, for this. I had been over a month at Thomas Joy's. I was quite wore out, patience and pocket.... But I hadn't nigh done yet. The Queen's bill was to be took to the Signet Office in Somerset House, Strand—where the stamp shop is. The Clerk of the Signet made "a Signet bill for the Lord Keeper of the Privy Seal." I paid him four pound, seven. The Clerk of the Lord Keeper of the Privy Seal made "a Privy-Seal bill for the Lord Chancellor." I paid him, four pound, two. The Privy-Seal bill was handed over to the Clerk of the Patents, who engrossed the aforesaid. I paid him five pound, seventeen, and eight; at the same time, I paid Stamp-duty for the Patent, in one lump, thirty pound. I next paid for "boxes for the Patent," nine and sixpence. Note. Thomas Joy would have made the same at a profit for eighteen-pence. I next paid "fees to the Deputy, the Lord Chancellor's Purse-bearer," two pound, two. I next paid "fees to the Clerk of the Hanaper," seven pound, thirteen. I next paid "fees to the Deputy Clerk of the Hanaper," ten shillings. I next paid, to the Lord Chancellor again, one pound, eleven, and six. Last of all, I paid "fees to the Deputy Sealer, and Deputy Chaff-wax," ten shillings and sixpence. I had lodged at Thomas Joy's over six weeks, and the unopposed Patent for my invention, for England only, had cost me ninety-six pound, seven, and eightpence. If I had taken it out for the United Kingdom, it would have cost me more than three hundred pound.[1]

John's extortionate tour through the English patent system was obviously something of an embellishment. Yet, like most of Dickens's work, it tapped into the latent public sentiment of the mid-nineteenth century, in this case the public's belief that the patent system was an exploitative one that catered to the rich and stymied the upward mobility of would-be inventors. Dickens's hidden critique of the English patent system was indeed prescient, anticipating the English patent reforms that would make it easier for workaday inventors like John to patent inventions with lower fees and less bureaucratic red tape.[2] Dickens himself also had a vested interest in IP. He had long been concerned with his own copyright protections and the unauthorized reproductions of his work that sprouted up everywhere, and there is probably a little

Marcus Stone, engraving illustration for "Poor Man's Tale of a Patent," as it appeared in Charles Dickens, *American Notes and Reprinted Pieces* (London: Chapman & Hall, 1896).

bit of a young Dickens in the character John, trying to fight for his inventive rights in a byzantine and adversarial system.

John, a provincial inventor who travels to the metropole to represent himself in the pursuit of a patent, is an attempt at a biographical sketch of a typical inventor around 1850. But that typified profile would change dramatically by the end of the century, making way for the inventor who would arrive at the patent office with a patent lawyer who, like a well-oiled machine, would guide his client through the patent process. This patent lawyer of around 1900 arrived at the patent offices in Washington, London, Paris, or Berlin ready to prosecute a patent with a suite of well-rehearsed arguments for the most important of all patent criteria: nonobviousness. Although the kind of supporting evidence (such as models and drawings) that the patent lawyer can bring has changed over time, the patent office was and still is the site of deliberation regarding the sufficiency of that critical argument about nonobviousness. A winning argument rewards the inventor with the privileges of exclusivity that a patent confers. However, more often than not, the initial application is neither an unqualified success nor a complete failure. Instead, a diligent patent examiner, having researched the inventor's prior art and surveyed inventions that are not cited, will distinguish between claims that are reasonable within that office's framework and those that are not, giving inventors a chance to revise the application and return when they are ready to resubmit it, typically with narrower, more tempered claims. Today, it is not unusual for a patent application to bounce back and forth between an inventor and an examiner a dozen or so times; as the reservoir of prior art grows exponentially, that trend toward more and more back and forth also grows.

The patent office is, as such, a place that has a distinctive role in the cultural and technological ecosystem of industrialized nations and yet is also shrouded in a great deal of mystery. The patent office is a civic space with which most people will never engage, unlike a courtroom, a post office, or a city hall. To most people, patents represent technical knowledge that seems abstract and distant. There are several ways to understand the patent office that may help demystify it as a place. One way is as the site where, more than any other place one might imagine, the nature of obviousness (and nonobviousness) is parsed. The setting for this parsing is not at all incidental. On the one hand, the patent office is fundamentally tethered to law and the legal system as a whole. However, it is not a place of juridical deliberation, as no one involved in the patent process has committed a crime, seeks to annul an agreement, or otherwise solicits recourse through the law (patent lawsuits do occur but they happen in courtrooms, as we will see). Rather, it is a place where the onus is entirely on the inventor to essentially audition for a privilege that is enforceable by law. While the nonjuridical character of the patent office may lessen the stakes, identifying nonobviousness—the most critical of thresholds for inventiveness—still entails a serious intellectual quest. It is important not to be fooled by the bureaucracy and paper pushing; patent offices are sites of intense philosophical activity that impact every aspect of everyday life.

A second way to understand the patent office is as a space of scrutiny. The certification of innovative ideas in the academy undergoes a similar process through peer review, where citations mirror the patent office's prior art. However, the scrutiny that takes place at the patent office is altogether different; its things and processes envision a concrete, material change in how industry—and in turn society—will work, no matter how minute that change is. Parsing inventions rather than ideas, patent examiners are the umpires of the playing field that is innovation. The sports analogy here is not simply an embellishment; it is invoked to signal what keeps innovation fair, orderly, and observant of the rules of the game—otherwise, it would be a melee.

Finally, the imperatives to demonstrate nonobviousness and to scrutinize are both activities that manifest the critical nodes of the American, British, French, and German patent offices and later the national and multinational offices around the globe that provide vital mediation between the state and the individual, between the corporation and its regulation, and between piracy and just compensation. To end this chapter, we will examine the spatial characteristics of these vital dynamics and how they have changed over time.

Nonobviousness

In everyday life, "obviousness" is the kind of trait that is measured by the gut, something that tends to follow the dictum that you will "know it when you see it."[3] The law, however, requires clear tools for recognizing what is obvious and what is not, particularly as this is so often the crux of an invention's patentability. The legal scholar Kevin Emerson Collins notes that a "functional good that departs from the prior art only in its aesthetics, such as a hammer with a wavy handle, is not functional innovation but rather aesthetic innovation in a functional good, and it is obvious under nonobviousness's functionality mandate. In contrast, a hammer whose innovative design makes it better at pounding nails may be nonobvious."[4]

The so-called "functionality mandate" of nonobviousness is a construct linked to Thomas Jefferson, the inventor, architect, and third president of the United States.[5] Jefferson, who was instrumental in shaping the American patent system, had feared that the patent system would merely be an index of fashions and trends if it did not rigorously apply metrics of newness, utility, and nonobviousness.[6] Jefferson was adamant that a functionality mandate was needed to ensure that an invention that was a "revolutionary alteration of only formal, aesthetic features without a concomitant functional departure from prior art" would not pass muster as a patentable invention.[7]

Newness is proved by cross-referencing prior art and the general marketplace and is, in the purest sense, unambiguous: something either exists already or it does not. Utility, the second threshold of patentability, is where interpretive metrics become necessary. All historical patent systems, by statute, have had some form of a utility doctrine, that is, a means for validating an assertion of functionality.[8] Ambiguity creeps into the assessment of utility when invented objects can theoretically have more than one distinct function. For example, a Brancusi sculpture, Ferrari Hardoy's butterfly chair, and Frank Lloyd Wright's Luxfer prisms could all be used as paperweights or doorstops if one wanted to do so, but that is clearly not their intended function. In addition, there are objects, such as wooden wedges used as doorstops, that are more suitable to the function because of their shape, material, and/or economy. In most utility doctrines, the utility that the inventor claims must not be a "throwaway" claim, or an incidental utility that is irrelevant to the design of the invention, such as using a sculpture as a paperweight.[9] This is just one of the many reasons that design—and knowing what it is and is not—is so critical to the patent process, and it is one of many moments when designers can claim an instructive role in thinking about patents.

Collins has described the measurement of these criteria—newness, utility, and finally nonobviousness—as a series of "screens" that have been encoded into law through various legal precedents over time. Although Collins's subject is US patent law, the concept of "screens" proves to be useful across global patent systems.[10] When one screen has been found insufficient to measure patentability, other screens are introduced through legal precedent to assist in measuring patentability. Just as all good jurisprudence relies on legal precedent, a mountain of case law dating back to the eighteenth century supports our understanding of the contours and mechanics of these screens. One example is the "functionality screen," which comes into play when an assessment of the utility threshold is insufficient to establish whether an instance of IP is patentable or copyrightable: "When the innovative feature is both expressive and functional and the feature's two attributes cannot be separated either physically or conceptually, the functionality screen excludes the innovation from the copyright regime, ceding jurisdiction over the innovation to patent."[11]

Another screen that is vital to patents in architecture and, more recently, patents of "dispositions of space" is the "aesthetic authorship screen," which "relies on nonobviousness to do [the] more difficult work ... of sorting innovation based on whether its departure from the prior art is functional or aesthetic."[12] This works as follows. The threshold of nonobviousness necessitates novelty, which in turn necessitates a perceivable departure from prior art. The procedure to establish this typically involves two steps, the first of which gathers basic facts about the invention's provenance and context and the second of which is the fictive act of conjuring a PHOSITA (person holding ordinary skill in the art) and *imagining* whether they would find the advance to be obvious or not. This is not as subjective as it may sound, as there are any number of precedents, as well as the "screens" developed from them, to look to as guides for such an assessment. One of those screens is Collins's aesthetic authorship screen, according to which "a mere change in form is obvious; the change in form must lead to a change in function in order to weigh in favor of nonobviousness."[13] That is, is the change a change merely in form or in form *and* function? Collins notes that both the "functionality" and "aesthetic authorship screens" "evolved from a practice to a statutory requirement to a judicially enforced, hard-and-fast rule to the purportedly advisory rule of thumb that exists today."[14]

What this meant in practice was that designs that did not impact functionality—such as the hammer with a wavy

The Patent Office 225

handle—were, even if new, of no concern to the PHOSITA and hence of no import for the measurement of obviousness.[15] As Collins notes, "mere changes in form are simply not variables in the calculus that the PHOSITA employs to identify nonobviousness. Satisfying consumer preferences for visual beauty does not count as doing something under the authorship screen because, if it did, patents would upset the competition-protection balance for aesthetic innovation that should be set by copyright law."[16] In other words, the functionality mandate was critical for keeping the regimes of copyright and patents sovereign and not interdependent. In European patent offices, the measurement of nonobviousness has often been called something a bit less mercurial: the measurement of an inventive step. In practice, however, the two are largely the same, with both demanding that explicit, critical knowledge related to prior art be included in an application.[17] Nonobviousness also tacitly frames "invention" as any kind of new step that was somehow not already entirely self-evident.

Obviousness also comes into play as an aspect of the patent disclosure itself. As Mario Biagioli has pointed out, the summary descriptions of patent disclosures "were not meant to enable the reproduction of the invention after the patents' expiration—a function that was typically taken up by provisions about the training of workers and artisans to build and operate the invention *in loco*—a 'disclosure' through bodies rather than texts."[18] Indeed, one of the most common mistakes made by patent applicants is disclosing too much information.[19] The imperative for an invention to be nonobvious or to contain an inventive step does not mean that the exegetical aspect of the disclosure of a patent (through drawings and writing) must itself be so exhaustive as to permit full reproducibility. This allows for a significant amount of strategic opacity in patent disclosures. For the specifically spatial patents that have emerged in the last five decades, patents whose inventive claim is primarily a "disposition of space," a number of thresholds and screens do not yet appear to be properly working.[20] This has led to a number of patents that are erroneous or redundant. One reason for this is that the "authorship screen," or more simply who is able to file a patent in the first place and how, is predicated on expensive patent filing fees that cannot be afforded by all, and another reason is a high level of multifunctionality, which dispositions of space tend to have.[21]

Johan Vaaler, Paper Clip or Holder, US Patent 675,761, filed January 2, 1901.

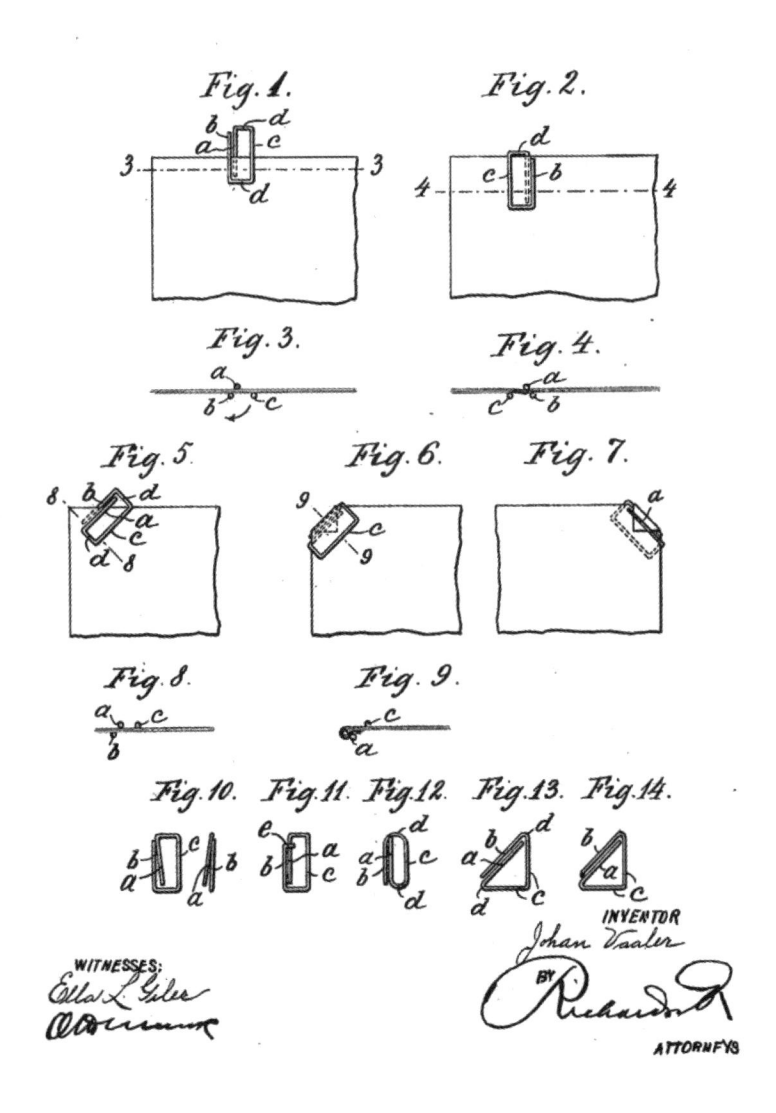

Let us turn to a heuristic consideration of these thresholds and screens. The humble paper clip is an invention couched in a great deal of lore (and not just a little uncertainty). The common origin story is that a Norwegian inventor by the name of Johan Vaaler saw the need for a simple device to bind sheets of paper together quickly and temporarily. Norway, which was a territory of Sweden at the time of the invention, did not have its own patent system. Vaaler's drawing for a paper clip was recognized by a special government commission, but he was encouraged to seek a formal patent in Germany (often the default patent system for European inventors who lived in countries that did not have their own patent systems). When Vaaler's invention reached the German patent office, it quickly passed the newness threshold because no identical invention existed, neither in the German patent system nor, to the best of the German patent office's knowledge, elsewhere. The utility threshold was a bit more complicated, as one could theoretically assert the paper clip's utility for purposes other than binding paper: as a clip for hair or, when extended, as a prod for cleaning out dirt and grime from small nooks and crannies. Here the functionality screen steps in, independent of what the inventor claims, to determine that, yes, while a paper clip could act as a hair clip or a prod, its functionality as a temporary binder of multiple sheets of paper was its most effective use, meaning that the other functions were "throwaway" functions for the purpose of patenting. Finally, there was the most important threshold of nonobviousness. Thin, bendable metal wire had long existed. So too had ribbon to tether loose sheets of paper together. Thus, there was sufficient evidence to prove that there was both an untapped material invention (the wire) and a need for binding loose paper (the ribbon). Valer's paper clip was rightfully patented, and the humble object took on special meaning in World War II when Norwegians would fix paper clips to their lapels to show patriotism and to annoy their German occupiers. The paper clip became such a strong symbol of resistance that one could be arrested for wearing it.[22]

Patent Examination

Patent examiners fill a largely unheralded role in the advancement of modern science, technology, design, and construction as they unwittingly influence and arbitrate design trends among consumers. As bureaucrats of the state who carefully and fairly deliberate over civilian claims, their role is not unlike that of a judge. However, they rarely have the kind of visibility a judge enjoys, which renders them largely anonymous (apart from having their name appear on successful patents) and seemingly interchangeable. Patent examiners were rare figures in the early days of patent offices, and the variety and slow pace of patent applications meant that patent examiners were often examining vastly different types of inventions, as Thomas Jefferson did as the first patent examiner of the US Patent Office between 1790 and 1793, when patent examination duties fell on the Secretary of State.[23] Ever since, the scope of an examiner's work has become exponentially more circumscribed, heightening the level of specialization required of the examiner—which, most agree, makes for a more rigorous review process while requiring a much larger work force in the Patent Office.

Clara Barton, later the founder of the American Red Cross, was the first woman to serve as a patent examiner when she joined the US Patent Office in 1854, a position that the US Patent Office ultimately withdrew from her because of her male colleagues' consternation, reducing her to the role of copyist by 1858.[24] As one article reported: "The Male clerks resented the regular employment of a woman, as bravely and gallantly as if they had been a lot of medical students resenting a woman's undertaking to study the healing art. They formed in line to stare and whistle at her as she passed by. They slandered her to the Commissioner. He told their spokesman to prove his charge by a certain date and Miss— should go; to fail with this proof, and the spokesman himself should go. Somebody went out of office, but it was not the woman clerk. She remained in office three years. Her books are still shown as models of good work."[25]

Albert Einstein, the only other patent examiner to make a major public name for himself, was more specialized, examining precision machines and timekeeping devices for the Swiss patent office in Bern between 1902 and 1909.[26] After graduating from the Federal Polytechnic University, Einstein had initially tried to obtain a job as a schoolteacher but, short on luck, settled for a job as a patent examiner that a friend, Marcel Grossman, had helped him identify.[27] By all accounts, Einstein found patent examination work to be tedious, but this did not preclude him from executing his own work. In 1905, while working as a patent examiner,

British comic strip pertaining to the fickle nature of Victorian consumers. It reads: "The public are capricious. What they want is the latest thing, not the best. If you want to make a fortune over patent articles, the thing is to be last in the field." May 4, 1887.

Einstein published his theory of special relativity. Einstein did this work in his off time as well as on the proverbial office clock, often quickly concluding his patent work and using the remainder of his time to tackle his own calculations. Einstein was nevertheless good at his job and was promoted in 1906, three years before finally leaving to teach theoretical physics at the University of Zurich. In his later years, as he reflected on his time as a patent examiner, Einstein referred to the patent office as "a worldly cloister" where he was able to hatch "his most beautiful ideas."[28]

The profession of patent examiner was one of the many new careers spawned by the Industrial Revolution, and by the end of the nineteenth century it was a well-regarded one, often attracting those who were interested in science and technology but nevertheless chose not to pursue the advanced degrees that were required to enter the academy.[29] In England, the profession received a royal charter in the late nineteenth century, recognizing it both as a profession and as a distinct personality of the legal system.[30] In Germany, patent examination had begun as a part-time job for professionals already in science and technology fields, making for a situation where examiners were often examining their colleagues' work.[31] The German patent office in the late nineteenth century had only three full-time employees, which allowed the patent office a certain flexibility but was ultimately deemed to be too conducive to conflicts of interest. By the early twentieth century, being a patent examiner in Germany was, as in England, a full-time and independent profession.

Jefferson, Barton, and Einstein all made most (but not all) of their major contributions to public life and public knowledge *after* their tenure as patent examiners, and this is no coincidence. Patent examiners on either side of the Atlantic were expected to keep low public profiles to avoid conflicts of interest. Some were even barred from submitting patents themselves. In most circumstances, examiners who had filed applications for which patents had not yet been issued prior to becoming an examiner had to drop their names from the application or, if they were the only person on the application, put the application under another person's name. They generally transferred the patent rights to a member of their family. (In the case of patents that had already been issued to an examiner prior to becoming an examiner, the examiner's name did not need to be removed or changed.)[32] It was and remains crucial that patent examiners maintain an arm's-length relationship

The Patent Office

with inventors and their attorneys, as repeat inventors working in a specific subfield will often prosecute patents with the same examiner several times over. The examiner must be reachable for questions, but answers to those questions are typically brief and purely factual, following manuals like the American Manual of Patent and Examination Procedures to the letter.[33] This also means that patent examiners do not typically feel that they are at liberty to discuss their work with the lay public as it is technically prohibited, a restriction that made interviewing patent examiners for this book a challenge.

One patent examiner, a specialist in HVAC systems who will be identified as "Raul," was willing to be interviewed and shared considerable insights into the mechanics of the examination process as it stood in the highly specialized patent landscape of January 2022, when the interview was conducted.[34] Raul has worked for the US Patent Office for 15 years and holds a bachelor's degree in mechanical engineering from a large state research university in a mid-Atlantic state. His interview for the position lasted approximately two hours and comprised several questions from the "art unit" managers who would become his supervisors about the kinds of classes he had taken.

When a patent application for an HVAC invention arrives at the Patent Office, it is first placed in something known as the "control repository," essentially a digital waiting room from which an analyst will ultimately direct the application to a specific "tree"—which in the case of an HVAC invention would be building systems. The building systems tree has several "classes" or "branches," and examiners are allowed to request which of the branches they will work in. One class is HVAC systems, for which Raul is a primary examiner. Raul noted that he felt qualified to examine virtually all of the cases assigned to him by his manager and would perhaps reroute at most one or two cases per year to a different examiner if he did not feel he had the appropriate expertise. The classes are hard and fast, so when Raul receives a case that seems to be both an HVAC invention and, let's say, an insulation invention, he must ask the applicant to split the patent application into two (or more) applications. The current application fee for a US patent ($1,800) allows for a guarantee of two days of Raul's time, and no matter how easy or hard the application is, he must complete it within that time to ensure equity between patent applications. These two-day slots always emerge from an application backlog that is typically between six and eighteen months, depending on a number of factors that can impact wait times: the number of applications in the class, the number of examiners in the class, and the overall caseload at the Patent Office.

Most of the HVAC inventions that arrive on Raul's desk are mediated by the inventor's patent lawyer, a mediation that Raul does not feel is actually necessary when the inventor and examiner are both working in good faith. He is nevertheless impressed with the expertise of the lawyers he meets, countering the assumption some might have that patent lawyers do not know very much about the science or mechanics of the invention they are representing. In fact, many of the patent lawyers also studied engineering in their undergraduate education, just as Raul had. Raul has found that patent attorneys representing HVAC inventions are well versed in technological jargon, but he has also noted a distinct practice of patent attorneys needlessly extending the time spent prosecuting a patent in order to maximize billable hours from their client, a burden that is also carried by the Patent Office itself as they spend time interacting with the lawyer. Fundamentally, Raul sees the examiner as the party with "narrowing" aims and the inventor and their attorney as the party with "broadening" aims.

Raul believes that many of his colleagues conduct examinations the same way that he does, reviewing the drawings first and then the text. Raul thinks that more than half the time, he has been able to fully grasp the invention's claims before reading the written portion of the application. When he does turn to the text, he often finds that the shortest texts are the best, while longer texts have greater deficiencies, in part because they have simply hamstrung themselves with wordy, complicated claims.

Although reluctant to admit it, Raul ultimately confessed that he is aware of the remarkable power he holds within the HVAC industry, determining more or less on his own what is inventive and monopolizable and what is not inventive and not monopolizable. Raul's understated and humble personality would seem to emerge from the very culture of the Patent Office itself, where, despite the myriad solitary decisions being made on a daily basis, individuals are trained to think of themselves as part of a larger whole. To an executive of an air conditioning or furnace manufacturer, Raul can easily seem like a mercurial gatekeeper with an immense impact on the success of their business. While Raul acknowledges that this may be the case, he makes a deliberate attempt not to think of his work in those terms, choosing instead to abide by protocol and minimize any trace of his own subjectivity. This is but one of the many reasons Raul felt that he could only give an interview after work hours on his private phone with assurances that his identity would not be revealed.

When asked, Raul was also reluctant to offer a critique of the patent system as it stands in 2022, but he ultimately did share some feelings about recent reforms to the US patent system. In 2011, under the Obama administration, the America Invents Act shifted claims to priority to the person who was first to invent a patentable invention, not the first person to actually file the invention. For example, if both Goodman Manufacturing and Carrier are working on solar-powered air conditioners, it is incumbent on the companies to document when they first started working on the invention internally rather than when they actually filed the invention with the Patent Office, which they could do at a later time. Raul noticed that this drastically reduced the filing of co-called "junk" patents in the HVAC sector, patent applications that are alternately described as "defensive" or "anticipatory." However, it also dramatically increased the importance of detailed and truthful internal documentation in the labs of corporate and individual inventors alike. Raul wonders whether the reforms have had the effect of creating an understimulated patent system, but insists that we still need to wait and see. Raul is also concerned about the sharp uptick of HVAC patents from China, which he estimates make up nearly half of his current caseload. He fears that the increase of foreign patents, combined with the decline of domestic patents, signals an America that is becoming less dynamic and less of a leader in HVAC manufacturing.

Raul also noted that, despite the ethos of collectivity at the Patent Office, there are sometimes wildly different rates of success in the patent process. Some of his colleagues approve as few as 10 percent of the patents that cross their desks, while others approve as many as 90 percent. On average, the odds are slightly in favor of the inventor. Raul comments that supervisors examine the decisions of their subordinates and direct them to reevaluate if the supervisor feels that they are being too harsh or too permissive. Part of the reason for reaching the correct balance is financial: while the application fees cover the expenses of the examination process, it is the annual renewal fees for patents that keep the overall office running, and with too few patents, that revenue will not materialize.

Historically speaking, the statistics show that successfully passing through the examination process and acquiring a patent was not particularly hard if the inventor paid attention to all of the thresholds, screens, and mandates described earlier. In Britain in 1906, for example, more than one third of the inventions submitted to the patent office were found to be entirely unanticipated by prior art, while about one half were found to be partly anticipated, requiring the inventors to revise their claims. In most cases, after reworking the claims for these applications, the inventors were successful in patenting the aspects of an invention that were unanticipated by prior art. The remainder were found to be fully anticipated by prior art and were thus rejected outright.[35]

In the majority of cases, patent examiners have responded to inventors with questions. Almost all of Buckminster Fuller's nearly three dozen patents began with a list of detailed questions from the patent examiner in charge, questions that would arrive on Fuller's desk by way of his lawyers. For his famed Geodesic Dome patent, for example, the patent examiner bristled at the grandiose language of some of the main claims, such as "a building framework of generally spherical form constructed of interconnected structural elements of substantially uniform length arranged as the chords of great circle arcs" and a "[b]uilding framework of generally spherical form in which the main structural elements are aligned with great circles of a common sphere."[36] Reading between the lines, it would seem as though the patent examiner was fearful that Fuller was attempting to make a claim over a broad swath of hemispherical structures that did or could exist, using the ostensible specificity of the verbiage "uniform length," "chords," and "arcs" to mask what was already a fairly commonplace method of assembling domes. The examiner pointed Fuller and his lawyers toward a claim that he believed was truly circumscribed, and not just masquerading as such: the dome's incredibly light weight. The examiner was quick to champion the claim that the main technical improvement of the patent was, in fact, the dome's reduction of the ratio of structural weight to weather-free square footage from 50 pounds in other structures to a mere 0.78 pounds in Fuller's structure.[37] The examiner had the responsibility of questioning the method and veracity of Fuller's calculations, an interrogation that the illustrious inventor was neither used to nor suffered with much pleasure. The examiners of Fuller's patents also commonly told him that his claims were too broad and needed to be narrowed, something that a universalist like Fuller balked at but with which he ultimately complied.[38]

Although Raul indicated that he finds patent lawyers largely unnecessary, it is clear that Buckminster Fuller's patents would not have come to fruition without the savvy of his patent attorney, Donald W. Robertson. Robertson, like all of the most serious American patent attorneys, was based in the Washington, DC area, which afforded him the luxury of quick drop-ins to the patent examiners with whom he was prosecuting patents. Robertson's ability to show up at the Patent Office to meet with the examiner in person was one way that he could soften his client's ire over rejected claims. In February 1954, amidst a great deal of back and forth in the Geodesic patent case, Robertson stopped in the Patent Office and was able to sit down with the patent examiner "to engage in conversation at a moment when he was undisturbed."[39] In a recap of the meeting for Fuller, Robertson noted that "[i]t seems that what was most needed was to swing the examiner from a prosaic to an imaginative approach to the invention. He appeared to have an open mind on the subject and even went so far as to say that if we could reduce the number of claims (which he thought was too great), leaving out the claims which did not bring in the great circle structure, he thought the case would be ready to be 'passed for issue.'"[40] Robertson's contention that the examiner was thinking "prosaically" and not "imaginatively" was a subtle way to stroke Fuller's ego, to be sure, intimating that examiners lacked the imagination of the inventors they analyzed. But it was also a nudge to Fuller, reminding him that straightforward prose— and prosaic thinking—was not a bad thing as they thought about what they needed to revise to push the Geodesic Dome across the finish line at the Patent Office.

National Offices

Before the creation of multinational IP organizations in Europe, Asia, Africa, and beyond in the last quarter of the twentieth century, national patent offices were the world's bulwarks of IP law and management. Each has its own history with its own vicissitudes.

Neither internationalization nor globalization altered the intrinsic character of these independent patent offices very much. In fact, the governance systems, policies, and processes of the national patent offices are remarkably more similar than they are for other systems such as education or healthcare, and there is little need to expand on their peculiarities and differences. Several specifics do matter, however, and these form a window for viewing the subsequent rise of multinational patent bodies.

In 1954, after the United States returned the hard copies of patents it had seized from the patent office in Berlin, West Germany required a new patent office and, in an effort to spread federal-level government offices across the country, established a new German patent office in Munich (on a site that had served as barracks, just north of the future home of the European Patent Office, which will be discussed shortly).[41] By the time the unremarkable structure designed by Franz Hart and Helmuth Winkler opened in 1959, German officials were already fretting that the building would not be sufficient for the enormous upswing in patent activity that the country was experiencing.[42] The rapid economic growth of the 1950s—the so-called *Wirtschaftswunder* (economic miracle) in Germany—pushed patent applications to and then beyond prewar levels.[43] The government made exceptions to a national hiring freeze to allow the employment of additional patent examiners, but the bureaucrats still found themselves overwhelmed by the workload. Although it was a good crisis to have, it was nevertheless a crisis, and it left Germany particularly eager to play a formative role in a supranational patent office that would not only help to unify Europe but also alleviate some of the burden placed on its own national bureaucracy.

The Italian patent office, long notable in the domain of design patents, also warrants reflection here. The measurement of novelty in the Italian patent system expressly included consideration of an object's readiness for merchandizing; in other words, its ability to fit into a market that was already exploitable.[44] Design patent jurisprudence goes so far as to make special accommodation for "crowded" sectors of the market: "Where a productive sector is particularly 'crowded' (so-called crowded art), even 'slight originality' can justify the patenting of a product, in that the degree of differentiation from pre-existing models is inversely proportional to the crowding of the merchandising sector in question.... [T]he more a sector is crowded, therefore, the fewer variations on the existing state of the art required when a patent is filed."[45] Another peculiarity of the Italian system was that designs that applied to different market sectors—say, an ornamental motif or pattern that applied to both wallpaper and clothing—were not considered redundant.[46]

Over the last two centuries, national patent offices have regularly attracted a fair share of detractors. These include those who find individual patent systems to be overly permissive or overly prohibitive, those who believe that patents stymie competition, and those who believe that patent systems give large corporations an oversized power that quashes the ability of small, regional, or individual outfits to compete. In Germany, for example, such opposition has long come from the Handelskammer, a union of industry-specific companies that regulate and monitor best practices in a manner similar to a chamber of commerce.[47]

In the specific realm of architecture and construction, other bodies have played a role in challenging the absolute authority of the patent office. One such example is the Building Research Station in the United Kingdom.[48] In 1917, anticipating the housing crisis that would follow World War I, the UK Department of Scientific and Industrial Research proposed creating a national organization that would test new and experimental building materials for housing. The organization, which came to be known as the Building Research Board, came formally into existence in 1920 and became the parent organization for a nationally subsidized laboratory for testing and analyzing building materials, based first in London and later in Watford. The Building Research Station studied standards and methods for bricks and brickmaking, analyzed the behavior of reinforced concrete and structural plastics, and researched issues such as building fire safety in far greater depth than had previously been done by national agencies.

The Building Research Station did not have an official relationship with Britain's Patent Office, but many building materials manufacturers came to see certification from the Building Research Station as the gold standard of achievement for building materials that inventors sought to bring to market. Beginning in the late 1930s, foreign inventors who joined domestic inventors in seeking a rigorous and neutral evaluation of their products would often send samples to Watford first for testing, and then, with what was hopefully a successful certification, would send a patent application to the Patent Office for patenting. Although there was no explicit proviso that certification of a new material or product by the Building Research Station guaranteed success in the Patent Office, there exists no record of an invention tested by the former that was not successful in the latter. Depending on how one looks at it, the Building Research Station was becoming an organization that was either competitive or symbiotic with the Patent Office.

A block construction system from Sweden, Siporex, provides a fine example of the ambiguous relationship between the Building Research Station and the Patent Office.[49] Executives from the company headquarters in Malmö sent samples of their concrete blocks to the Building Research Station in 1939 (along with a payment for the testing) with the hope of obtaining glowing reviews from the well-respected agency. Even though the company *already* held patents for the product in both Sweden and the United Kingdom, its executives believed that the qualitative assessment of the Building Research Station would go one step further in making Siporex the go-to concrete block for mass housing construction.[50] Furthermore, with another war unfolding, the need for mass housing was once again crystal-clear. Yet, to their dismay, the company did not get the results they wanted: over the course of two sets of tests, the technicians at the Building Research Station found that the concrete was too prone to shrinkage after it became wet and was thus not certifiable. Siporex executives panicked and pointedly questioned the Building Research Station's testing methods, claiming that no such results had occurred in their own labs. The Building Research Station could not be pressured and insisted that all of the correct controls and methods had been in place for the testing. The company found itself with a patent for a product that was, at least according to the most well-respected building materials agency in the world, suboptimal; and without the certification, the patent was essentially meaningless. The Building Research Station's archives also reveal a troubling pattern of writing off inventors from countries (Yugoslavia, Turkey, Poland) that they felt did not have strong inventive traditions before even testing the products in question.[51]

Some unconventional building materials did, however, get the Building Research Station's seal of approval. One such example is cannetis (both the company and the product name), a rigid sheet-like building material comprising woven cane, which was sent to the Station by its French manufacturer in 1946.[52] Cannetis's success is an excellent example of how patents were not always

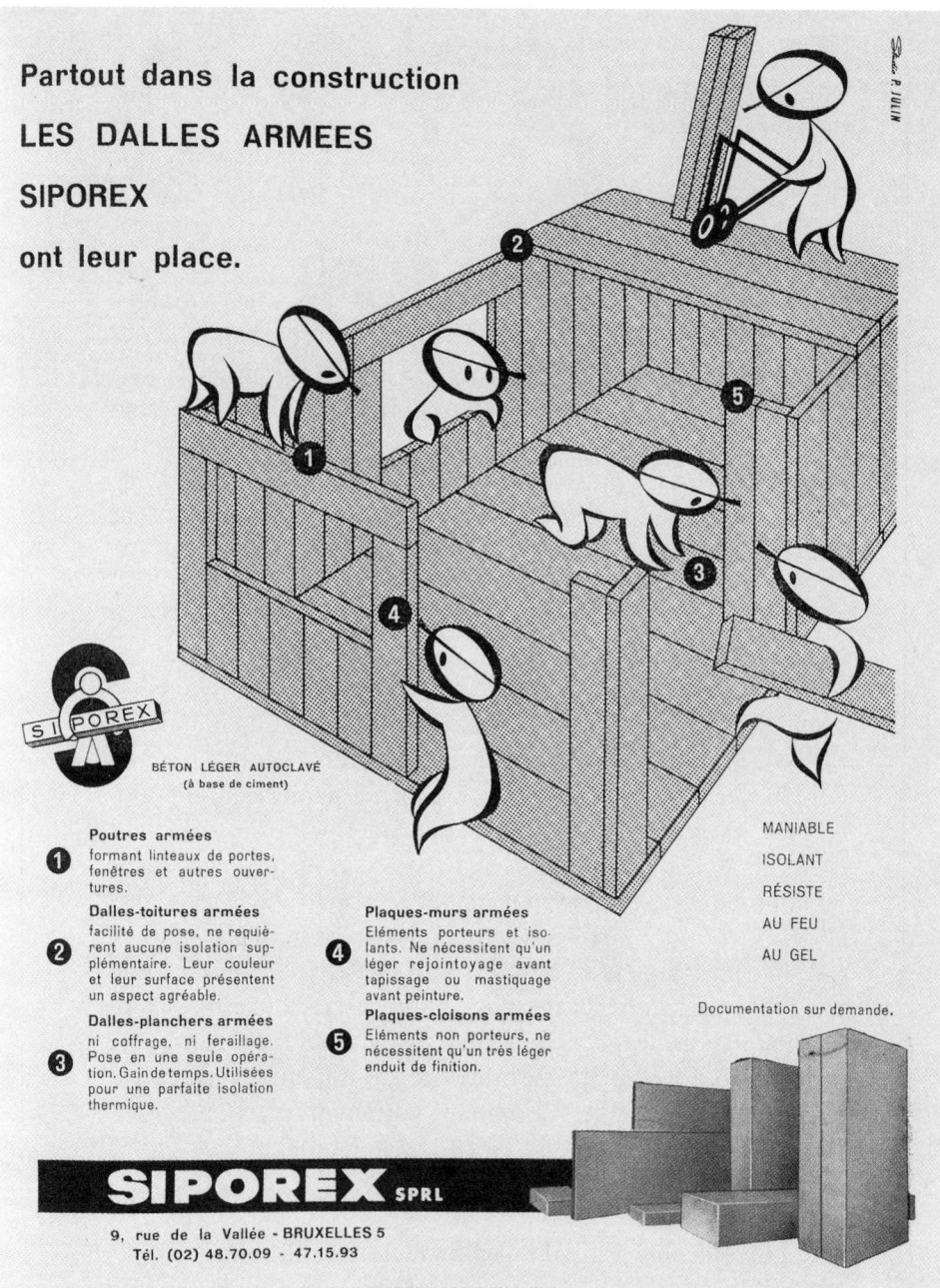

Advertisement for Siporex in *La Maison* 16, no. 3 (1960): lii.

Postcards depicting cannetis, a building material made from split plaited reeds, 1946. National Archives (UK) DSIR 4/2495. Also color plate 10.

necessary for advancing building materials. A building material and technique common to the Mediterranean littoral since antiquity, cannetis was essentially unpatentable because it had been around in more or less the same form for centuries and thus, like a simple brick or a two-by-four, could not be proven new. However, patents do exist for manufacturing accessories for cannetis, such as Paul Badosa's 1919 patent for a machine that facilitated the manufacture of cannetis by the disabled.[53] The manufacturer was essentially seeking the Station's certification of their weave of the cane, which in turn reflected the quality of their weaving machinery and their crops. Although they could not claim a monopoly on the material, the Cannetis company could use the certification as a marketing tool, reinforcing the status of cannetis as an excellent and low-cost substrate for plaster and concrete, whether arranged horizontally or vertically. The Building Research Station noted that, in addition to its high performance in lab tests, cannetis's "possible use for the cheap erection of efficient houses in the colonies is of some interest to us, as our advice on such matters is sometimes sought."[54] Indeed, the manufacturer's letter includes a postcard with a picture of a modest house built with a cannetis substrate for the Colonial Exhibition of La Rochelle in 1927.[55]

International Offices

In the twentieth century, one of the greatest challenges for the global patent system was the unification of Europe's multiple national patent systems into one streamlined office, an imperative of the greater project of post-World War II European unification.[56] Today's European Patent Office in Munich, the unified patent organ of the European Union, originated in the Council of Europe, where Henri Longchambon, a French senator, proposed its creation.[57] Longchambon's proposal, known as the "Longchambon plan," was one of numerous efforts in the wake of World War II to create an insurance policy of sorts against another world war through continental unification, efforts that led directly to the creation of the European Coal and Steel Community, a forebear of the European Union. Unification and cooperation between European states was, it should be noted, not merely a project of enlightened, anticonflict politics; it was also believed by many to be necessary for the relatively small economies of the European nation-states to be able to remain competitive in a global postwar economy that included the United States and emerging markets, particularly those in Asia.

Longchambon's plan was ultimately deemed impracticable, but it nevertheless led to several major multinational conventions that opened up channels of communication between the various national patent offices existing in Europe at midcentury. These efforts to foster cooperation between separate patent systems culminated in the Strasbourg Patent Convention of 1963, which created certain unified criteria of what constituted patentability within national patent systems, bringing greater harmony and trust to international patent policy as well as transferability.[58] Proponents of a single system, Germany's Kurt Haertel and France's François Savignon foremost among them, maintained pressure on individual governments, ultimately leading to the Munich Diplomatic Conference for the Setting Up of a European System for the Grant of Patents (the Munich Convention for short) in 1973.[59] The convention was ratified, and by June 1, 1978, when the first European patent (for a heart attack drug by BASF) was filed, eight of Europe's largest economies had formally adopted the unified patent system.[60] By the fall of the Berlin Wall, the European Patent Organization (EPO), as it came to be known, was a robust force in global patent governance, with strong connections to the World Intellectual Property Organization and collegial exchanges with the United States Patent and Trademark Office and the Japanese Patent Office as well. The EPO adopted quite similar policies to the US office on nonpatentability, although it allows animal patents as long as they don't contravene *ordre public* (in other words, morality).[61] Today, including the states with extension and validation agreements (including countries such as Cambodia, Tunisia, and Morocco on other continents), there are 44 participants in the system. The body has since served as a model for other global spheres of patent coordination in Eurasia, East Asia, Africa, and Latin America.

For European inventors, the greatest changes to the process of obtaining a patent lay in the new notions of novelty and nonobviousness that came with the unified European system.[62] Inventors from countries that tended to be more permissive suddenly needed to clear a higher bar, whereas inventors from countries that tended to be less permissive were suddenly able to make broader claims and patent a larger array of inventions. Either way, supporters of a unified system argued that, because of the variations from one nation to the next, national patent regimes violated principles of free and fair trade and thus merited harmony or, at the very least, a form a reciprocity.[63] As Pascal Griset explains, in the new system, "an international filing could have the same status as a national application in all the countries it designated . . . it provided for a single filing, resulting in a single patent."[64] Competition was, of course, also part of the hard-won battle to internationalize: "To face up to industry outside Europe in general, and in particular in the USA, Europe put a new guard at its doors. . . . A political reading of the inception of the Convention would therefore probably see it as a European response to the challenge of the global—and particularly transatlantic—competition and trade, while at the same time respecting member states' national sovereignty in a field affecting industrial policy."[65]

The creation of the EPO necessitated new supranational headquarters. Since the Strasbourg Convention, multinational patent offices had existed in The Hague, Luxembourg, and Munich, all of which would compete for the main seat that the ratification of the 1973 convention necessitated. Fresh from hosting the 1972 Summer Olympics, Munich, with its central location in Europe, excellent infrastructure, and strong history of innovation, won out.[66] The team promoting Munich argued that even though "[m]any cities feel as though they are 'in the heart of Europe' . . . [t]he heart does not have to be the geometric center. . . . Munich is not a cosmopolitan city like London, Paris, New York or Rome. It is no super city . . . [but] a city of . . . over 1.3 million inhabitants . . . a city where people defend their individuality and their way of life more stubbornly than anywhere else."[67] A purpose-built central patent office followed, with Gerkan, Marg & Partners, a German architecture office based in Hamburg, completing the prominent new headquarters in 1980.[68]

The building comprises two volumes that converge on a roughly east-west line bisecting the site. The building's exterior detailing is clearly indebted to the German architect Egon Eiermann, among whose numerous accomplishments in the very impoverished state of immediate postwar German architecture were his creative interpretations of the German building code, including wrapping taller structures in sylphlike balconies running the horizontal course of the building, like a series of rubber bands breaking up the monotonous glass curtain walls beneath. In

Meinhard von Gerkan, rapidograph drawing of the European Patent Office as seen from Museumsinsel, Munich, looking south, c. 1979. Bundesarchiv Koblenz Akte B 141/47800.

Typical Office Floor

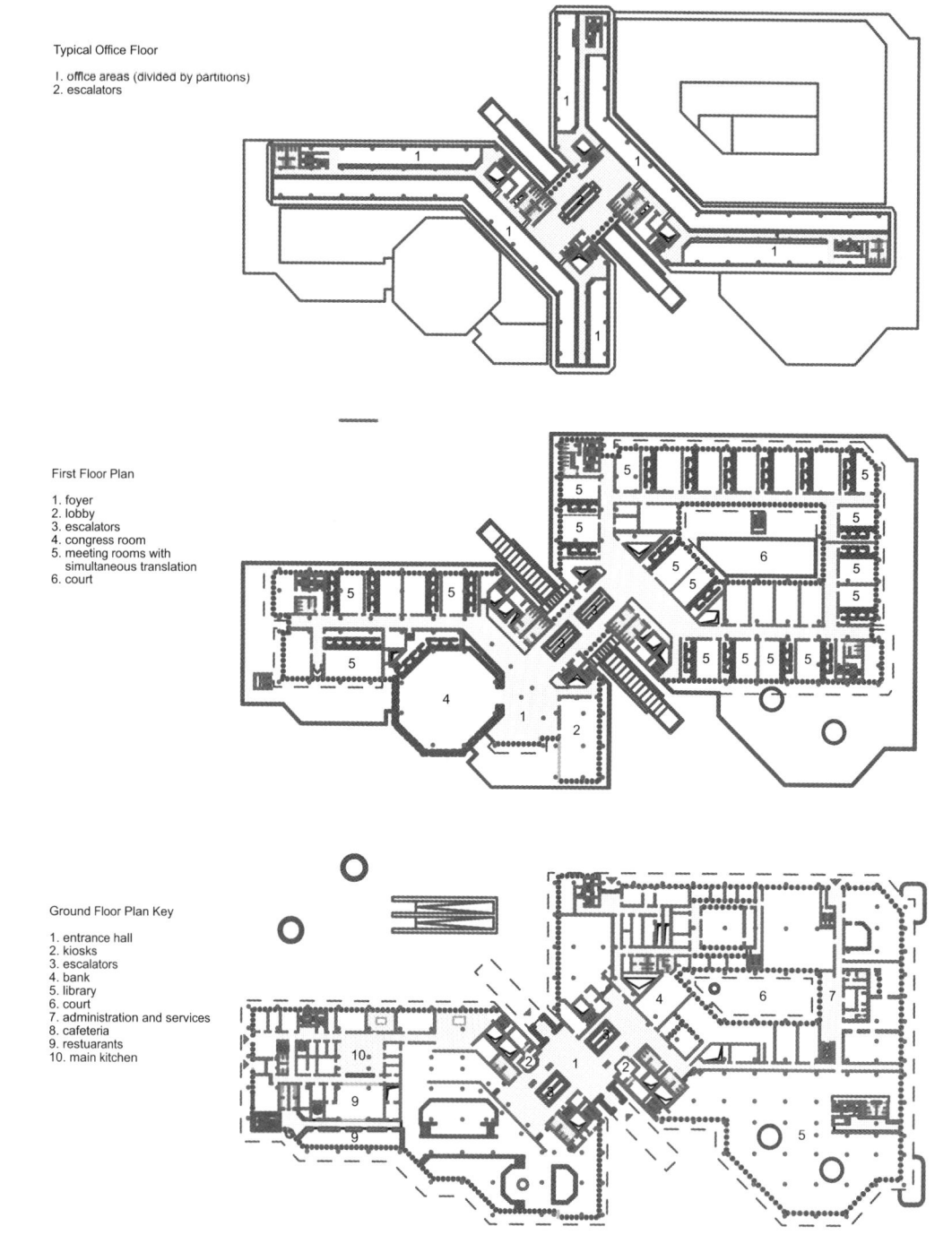

1. office areas (divided by partitions)
2. escalators

First Floor Plan

1. foyer
2. lobby
3. escalators
4. congress room
5. meeting rooms with simultaneous translation
6. court

Ground Floor Plan Key

1. entrance hall
2. kiosks
3. escalators
4. bank
5. library
6. court
7. administration and services
8. cafeteria
9. restuarants
10. main kitchen

Plan of the European Patent Office.

the EPO, this motif is reinforced by an array of slender vertical steel mullions, creating a lattice-like facade that appears as a light but still cage-like shroud that covers the volumes beneath, all the while avoiding a fortress-like appearance. The larger, lower volumes act as a kind of pedestal for the two towers rising above them.

Overall, the complex accommodates approximately 2,500 civil servants. The headquarters, located along the leafy left bank of the Isar River and directly across from the Nazi-era Deutsches Museum, a museum of German technology and engineering, became "the heartland of technology in a Germany that had committed itself to the project of European reconstruction."[69] It remains common for EPO patent examiners to use the library of the Deutsches Museum for prior art research, signaling the real synergy between the two institutions.[70]

The programmatic needs of the EPO were complex. In addition to providing the regular inward (staff offices, prior art research facilities, a staff cafeteria) and outward (meeting rooms for inventors and patent attorneys prosecuting patents) features of a typical patent office, the EPO had to account for the added challenges posed by the interpretive needs stemming from the multiple languages spoken by both its staff and the inventors. Many meeting rooms, for example, were equipped with simultaneous interpretation booths that sought to lessen the hurdles for inventors and their representatives who might not speak English or German, the official administrative languages of the EPO.[71] The EPO had self-imposed mandates for a diverse, multinational staff, one that could reflect the multinational body it represented proportionally to how much money each member country contributed to the organization.[72] This meant luring top patent bureaucrats from now defunct or otherwise redundant national patent offices to Munich.[73] A complex pneumatic tube system by the Telelift company (a European patentee) was almost integrated into the building to ease communications between the multinational staff, but ultimately abandoned.[74]

A sculpture entitled *Blauer Ritter* by the Swiss sculptor Bernhard Luginbühl was prominently installed adjacent to the escalators and the conference area on the first floor.[75] *Blauer Ritter* is an assemblage of found machine elements, many of which Luginbühl collected from scrapyards.[76] The blue elements of the sculpture lend the sculpture its name ("Blue Knight"). This particular piece is not kinetic, but many of Luginbühl's pieces, like those of his close

View of a patent litigation meeting at the European Patent Office, Munich, with synchronous translation booths. Also color plate 12.

Brochure photograph showing an office employee using the Telelift pneumatic tube system, 1978. The system was proposed for the European Patent Office but ultimately excluded from the final design. Bundesarchiv Koblenz Akte B 141/47802.

View of Bernhard Luginbühl's sculpture *Blauer Ritter*, as installed in the European Patent Office, 1976.

friend Jean Tinguely, were.[77] The selection of Luginbühl for this commission could be viewed as either inspired or facile, depending on one's perspective. Luginbühl's three-dimensional machinic collage immediately calls to mind the technical and industrial ethos of patent culture writ large, and the sculpture could easily be a large patent model for some unknown invention selected at random. At the same time, it pushes the technocratic idiom of the building's exterior and the neighboring Deutsches Museum further into the life of the building, when it could instead have provided a moment of relief and humanization.

The EPO's move into its new headquarters on the Isar coincided with the advent of computing. The EPO administrators had to decide on the extent to which the new architecture and spatial organization of the office should reflect changes that were still largely projective in nature. Dutch patent administrators had already instituted IBM computing systems in The Hague for file administration and documentation.[78] The EPO was also working with engineers to explore the potential of computing in what was probably the process most in need of automation: prior art searches.[79] At the time of the new facility's opening, EPO administrators had taken the bold step of extending the digitization processes in place at The Hague to a number of new functions, including a patent publication system, the patent register, patent applications, application process monitoring, renewal fee reminders, and accounting.[80] The centralized computer and its 65 terminals across the building were a massive expense and presented any of a number of risks should the computer fail. For almost two decades, anxiety over the reliability of computing systems meant that the EPO was duplicating most digital work in an analog manner, including the patent filing system that many traditionalists preferred.

Critical reaction to the building was predominantly unfavorable, citing the incongruence between the building's technological rationalism and large scale, on the one hand, and the subdued landscape around it (a criticism that rings very ironic when one considers that the mammoth scale of Gabriel von Seidl's Deutsches Museum was also at odds with the setting when it opened to the public in 1925).[81] The architectural critic Peter Davey, who saw some good in the project, put it this way:

> The German architectural climate has changed radically in the last decade. Ten years ago, the emphasis was on

An official during his work at the Reich Patent Office in Berlin at the index of patent specifications, c. 1910.

View of an EPO employee obtaining patent records for research, c. 1980.

technology, industrialization, the isolated statement derived directly from programme; energy was of little account. Now the focus has moved to art, handwork, contextualism and energy consciousness. The Patentamt is a dinosaur, but a remarkably elegant one.... Every detail is perfectly thought out: everything meets impeccably; every change in direction, texture, material and finish is logical; everywhere the technological effect is attained (paradoxically) by the highest standard of craftsmanship and fitting impeccably supervised.... However unfashionable the Patentamt may be at the moment, historians will surely say that it represents the late apogee of a very fine German tradition of (mostly commercial) building that flowered in the '60s and early '70s. That is why the building has a place in this issue [of *Architectural Review*]. We no longer need to attack it, for its like will probably never again be seen in our lifetimes. Let us accept it for what it is.[82]

Davey's take on the EPO is emblematic of the critical charity one needed to extend to buildings forged under the auspices of globalization and European unification. Design by committee rarely works, but Gerkan, Marg and Partners had done a good job in expressing a technological moment tectonically.

6   *The Courtroom*

"It is not my custom to write 'fan' letters," Robert Fiddler wrote to Seymour Melman, a widely acclaimed professor of industrial engineering and operations research at Columbia University in the summer of 1958. "However in view of the furor which your study has created among the patent profession, I feel you are entitled to whatever small support this letter will provide."[1] Fiddler, a young law practitioner fresh from a career as patent examiner, was referring to a report that Melman had issued to the US Senate Subcommittee on Patents, Trademarks, and Copyrights entitled "The Impact of the Patent System on Research." As the report summarizes, "the patent system, whatever its past contributions and its value and virtues in other respects, contributes little to the progress of science and useful arts," a conclusion that, "without doubt, will be greeted with skepticism by some and vigorous disagreement by others."[2] Noting his agreement with Melman's position on the patent system, Fiddler expressed frustration over the years he had given to his country as a patent examiner, contributing to a system that he felt had become increasingly futile:

> Like many of my brethren at the bar, I have often wondered at the utility of our activities. The wonderment, however, is often obscured by the maze of technical details involved

in carrying on these activities, so that one rarely stops to question the basic premises upon which these activities are grounded, and a variety of myths are generated regarding the technological and scientific spur provided by the patent system. Myths, of course, die hard, particularly those which are treasured as presumably providing economic sustenance. Though none of my colleagues has dared to voice any fears that the threats posed by your study are directed at his pocketbook, these fears seem quite implicit in any patent lawyer's defense of the patent system. Whatever merit there may be in some of the criticism leveled at your report, my colleagues' analyses seem to me extremely nearsighted as failing to view the broader implications of your study which, as I see it, merely indicated the archaic nature of our present patent system, and its lack of fulfillment of its alleged objectives in our present state of development of science and the industrial arts. I, for one, am pleased to find our basic tenets being publicly questioned, and I hope this may provide the seeds for a revision and development of the patent system into a body of law which will truly serve its avowed purposes.[3]

Fiddler's private "fan mail" to Seymour Melman provides a rare glimpse into the views of a disillusioned patent system insider; his call for reform sheds light on the complex relationship between the patent office and the judiciary. As we move out of Fiddler's domain of the patent office and into the domain of the courtroom, we can benefit from an analogy. Imagine the patent office as a kind of maternity ward, a place where inventors (the mothers of invention, if you will) give birth to a new generation of technological objects. As in any good maternity ward, the inventions are monitored for their health and soundness before being brought into the world. For an amount of time similar to the mother's, the inventor is then the custodian of that invention and has the right to rear it as the inventor sees fit. Then, once that custodianship has legally expired, the invention becomes part of the commons, on its own, if you will. The courtroom and its allied spaces are not unlike the classrooms where the child's identity is expanded, refined, remediated, and perfected based on cultural mores, all while still under the custody of a parent. This analogy, however imperfect, helps us imagine the courtroom beyond its narrow identity as a space of adjudication. Like the classroom, the courtroom will be a place of success and disappointment, of maturation, and of conflicting philosophies that act on the invention and continue to shape it even after it comes into the world as an object. Just as pedagogy changes over time, so too do judicial philosophies, and patents mirror the vicissitudes those philosophies produce.

What follows is an attempt to portray the dynamism of the judicial sphere with respect to the very nature of patents through a reflection on three key domains: design as it is manifest in law; infringement and legal deviance; and reform that reflects social, cultural, and philosophical change over time. These three domains offer important insights into the ways that patents and the nature of invention exist in a kind of constant flux, shaped not only by national and international laws as they are written but also by the clarifications, challenged assumptions, and refinements made through adjudication, case law, and evolving legal philosophies.

Design in Law

While the development of the patent system grew out of the early modern tradition of grants and privileges in most European nations, in the United States and other nations formed from the late eighteenth century onward, IP was enshrined ex nihilo in new national constitutions. The intellectual property clause, for example, is part of the very first article of the US Constitution, passing with unanimous approval at the Constitutional Convention in Philadelphia in 1787.[4] Officially, it stated that Congress has a mandate "to promote the Progress of Science and Useful Arts, by securing for limited Times to Authors and Inventors the exclusive Right to the respective Writings and Discoveries."[5]

The modern patent system as an inalienable right and "democratic" institution may have been one of the first great intellectual exports from the United States back to the old world.[6] Indeed, the first decades of the nineteenth century saw a number of European countries enshrining their own patent systems in their constitutions: Russia in 1812, Prussia in 1815, Belgium and the Netherlands in 1817, Spain in 1820, Bavaria in 1826, the Papal States in 1833, Sweden in 1834, Württemberg in 1836, Portugal in 1837, Saxony in 1843, and so on.[7] The American system of promoting industry and the arts by rewarding the inventor with a temporary monopoly was a major source of inspiration for these new patent regimes, but three other major models were available. There was the English system, which deprived the state of the power to grant monopolies *except* in the case of inventions, the French system, which was framed simply as a property right of limited duration, and the Habsburg system, which restricted "the right to copy inventors for reasons of social utility."[8] Variations on a theme, one could say, but these fundamental frameworks guided parliamentarians and constitutionalists as they decided where to place their stake in the ideological ground of IP law.

To understand why these systems and their offspring were later reformed, as well as how legal disputes in IP arose, it is important to first have a grasp of the myriad ways that design and construction as practices, with all of their particularities, have come to be inscribed within extant law in the countries where patent systems first resided. As discussed in chapter 1, women were technically allowed to submit patent applications in the American and most European patent systems of the nineteenth century, but this did not spell utopian equality. The hurdles to women's full participation in this arena were social, cultural, and economic, and not explicitly legal. The reason for the qualification "explicitly" is that there were nevertheless ways, through words or processes, in which law pertaining to design was still highly gendered and tended to diminish, if not entirely discourage, inventions associated with women and womanhood. The legal scholar Charles Colman's research on this question is most insightful, and his key findings bear careful consideration.[9]

Colman has demonstrated through exhaustive examples how words like "ornamental" and "decorative," which are crucial to design patent case law in America, became stigmatized despite the congressional mandate to "promote the progress of the decorative arts."[10] The gendering of design in patent law was also alive and well in Europe, signaling a transatlantic phenomenon. The root cause of the stigmatization of the very idea of "design" in law was that the "creation, appreciation, and consumption of design 'for its own sake' grew increasingly intertwined with notions of decadence, effeminacy and sexual 'deviance.'"[11] Because case law marks precedent, earlier instances of gender stigmatization in design patent law were there to be echoed and further entrenched as time went on. The result, Colman argues, is that design patents came to hold very little legal standing as a viable form of IP for much of the twentieth century.[12] A court decision about a design patent infringement case relating to designs for men's ties in *Franklin Knitting Mills vs. Gropper Knitting Mills* indicates just how gendered and sexist design patent law was: "[Men's ties] are bought not only because of their utility to the wearer and their attractiveness to others when worn, but also because of the appeal, as novel, ornamental and pleasing, that the design makes to the aesthetic sense of the purchaser, ofttimes the wife, sweetheart, or female relative of the man who is to wear it."[13] There is also this rhetorical flourish from a circuit court: "Can it conceivably involve patentable novelty to draw a few spaced apart parallel lines on a gown, a parasol, a shirt, a shawl, a rug, or the many other articles made up of textile fabrics? To so hold would undignify the whole theory of invention."[14] Yet sharper was a critique of domestic design and the frivolity of patenting it as being an "excessive and effeminate concern with taste and home decoration, self-absorption at the expense of wider issues, and associations with decadence."[15]

Another common strategy judges used to diminish the standing of design patents was evasiveness. In their dismissal of cases pertaining to design patent law, judges would commonly cite the purported ambiguity of certain words like "ornamental," "decorative," "pleasing," "shapely," "well-constructed," "well-designed," "balanced," and so forth, to perform ignorance or deference, to offer editorializations, or to justify irregular procedural moves or evidentiary approaches as means of evading the deeper discursive context that all of these words have held in both case law and legal philosophy.[16] Colman describes this as one of many manifestations of so-called "judicial distancing," a practice that typically followed four steps to diminish and gender design patents well into the twentieth century: "1. Evaluation of visual material through an emphatically 'commercial' as opposed to 'aesthetic' lens; 2. Disavowal of any ability (as 'respectable' men) to make informed or nuanced evaluations of design—except, perhaps, where indisputably 'masculine' objects were in dispute; 3. Summarily dismissing as uninventive and unappealing most patented designs appearing before the court; and 4. Selectively invoking utility patent principles in a manner that effectively rendered design patents a dead letter."[17] In other words, design patents, a tremendous well of opportunity for architects to protect the more aesthetic dimensions of their work, were under regular assault from the courts, which diminished and feminized the very idea of design. Indeed, for architects, the courts' diminishment of the design patent and its connotative association with decadence, women, and homosexuality likely played some part in discouraging architects from patenting in the first place.

That was probably the case for the autodidactic American architect James Renwick, the architect of St. Patrick's Cathedral in New York City and the Smithsonian Institution Building in Washington, DC, among other prominent buildings.[18] Renwick was very close to his brother, Edward Sabine Renwick, an inventor and well-known patent expert; the two exchanged hundreds of letters over the course of their careers, several of which discuss the issue of patenting elements of architecture.[19] James Renwick was aware, unlike many of his contemporaries designing prominent projects, that patents for the design of buildings and building parts were commonplace, and he had regularly pondered whether some of the architectonic inventions of his practice—such as the interlocking masonry unique to Gothic revival building elements—were patentable and, if so, whether the effort was worth it. Naturally, James turned to his brother for advice, and time and again Edward discouraged him from pursuing patents. Edward Renwick was well aware that the courts' opinion of patents in design was diminishing by the year, and he believed that any effort on his brother's part to pursue patents could easily end in both legal fruitlessness and an association with the frivolity and gendered domain of design patents. Edward appeared to believe that architecture should be as widely patentable as any other industry, but he was concerned about the lack of fixity associated with the terms defining design patent case law. This sentiment was perhaps best articulated in his most important publication, simply titled *Patentable Invention*: "[T]he decisions of the courts," he said, "upon the question of what constitutes invention have been so contradictory, and of late years have been so frequently at variance with earlier decisions, that at the present time unless the subject-matter of a patent is wholly new, it is practically impossible to presume whether it is likely to be regarded as an invention in the estimation of a court, or is not."[20]

Although much of the diminishment of design patent law rested on its gendering well into the twentieth century, there were also other forms of bias that can be seen in between the lines of patent law. This is particularly evident in the slippages between national patent laws when a patent had some kind of transnational significance. German and Japanese patent laws, for example, have both hinged largely on the concept of an "inventive idea," which marks the successful passing of a patent through the nonobviousness threshold. In Germany, however, an "inventive idea" extended protection to obvious modifications of that idea, whereas in Japan it did not. Imagine, for example, a patent for a manual device for shutting or opening the blinds of a sunroof that could not otherwise be reached with the arms. Now imagine that the blind is electrically wired and can be opened or closed with the press of a button. In Germany, the latter system would likely come under intense patent scrutiny regarding its inventiveness because there already exists a means for remotely closing the blinds of a sunroof, and the electrical wiring and button are simply an obvious modification contingent on technology. It is quite possible that this patent would fail. In Japan, however, the intrinsic differences between the manual and electronic systems would automatically confer the possibility of a patent, as the systems do the same thing but by different means. For Japanese inventors, this understanding of the "inventive step" has made for a significantly more permissive patent culture, an understanding that was the backbone of much of Japanese patent invention from the 1950s through the 1990s.[21] This difference between German and Japanese patent laws, representing perhaps the extremes of the interpretations of permissibility in design and construction patents, hinged essentially on their different interpretations of the same aspects of patentability. However, as Asian nations, particularly Japan, fortified their patent systems and competed on the global marketplace, this permissibility was characterized not as semantic but as cultural, simplistically (and falsely) implicating the Japanese (and later the Chinese) as a culture of copyists lacking the scruples that came with Western notions of IP.

Another important variable that is particularly germane to the realm of patents in the building sector involves the ownership of workplace knowledge. As we have seen, patents in this sector have come from both private inventors and corporations like BASF and DuPont. The nature of what an employee does and does not own while working toward an invention under corporate tutelage looked very different in the nineteenth century than it did in the later twentieth century, when corporate patents in the building sector became dominant. Nineteenth-century legal culture on either side of the Atlantic tended to preclude the idea of ownership of any kind of knowledge, including embodied knowledge learned on the job.[22] As such, firms had very few ways to legally (or conceptually) control the creative products of employees. This is why trade secret law was created. As Catherine Fisk explains, "Workers considered themselves at liberty to make the most of whatever knowledge they possessed. Most employers believed they had few ways to prevent the loss of information entailed in employee mobility because knowledge and skill were widely considered to be attributes of skilled craftsmen rather than assets of firms. Of course, there was a market for ideas, but inventors and authors (and not their employers) were envisioned as the primary entrepreneurs in that market."[23] As we have seen in chapter 3, with the rise of the power of private corporations, this marked autonomy of individuals was ultimately suppressed, not only by trade secret law but also by other legal apparatuses such as nondisclosure agreements.

It is also important to keep in mind the perennially ambiguous state of architecture in these affairs, existing as it does well within the framework of copyright law but more tenuously as an inventive art that proffers patentability. To return to Collins, this is largely a situation framed by thresholds and screens: "Architectural innovation occupies a contested zone on the copyright-patent boundary: it is widely recognized as copyrightable subject matter, and ... is routinely patentable. Furthermore, the same features of architectural designs embody both aesthetic and functional innovation, making it impossible to tease the aesthetic and functional apart."[24] Today, architects must contend with the additional ambiguities posed by patents for disposition of space and the altogether new regime of trade dress.[25] Retaining effort and individuality as the essence of IP law is critical to continued debates on the advancement of patent law in design and construction.[26]

Reform

Seymour Melman's radical suggestion in 1958 that the US patent system be dissolved is one extreme in a spectrum of reforms that inventors, IP lawyers, legislators, corporations, and countless others have advocated across history and geography. Indeed, reformist movements for the patent system are as old as the system itself, and it would be fair to say that the vitality of the intellectual debates around the patent system and its merits (or lack thereof) demonstrate that the very existence of this legal subfield is a matter for discussion. Even when we limit the cone of our historical vision to reform movements in architecture and design patents, there still remains a robust and contentious landscape.

We can go back to the eighteenth century to chart the desire of those wishing to reform the patent system. In England, the attorney general or solicitor general reviewed applications for the exclusive manufacture of certain items until 1713, when Queen Anne, objecting to this procedure, ordered that the Royal Society instead be the examiners of patent applications, a reform that tamped some of the capriciousness associated with the early modern privilege system.[27] The real watershed moment for patent reform in Britain came in 1852 when the British Patent Office was formally established, moving the patent process out from the immediate aegis of the Crown. A universal patent for all countries within the kingdom was the strongest move toward a streamlined, modern system. The new system also altered the rates at which different industries sought patents. For sectors where the initial patent research was done in secret, such as in mining and chemicals, the propensity for patenting shot up dramatically, while in manufacturing and the building sectors it went down.[28] The reforms also inaugurated the caveat system, in which inventors could submit a form, the caveat, to a particular branch of the Patent Office requesting that they be notified of any petition submitted to that office that was applicable to the branch of patents covered by the caveat.[29] This helped inventors in the process of preparing a patent to keep abreast of concurrent patent activity unfolding at the Patent Office. Thomas Edison, to reiterate, used caveats extensively as a means not only to remain informed about competitive inventions but also to keep his pulse on the state of the art, so to speak. Coinciding with the International Convention for the Protection of Industrial Property a few decades later, the Patents, Designs, and Trade Marks Act of 1883 brought into being the office of the Comptroller General of Patents, which coalesced a staff who examined patent applications, a process that was meant to ensure that the descriptions of inventions were accurate, although no investigation into novelty was carried out.[30] That ended with the Patents Act of 1902, which introduced a process for examining the novelty of inventions. This extra work necessitated a major reorganization of the Patent Office, including hiring 190 additional examiners to assist the existing 70.

The story of the British patent reforms that began in 1852 would not be complete without a discussion of the very considerable patent abolition countermovement that existed alongside these reforms.[31] The countermovement was actively supported by a number of prominent inventors who had never participated in the patent system and saw no value in it, including the engineer Isambard Kingdom Brunel, the arms magnate Sir William Armstrong, the sugar trader Robert MacFie, the electrical engineer and jurist William Robert Grove, and many of Britain's top legal officials who published anti-patent sentiments with the London publisher Longmans, Green, Reader, and Dyer.[32] These men led abolition efforts over the ensuing two decades, prompting the introduction of new and ultimately unsuccessful legislation almost every year, at times coming very close to reaching their goal of totally eradicating the patent system. In 1869, MacFie, for his part, compiled a volume of speeches, discussions, editorials, and parliamentary proceedings covering the United Kingdom, France, Germany, and the Netherlands that laid out myriad arguments that were contemporaneously unfolding on patent abolition, with the hope of winning hearts and minds for the abolition cause across borders.[33]

Armstrong's arguments were some of the most compelling. With regard to monopolies, which he saw as the most treacherous effect of the patent system, he noted:

> Nothing, I think, can be more monstrous than that so grave a matter as a monopoly should be granted to any person for anything without inquiry either as to private merit or public policy—in fact, merely for the asking and the paying. Amongst other evils of this indiscriminate system is that the majority of Patents granted are bad, and yet such is the dread of litigation, that people submit to a patent they know to be bad rather than involve themselves in the trouble and expense of resisting it. So that a bad Patent, in general, answers just as well as a good one.[34]

Yet Armstrong, himself a successful capitalist, made something of an about-face when he admitted that the patent system was not the protector of poor inventors it purported to be:

> One of the most common arguments in favour of Patents is, that they are necessary to protect the poor inventor, but

> it is manufacturers and capitalists, and not working men, who make great profits by Patents, and that, too, in a degree which has no reference either to the merit of the inventor or the importance of the invention. One rarely hears of a working man making a good thing of a Patent. If he hits upon a good idea, he has seldom the means of developing it to a marketable form, and he generally sells it for a trifle to a capitalist, who brings it to maturity and profits by it. He could sell his ideas just as well without any Patent-Law.[35]

Armstrong also highlighted the roadblocks to creativity writ large that were put up by the patent system:

> As to the cost of the system to the public, I don't see how it could be calculated, for it consists not merely of the licence fees, but also of the loss resulting from the stamping out of competition, which would cheapen production and, in most cases, lead to improvement. My great objection to our indiscriminate Patent system is, that it is scarcely possible to strike out in any new direction without coming in contact with Patents for schemes so crudely developed as to receive little or no acceptance from the public, but which, nevertheless, block the road to really practical improvement.[36]

As Adrian Johns contends, had the likes of MacFie and Armstrong succeeded, "Industrial creativity would have fallen to free trade. And subsequent scientific, industrial, and economic history would surely have looked very different."[37] However, by the 1870s, and even more so the 1880s, the patent system was operating very smoothly and making many men very rich, and the effect of this success and monopoly-driven wealth ultimately tamped down abolition efforts, transforming them into a minor (although still extant) fringe movement. In Germany, the nongovernmental German Patent Association (Deutscher Patentschutzverein) played a major part in contributing reform-minded ideas in the passage of a new German patent law in 1877, advances that were generally seen as more democratic and less costly to everyday inventors.[38]

Before the late nineteenth century, design patents in both the European and North American patent systems had been an obscure corner of patent law that did not receive much attention from either inventors or legal scholars.[39] A good deal of this obscurity was probably the result of semantic confusion over several of the words that actually defined what a design patent is in relation to a utility patent. Around the turn of twentieth century, most of the design patent law on the books had some variant of the word "utility" as a requirement of a design, which was the source of much of this confusion because it failed to distinguish, as the trailblazing patent scholar Kelsey Martin Mott has explained, "between usefulness of a design and usefulness of the article embodying the design—between utility of appearance and utility of function."[40] Legal precedents began to accommodate the idea that utility of appearance could embody the notion that a certain design could promote pleasure, refinement, or comfort, which was in turn a utility that redoubled any practical utility an object might have.[41] As one judge put it: "I decide this case upon the broader ground that patents for design are intended to apply to matters of ornament, in which the utility depends upon the pleasing effect imparted to the eye, and not upon any new function.... Design patents refer to appearance, not utility. Their object is to encourage works of art and decoration which appeal to the eye, to the aesthetic emotions, to the beautiful.... The term 'useful,' in relation to designs, means adaptation to producing pleasant emotions. There must be 'originality' and beauty. Mere mechanical skill is not sufficient."[42]

Another telling case with considerable case law influence in the decades to follow was *Forestek Plating and Manufacturing Company vs. Knapp-Monarch Company* in 1939, in the US Court of Appeals' Sixth Circuit.[43] The subject of the case was US utility patent 2,040,369, an electric sandwich griddle resembling a panini press, and a subsequent, correlative design patent, D96,481. The designer, Andrew Knapp, and the manufacturer, the home appliance specialist Knapp-Monarch Co., were sued by Forestek, which believed that all of the design's key elements—two encased heating elements, a tray, struts that did not conduct heat to the surface below, and a multipoint hinge that allowed accommodating the depth of the sandwich—infringed upon existing patents, some of which were held by Forestek. The court's decision acknowledged that these elements were patented and that Knapp-Monarch had not been thorough in its disclosure of prior art, but affirmed that the invention remained patentable as these elements had not been combined in this manner before. This meant, by extension, that recombination in design itself was just as patentable as new inventions. Moreover, the court acknowledged that the streamlined, stainless-steel design of the encased heating elements was itself inventive in that it departed from the boxier precedents in question, and this in turn acknowledged that style—in this case, the machine aesthetic of art deco industrial design—was also an element of aesthetic power in law: "Each design must be measured at its own particular time and in its own special field. The stream-lined train design was rejected as early as 1874 but was universally accepted in 1934. In our own time, original work has been done outside of the official province of art. Technical progress has been made in all branches and new forms have been invented in the sphere . . . of articles of daily use of all kinds. . . . The use of the machine produced a different type of designing genius. Its aesthetics required precision, calculation, flawlessness, simplicity, and economy."[44] In her analysis of the case, Mott contends that what the court was in effect doing was tethering tacit ideas about beauty and style to the more prosaic usage of the term "ornamental" in design patent law. One of the strengths of the Knapp-Monarch griddle case was that it was able to assume this guise of inventiveness through its association with both a design and a utility patent at the same time.

The US Commissioner of Patents in 1902, Frederick Innes Allen, understood the importance of an amendment to clarify the divergent notions of "utility" as well as the screen between utility and design patents. He recommended an amendment to Congress that included substituting the word "artistic" for the word "useful" in design patent law. Allen believed that this would steer design patents to their "proper philosophical position" within the realm of invention. The Senate Subcommittee that dealt with patents agreed with Allen's assessment that a clarification was necessary but disagreed that the word "artistic" was the solution. Instead, they used the word "ornamental," which they believed would create less confusion between design patents and copyrights and thus function as a better screen between the patent and copyright regimes.[45] About a year after the amendment was enacted, Allen reported a significant drop in the quantity of design patent filings and a sharp rise in the suitability of the applications that were filed in terms of matching the intended "philosophical position" of a design patent. Allen ardently maintained that the major reform that resulted from changing a single word in the law did not represent any shift in policy at the patent office but was rather

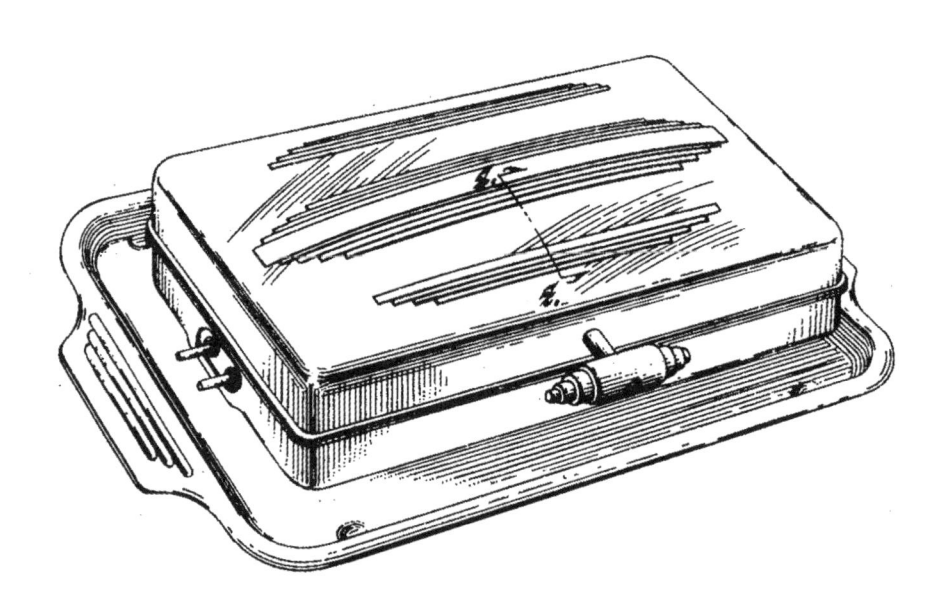

Left: A. S. Knapp, Combined Sandwich Toaster and Tray, US Patent D96,481, filed May 21, 1935. Right: W. H. Fischer, Electrical Appliance, US Patent 2,040,369, filed June 12, 1935.

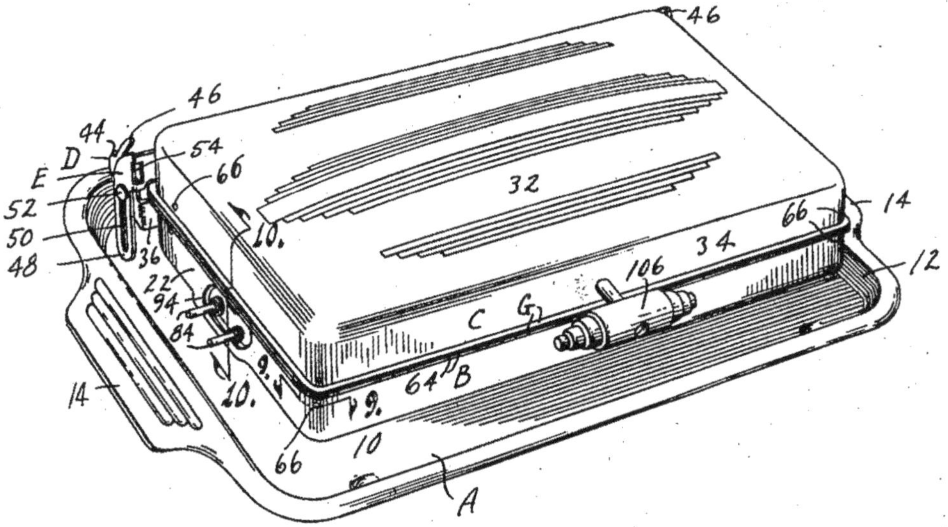

46
46
44
22
D
E
52
50
48
54
66
32
66
14
34
12
106
36
10.
22
94
84
C
G
10.
9.
64
B
9.
14
66
10
A

a mere fine-tuning of the law to match the trajectory of inventive activity as it unfolded in real time. Both before and after the amendment, Allen insisted, "design refers to appearance and not to mechanical utility."[46]

Japan instituted its own major reforms with its Patent Act of 1921.[47] This act included a progressive workers' rights provision stipulating that corporate employees, and not their employers, were automatically the owners of their own occupational inventions but that they also had the obligation to transfer their inventions to their employers with "equitable remuneration," which meant that Japanese corporations were constantly in negotiation with their own employees regarding the monetary value of their patentable inventions made on the job.[48] In the same year, Japan also adopted a screen separating its very robust design patent system from its system for utility patents in a manner similar to that in the United States.[49]

The middle of the twentieth century was a period of robust legal activity for patent systems in the Global South, many of which of were forming in the wake of postcolonial independence. Under British rule, India had a nominal patent system that accommodated patents by those living under the Raj (British patents applied to India, but Raj patents did not extend to the United Kingdom).[50] In 1950, a Patents Enquiry Committee issued a report known as the Justice Bakshi Tek Chand Report, which outlined a plan to establish a fully independent patent system based partly on the 1949 UK Patents Act and incorporating many suggestions on how to stymie the many abuses of the Indian patent system that had existed prior to independence.[51]

The single most impactful reform effort pertaining to architectural patents was probably the successful push for a pair of bills in the United States, the Architectural Works Copyright Protection Act (H.R. 3990) and its companion bill, the Unique Architectural Structures Copyright Act (H.R. 3991), both in 1990. These bills are so important that they, along with the synchronic rise of digital design, mark the main reason why the later bookend of this study, around 1990, is placed where it is. To be sure, digital design is the defining motif around which most reform efforts are focused today, and once some historical distance has been acquired, that too will constitute a subject worthy of being placed in historical context. Nevertheless, if a designer today is familiar with any watershed legislation related to IP, it is most likely H.R. 3990.[52]

H.R. 3990 is not actually about patents per se. It is a law about copyright, but it came to define and demarcate numerous terms used in both copyright and patent regimes. The bill had bipartisan support, its two congressional sponsors being Robert W. Kastenmeier, a Democrat representing Wisconsin's 2nd Congressional District, and Sherwood Boehlert, a Republican representing New York's 25th Congressional District.[53]

Even after the Copyright Act of 1909, which protected "published" works, the status of architecture itself and its representations in drawings and other media as a form of IP had remained vague.[54] The issue of IP protection in architectural design was elucidated to a certain degree with the Copyright Act of 1976, which expressly articulated that an architect's plans and drawings were to be included under the rubric of "pictorial, graphic, and sculptural works"; yet the law was still ambiguous regarding which protections architectural form, as built, did or did not enjoy, noting that "the extent to which protection would extend to the structure depicted would depend on circumstances."[55] By 1976 an architect's work was easily infringeable as emerging economies continued their headlong journeys into development and globalization, places where locally trained architects were often in short supply and where mass media could rapidly disseminate imagery of architectural works. Many of these architects were frustrated with the ambiguity of copyright law, even in the United States, where the effort to offer architects IP protection was very robust. Indeed, most courts ruled that the Copyright Act of 1976 simply did not offer legal cover to actual structures. Architects owned the copyright for their architectural drawings but not for the buildings that could be made from them, and this was clearly a legal failure that required legislative intervention.

This revealed for the first time, in no uncertain terms, that architecture straddled the artistic domain that copyright sought to protect and the utilitarian functions protected by patents, an overlap that architects themselves have understood since the profession was created. The problem, it seemed, was that the "screens" of copyright law excluded objects with utility from copyright protection. In other words, a plan or a section was only an artistic object in the eye of the law, and not a document describing how to build a functional object, even though it is clear that it often is both. This siloing of architectural copyright was only further aggravated by the idea-expression dichotomy that barred the copyright of "ideas" or, in this case, things that artistic works might prescribe or inspire.[56] The closest legal analogy to this dilemma is probably culinary recipes, which in and of themselves cannot be copyrighted (as the recipes are not themselves art or literature) or patented (as, being in list form, they have no utility per se), that is, unless they are compiled into a book, in which case the book as a whole is copyrighted.[57] Although the topic was copyright, patentability with its contours was incredibly relevant to this discussion because it ostensibly offered the protection that the utility doctrine in copyright stymied. But there was also no drive to push architecture into the mainstream of the patent system, because architects rightfully enjoyed the artistic connotation associated with copyright. What's more, they were typically opposed to the mass reproduction of their designs, an intention that was tacitly associated with filing patents.

Some clarity on this issue emerged from the US adoption in 1989 of the updated Berne Convention for the Protection of Literary and Artistic works, an outline of international IP protections initiated by the World Intellectual Property Organization, which had an impressive roster of 179 signatory nations covering all but a handful of parts of the governed Earth.[58] Architecture was one of the topics that were updated in the convention: completed architectural works should not be infringed upon, a seminal statement that stood to bridge the copyright-patent divide in built works. When the convention reached American shores, the details of this bridging required a great deal of fine tuning and, perhaps more than anything else, the codification of terms that had to date generated more ambiguity than guidance on the matter. This is what the subcommittee hearing for H.R. 3990/3991 set out to accomplish in 1990.

As summarized by the American Institute of Architects (AIA), under the act, an architectural work is statutorily defined as "the design of a building as embodied in any tangible medium of expression, including a building, architectural plan, or drawings" and "includes the overall form as well as the arrangement and composition of spaces and elements in the design, but does not include individual standard features, such as common windows, doors, and other staple building components."[59] In effect, this automatically protected all works in any form of "publication," including in the form of a completed building, and it also did so for all works retroactively for 75 years after the death of the architect.

The hearing for H.R. 3990/3991 took place on March 14, 1990, presided over by Kastenmeier.[60] Some notable architectural personages were in attendance to offer testimony, including Richard Carney, the chief executive officer of the Frank Lloyd Wright Foundation, the architect David Daileda, representing the AIA, and the architect and Princeton professor Michael Graves. Kastenmeier opened the hearing with an homage to the art of architecture that was rare for a political setting:

> The hearing this morning combines two special subjects: architecture and law. Both serve as tests of civilization and gauges of societal development. The structure and groundwork of our society are found both in our architecture and our laws. Architecture is an art form that performs a very important public and social purpose. We rarely see works of architecture alone, but instead tend to view them in conjunction with other structures and the environment at large, where they serve to express the goals and aspirations of the total community. Frank Lloyd Wright, a proud son of my home State of Wisconsin and, indeed, of my congressional district, aptly observed: "Buildings will always remain the most valuable asset in a people's environment, the one most capable of cultural reaction."[61]

Kastenmeier's lofty introduction gave way to comments, questions, and input on the topic from all present as well as in the form of documents submitted for the record from those not in attendance. Carlos Moorhead, a Republican representing California's 22nd Congressional District, commented that legislation from Congress needed to be sensitive to customs of the profession of architecture, on which Congress would need edification: "We must be sensitive to long established practices and traditions among architects and others in the building industry that may be greatly affected by a change in the present law."[62] Written testimony included testimony from Columbia law professor Jane Ginsburg, who had workshopped the issue of architectural copyright with her students, performing a number of so-called "fact-pattern tests" for the proposed law.[63] The method Ginsburg and her students used was to treat architecture not as an ex nihilo medium, but rather comparatively to other forms of art. Ginsburg said in her testimony that "in copyright law generally, the 'work' exists

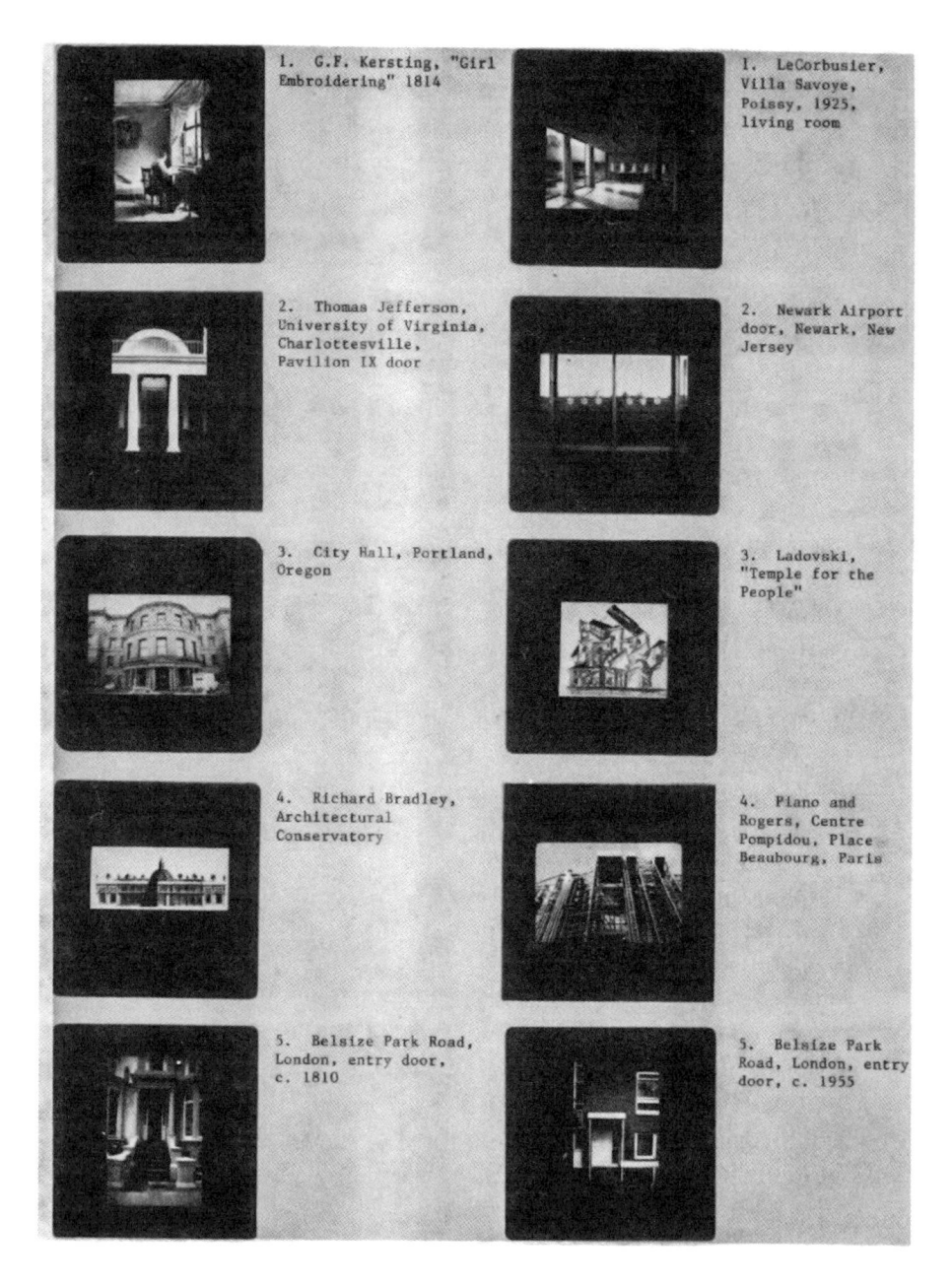

A reproduction of a portion of the slides utilized by Michael Graves in his congressional testimony on the issue of copyrights in architecture, part of *The Report of the Register of Copyrights on Works of Architecture* (Washington, DC: US Copyright Office / Library of Congress, 1989).

independently of the form of its fixation: A song is a copyrightable musical composition whether it is fixed in sheet music or on a sound recording."[64] What was needed was a new definition of the very term "architectural work": "An architectural work," Ginsburg proposed, "is the design of a building or other three-dimensional structure, as embodied in the building or structure or in the architectural plans."[65] As Hirschman explains: "This small change shifted the entire media-based paradigm for architectural copyright and paved the way for a broader notion of 'architectural works' that avoids treading into dangerous moral rights territory. Ginsburg's proposal recognizes design outside of and separate from the artifact, much like Alberti might have done. If the work of architecture is a design that is only fixed in a building or a drawing, the architect then is no more than a contractor or a draftsman."[66]

David Daileda's comments as a representative of the AIA touched on some key issues the others did not; for example, he urged the committee to remove any test of artistic merit from the bill.[67] His comments were perhaps most enlightening for how they challenged the circumscription of architectural drawings as primarily objects of art: "The AIA position," he argued, "is based on the premise that the value of architectural plans is in their execution, which is not protected under current law. The numerous small architectural firms in the United States might benefit the most from this increased protection, since their works are most likely to be copied."[68] Daileda did, however, gesture to some of the complexities of this matter for the practicing architects the AIA represented: "It [is] worth noting that the AIA view is not shared by all its members, some of whom feel that buildings should be protected in and of themselves. The reason for this split within the profession seems to be that some architects want their designs to be copied, in the belief that imitation manifests professional approval and respect."[69] Ralph Oman, the Register of Copyrights and Associate Librarian of Copyright Services at the Library of Congress, pointed out something that Daleida would also know: regarding alterations and renovations, architects are generally interested only in disavowing and not protesting major alterations.[70] Oman also stated his belief that representatives from the real estate and construction sectors should also participate in such a discussion, something the architects in the group would likely have disagreed with.[71]

To their credit, the subcommittee was cognizant of the retrospective as well as the retroactive importance of the law to figures in architectural history who were no longer practicing. It was for this reason that Richard Carney was invited to represent the estate of Frank Lloyd Wright, certainly one of the few architects that many of the lawmakers involved in this decision would know by name. Carney testified to how the licensing of images of Wright's drawings generated steady revenue for the Wright Foundation and that he, like Ginsburg, was supportive of a definition of the term "architectural work" that was not circumscribed within a particular medium. Implicit here, in the juxtaposition of revenue from the licensing of archival drawings, on the one hand, and support for a more expansive term, on the other, was the suggestion that the Foundation could be generating even more money if Wright's buildings were themselves protected.[72]

The subcommittee had fashioned a marquee slot among the witnesses, a slot for a witness who would edify the committee on the very nature of creativity and authorship in architectural practice and who would be most directly prepared to speak to the philosophical questions at the core of the bill. This slot went to Michael Graves, who, in addition to being a prominent American practitioner, came with extra pedagogical cachet as a professor at Princeton University. For his presentation, Graves deployed the well-trodden convention of the two-slide projector method, a common strategy in the teaching of art history that was meant to encourage thinking about form relationally through juxtaposition. The many sites Graves showed to the committee included the Villa Savoye, Thomas Jefferson's quadrangle at the University of Virginia, Newark Airport, the Centre Pompidou, the Villa Lante in Rome, and Walt Disney World, among many others.[73] Considering the relatively quick timetable with which the hearing was put together as well as how polished the presentation was, it is likely that the subcommittee was getting much the same presentation as Graves's architecture students at Princeton.

Graves, an ardent postmodernist, was really the perfect person for this job, as he could demonstrate how the use and reuse of symbolic form in historical architecture was an inherently creative, rather than derivative, practice. Even when discussing high modernism, Graves was able to offer a convincing argument that would disabuse any skeptics of the notion that any kind of architecture was *purely* functional, like a utility patent. Hirschman describes

Graves's contribution to the hearing as one that "complicated the notion of function on which American copyright law ostensibly hinged. His reasoning supports a reading of architecture as an overall work, a compilation, rather than individual variously protectable elements."[74] Oman recognized the impact of Graves's words on the committee in his own testimony, citing excerpts from Graves's essay "A Case for Figurative Architecture": "Internal language is 'intrinsic to building in its most basic form—determined by pragmatic, constructional, and technical requirements.' Poetic language is 'responsive to issues external to the building and incorporates the three-dimensional expression of the myths and rituals of society.' . . . The success of Mr. Graves's poetic language is evident to all of us in this hearing room today."[75]

Infringement

As with any form of property, IP's prime function from a legal perspective is to ensure that others cannot plagiarize, steal, or otherwise infringe upon what is rightfully yours. In the Lockean sense of property, trespassing on real property (such as entering someone's house without their permission) is a crime because it violates the autonomy and personal liberty that the act of purchasing or renting property is intended to afford. Trespassing and patent infringement share many similarities in that they are property violations, but IP is also fundamentally different from real property. The primary difference is that IP is something that ostensibly emerges from one's own body—be it the brain or the hands or both—and since one is intrinsically the owner of one's own body in modern legal systems, these creations are by extension automatically one's own, whereas real and personal property represent things one can acquire through a purchase or other form of transaction. IP represents a kind of centripetal form of property, distinct from the centrifugal property that we are most familiar with. As such, patent infringement represents a fundamentally different kind of transgression, one that typically violates not possessions that already belong to someone (as with a burglar who trespasses and subsequently steals) but rather possessions (especially money) that could potentially materialize from IP. Moreover, although patent infringement is a common problem, it is also very much a theoretical problem insofar as it represents losses that might occur as well as losses that have not yet occurred.

For those interested in the philosophy of law, this makes patent infringement cases a particularly stimulating domain, and numerous infringement cases adjudicated in the industrial period speak to the intellectual complexity of patent infringement in the domain of architecture, design, and construction. Yet another thing that patent infringement cases reveal is the power dynamics at play in the patent system, including David-and-Goliath battles between sole inventors and massive corporations, between designers and those who manufacture their wares, between the legally informed and uninformed, and so on.

Anyone studying the history of patent infringement cases will immediately notice a profound asymmetry between how utility patent disputes have been treated and how design patent disputes have been treated. Utility patents have long enjoyed serious review in European, Asian, and North American courts, generating a great deal of case law that has made the field of utility patents as complex as it is refined. Design patent cases, however, were not taken very seriously on either side of the Atlantic until well into the twentieth century. One statistic can stand for the rest: between 1926 and 1959, the United States Second Circuit Court of Appeals, whose territory includes New York, did not uphold the validity of even one contested design patent.[76] Indeed, as Colman explains, after several decades of design patents being adjudicated thoroughly and with considerable deference to the designer, the US Supreme Court would "never again agree to hear a single appeal squarely posing an issue of substantive, design-patent doctrine after 1895 . . . the court adhered to its practice of blanket denials of petitions in design patent cases even as it continued to hear fairly frequent appeals concerning the validity or infringement of utility patents."[77]

In May of 1951, the Argentine architect and industrial designer Jorge Ferrari Hardoy was distressed when a friend sent him an advertisement from the California furniture manufacturer Modern Color Inc.[78] The advertisement depicted an uncanny replica of the widely known "butterfly chair" or "BKF chair" he had designed with Antonio Bonet and Juan Kurchan, marketed by Modern Color as the Style 100 African Campaign Chair. Ferrari Hardoy, who at this point was clearly rather naïve about copyright and patent protections, wrote to Modern Color with a suggestion:

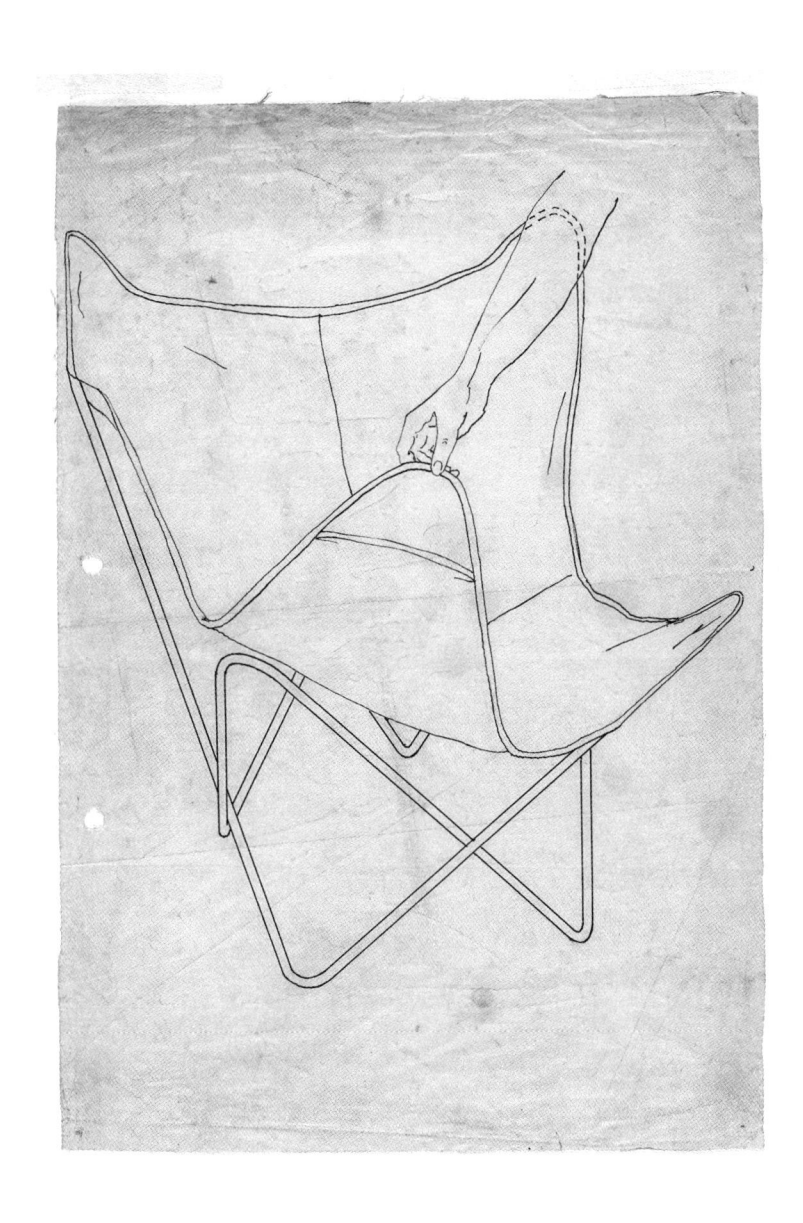

Jorge Ferrari Hardoy (in conjunction with Antonio Bonet and Juan Kurchan), iteration of an axonometric drawing of the BKF chair, design 1938, drawing c. 1941. Harvard University, Loeb Library Special Collections D083a. Also color plate 6.

A furniture store in Kyoto in May 1934. Glass slide published by the New York State Education Department.

> This chair, of which, as you know, I am the designer, and has been published in every part of the world with my name, must be manufactured in the United States by Knoll Associates. However, and for ending this situation, regarding that I have no definitive contract with Knoll, I shall ask you to propose me some base for an agreement to make legal the actual situation. Perhaps, if we could do it, you could remain as mine [sic] representing in California and Knoll in the rest of the country. That is simply a suggestion I want you to think about to propose me conditions to draw up a contract. On the contrary [sic], I think Knoll will carry out a legal action against you.[79]

Ferrari Hardoy was in for a tough lesson on the slippery maneuvers of manufacturers when he heard back from Dorothy Schindele, a fellow designer and the proprietor of Modern Color, three weeks later:

> We consider the plight of the designer in this country (at least) very tragic. The particular design you mention, and its variations, is being made in many, many blacksmith and metal workshop shops in this country. Part of the reason for this situation was that the design was not originally produced at a price the general public could afford to pay. And even when they were willing to pay the very extreme price, they could not obtain the article for six months or more. When a manufacturer takes a particular design for exclusive production and distribution he must face the possibility that it can be a tremendous success and be able and willing to stand back of it and produce and deliver in great quantity. Before producing any of the African Campaign Chairs (our trademark name), our attorney corresponded with Mr. Knoll and received the reply that Mr. Knoll had exclusive rights to the term Hardoy chair only, a name which we have studiously avoided for that reason and never permitted our dealers to use in reference to our chairs. We mention these various points not in argument at all, but merely as occurrences that you might find of interest. . . . We were interested in your suggestion that we handle the western distribution. . . . Do you have a design patent or mechanical patent with which we can eliminate by legal

means the very poor imitations currently being sold? Do you have signed clearances . . . to total rights of the design?[80]

There is an unmistakable tone of one-upmanship in Schindele's response to Hardoy. Politely, she points out how her manufacture of his designs was entirely legal because he did not own a patent for the design, roundly eschewing any ethical culpability for the plagiarism of his design. Schindele also attempts to divert Hardoy's attention away from Modern Color toward the large number of lower-quality plagiarists in California, where the chair was immensely popular. Schindele's final manipulation is her proposal to become the US distributor for Hardoy's designs, after she asks him whether he has any legal claim over the design, already knowing full well that he does not. Hardoy, unprepared to tackle the issue with the American legal system and given no assistance by Knoll, ultimately filed for a patent as a result of the situation in his home country of Argentina the following year, working with the Agencia "el Mundo" Consorcio Argentino de Patentes de Invención y Marcas de Fábrica, a Buenos Aires firm specializing in global patent and trademark protection.[81] Ultimately, Hardoy declined to work with Modern Color, and the butterfly chair, although patented in Argentina (where international enforcement was much more lax), continued to be copied around the world for decades to come.

Across the globe, the 1950s and 1960s were a time of intense tightening of IP protection, in no small part due to the correlation between intensifying globalization and copyism. Many nations added "worldwide" as a modifier to novelty requirements for patent claims, requiring inventors to conduct prior art research that was global rather than national in scope.[82] This changed the nature of many national patent systems. Prior to the 1950s, for example, Japan had long been viewed by Western counterparts as a place where unauthorized reproductions of goods and violations of copyright and patents ran rampant. The German Wilhelm Plage was famously despised in Japan for his numerous interventions in the Japanese entertainment industry to thwart unauthorized performances of European plays and music.[83] But amendments to Japanese law in the 1950s, accompanied by public awareness campaigns against piracy, worked to jettison Japan's global reputation for IP theft. Indeed, by the late 1960s, Japan had already transformed itself from a so-called copycat nation to a nation that was itself being copycatted, particularly by its neighbors in China, South Korea, Taiwan, and Hong Kong.[84]

A number of cases falling after the chronological scope of this book prove that patent and copyright infringement in architecture, design, and construction continue to be a robust arena of legal contention.[85] Clarifications of thorny questions of IP in architecture and design that have come out of legislation and case law have, in all likelihood, not kept pace with the unforeseen questions that have emerged from the proliferation of new modes of technology and dissemination. Design ideas manifested as digital files are easy to disseminate and reproduce, 3D printing has turned individuals into their own manufacturers, and mass customization has challenged the convention that mass production inherently manifests sameness. These developments present distinct new challenges for designers, patent attorneys and examiners, and consumers, challenges that are debated and acted out in the so-called "commons" of public resources, such as the Internet.

7 *The Commons*

For all of their affordances, patents have not been without drawbacks, and legal disputes are only the tip of the iceberg. Ever since the institution of state patent systems in the eighteenth century, patents have spurred many problematic ways of thinking about invention. The hagiography of the genius inventor at the expense of collaborative forms of inventive activity is one. The role of of colonialism in international intellectual and creative activity is another. And the denigration of the intellectual commons writ large, at the cost of the progressive character of industrial society, is yet another, not uncontroversial, charge that has been made against patent systems: as Adrian Johns puts it, "Patentees were the equivalent of squatters on public land—or, better, of uncouth market traders who planted their barrows in the middle of the highway and barred the way of the people."[1] These accusations against patents, which were and remain central to the concerns of patent abolitionists, demonstrate the myriad ways in which the "commons" is the final theater in which patents figure as a matter of public concern. At the risk of making an imperfect analogy, we might consider the intellectual commons that patents do or do not afford as something akin to the opportunity for public life that urban and architectural design does or does not afford. This chapter extends that analogy as a heuristic exercise.

What exactly is the "commons"? The most generous descriptions tend to see it as the land or resources belonging to and/or impacting the entirety of a community. The entire human population is now often described as sharing a commons, particularly within the context of shared fates, such as the climate crisis, whose effects will not discriminate based on national borders, race, or class. As a legal concept, the intellectual origins of the commons are inextricably attached to England's 1217 Charter of the Forest, a document that assured all free men the right to access the royal forests for their natural resources (for hunting, timber, etc.), a right that had been chipped away under the rule of William the Conqueror and his heirs.[2] The Charter of the Forest is a companion law to the more widely celebrated Magna Carta. In his study of the Charter of the Forest, Guy Standing argues that the commons, understood with fidelity to the charter, "refers to our shared natural resources—including the land, the forests, the moors and parks, the water, the minerals, the air—and all the social, civic, and cultural institutions that our ancestors have bequeathed to us, and that we may have helped to maintain or improve."[3] This book has featured patents from the fields of architecture and industrial design, but land, ambiance, and "the disposition of space" have also become the subject of controversial IP regimes.[4] From patents for geogrids and turf to corporate "trade dress" and "dispositions of space," these examples show just how prescient the spatial analogy truly is.

The term "commons" includes the knowledge that we possess as a society, built on an edifice of ideas and information constructed over the centuries. The commons that is regularly conjured in connection with patents concerns the edifice collectively constructed by generations of inventors and the knowledge they have bequeathed to us. The ambiguity of patents as a fundament of the intellectual commons, however, stems from our collective international and intercultural differences in designating when, exactly, knowledge enters shared ownership—when that knowledge must emerge from a monopoly, if a monopoly is warranted at all.

Needless to say, these matters quickly involve weighty philosophical questions about private rights and the public good, about international cooperation and sovereignty, and about ethics. The analogy between the commons and the design of public and private space presents us with an incentive to become more specific with these questions and to consider what, exactly, patents for our shared built environment mean. The analogy also begs us to ask

Hunting in the commons, scene from Gaston Phoebus, *Le livre de la chasse*, c. 1388, Bibliothèque Nationale de France Ms. fr. 1295 / Ms. Fr. 12398.

how they have acted as the main link between architecture, the public good, and private benefit and yet, to date, have gone almost entirely unexamined. This chapter considers the commons as it is addressed by architectural patents through a handful of different spheres of interaction: the mediating actors, that is, patent scholars and patent attorneys; the international landscape, particularly as it applies to countries that have forged patent systems more recently; and the philosophical arena in which conceptions of the commons and IP are debated and refined.

The Actors

My examination of the questions surrounding the intellectual commons in general, and in architecture specifically, has been enriched by interviews with legal scholars and practicing patent attorneys. Robert Merges, a prominent American legal scholar, has memorably equated the public domain with nature, an analogy that makes clear his strong leaning toward a less restrictive IP regime.[5] Drawing from Locke and Kant, Merges also makes clear his belief that IP monopolies are only moral in societies where economic and scientific activity meets a certain threshold of robustness. Two prominent examples of the socioeconomic conundrum presented by IP monopolies include pharmaceutical patents on life-saving medicine and vaccines and copyrights of educational material, such as textbooks. Both examples comprise desirable creations that are simply too expensive for most consumers in the Global South. Merges wrestles with the critical question of whether it is ethical to withhold life-saving medicine or educational resources from poorer people, and his answer is, by and large, no. This is where something known as the Lockean proviso or charity proviso comes into play. The charity proviso of John Locke's labor theory of property states that while individuals have a right to develop private property from nature through their own labor on it, they can only do so "at least where there is enough, and as good, left in common for others."[6] This originates in a moral imperative for fairness and equanimity. Merges distinguishes between vital intellectual property (e.g., medicine- and food-related IP) that is needed for sustenance and cultural intellectual property (e.g., textbook copyrights), with the physical growth associated with the former being more important than the nonphysical growth associated with the latter; hence vaccines rank above textbooks as shared property of society.

I asked Merges how to consider a patent for a low-cost, rapid construction system for simple houses filed by someone in the Global North who intends to implement the design in the Global South. The assumption behind this question was that shelter (or architecture, if we are to abide by the widest definition thereof) would seem to be part of the realm of physical growth and hence part of the realm of the "vital" component of his interpretation of the charity proviso. The patented system would, of course, also contain cultural material, i.e., conceptions of how to construct buildings in modules, the materiality of the system and its relationship to cultural traditions, innovations in rapid manufacturing and shipping, the ultimate form and design of the home, and so forth. I asked Merges how he disentangles the nonphysical and physical assets of a design object (and hence of a design patent) and to what extent, if at all, Western designers would be expected to forgo their patent rights in poor places that want to use their designs.[7] Merges responded, "There will always be a problem drawing lines with something like the charity proviso. The overall idea is to keep it limited to the 'preservation of life' core. So aesthetic elements of an architectural design might not be covered in the sense that a needy person has no right to claim these under the proviso. However, if the aesthetic is inseparable from the concrete utilitarian aspects of a building design, I'd say the aesthetic elements should run along with the concrete aspects and therefore be exempt from IP claims if necessary to preserve life."[8]

I also asked Merges for his thoughts about cultural material generated in places with so-called "charitable" conditions that arrives in the Global North without IP protection, e.g., a recording of traditional folk music in the Amazon and its dissemination in North America, or the repetition of a traditional building method from sub-Saharan Africa in an architectural practice's work in Europe. Do we assume that the IP expectations ought to work the same way when moving from places where IP is not formally secured? Or should we see the absence of IP protection in the place of origin as an assertion of IP's insignificance?[9] Merges responded, again, by emphasizing the specific contours of a given situation:

> I personally am in favor of traditional cultural expression being covered by a "group-oriented" IP right. That is, I like

the idea of exclusive rights, to protect a traditional form of culture (music, dance, textile design, etc.) from being exploited or twisted or used in ways that harm the creators and stewards of the cultural expression. I don't want white supremacists using Hopi music or Navajo rug designs in a pejorative way. Also, I think some compensation to a creator group (tribe, band, People, etc.) is in order. I'd copy the Amish here; they have one price for dealing inside the "community" (with other Amish), and another price when dealing with "English" people (everyone else). So, I'd make use within the originator group royalty-free, and only assert IP rights for commercial use outside the community.[10]

Merges's purview as a scholar has allowed him to focus on vital questions such as how the global commons relates to patents and other forms of IP. However, the vast majority of the legal minds shaping the practice of patents are patent attorneys who work for paying clients—an altogether different context. A prominent retired US patent attorney, whom I will call "Carmen," has taken patent cases to the US Supreme Court, has worked at two of the most prominent IP law firms in the United States, and holds a refreshingly clear-eyed view about the pros and cons of the patent system as a commons (or an "anti-commons").[11] When asked who she believes benefits most from patents, she does not hesitate to say that it is corporations, followed by corporate inventors and, as a distant last, the general public. Carmen maintains that patents generally incentivize the STEM fields as a whole and that the charity proviso is important. At the same time, she believes that the US patent system is becoming overly bloated and bogged down with worthless patents that stymie inventive activity. Carmen believes that a a more rigorous epistemological framework for knowing what constitutes a sufficiently inventive step is necessary to prevent patent systems from becoming unwieldy and counterproductive. While Carmen respects the work of patent examiners, she is frustrated by the wildly different standards that are applied to analyses of nonobviousness at the US Patent Office. Carmen amusedly recalled absurd patent proposals that some examiners seemed to take seriously, such as a proposal for a rosary, an alternative way to swing a swing (sideways instead of from behind), and a premade peanut butter and jelly sandwich. Carmen likes to posit questions regarding nonobviousness to her clients by asking a question in the past conditional tense: "Would that have been obvious if ...?," which she says is a way to veer the conversation away from the vagaries of philosophy and into the realm of the analysis of cold hard facts.

A prominent Canadian patent attorney whom I will call "Dave" demonstrated a markedly less cynical tone when I interviewed him.[12] Dave has an impressive roster of corporate clients, including many who hold patents for production lines in the construction sector. Unlike Carmen, Dave believes that patents benefit the private and public sectors in equal measure, and he adamantly rejects the idea that the patent system is bloated. Dave evoked the adage, sometimes attributed to Charles Duell, the US Commissioner of Patents in 1899, that "anything that could be invented already has been" (a favorite among patent attorneys) to assert that invention and inventive activity are perpetual, sedimentary, and infinite cultural phenomena.[13] In fact, Dave believes that the present represents a third great industrial revolution and that the patent system is integral to that revolution's ultimate success. Dave suggests that patent law has become too restrictive toward the defense of patents; injunctions, for example, have become more difficult to pursue in recent history, and the scope of what is patentable and the "doctrine of equivalents" have both contracted, he believes these changes originate in anti-patent ideology, not in sound legal reasoning.[14]

Another patent attorney in Germany, whom I will call "Hilde," describes herself as a specialist in "patent modernization," a position she calls a form of activist capitalism.[15] Hilde works with inventors on patents of inventions that some might see as difficult to patent and tries to incrementally expand the scope of what is patentable in prosecuting patents in both the German and the European patent offices. Hilde describes a patent as "nothing more than a piece of paper which allows an inventor to litigate." Indeed, Hilde does not feel the patent system is intellectually or morally sacrosanct; rather, she believes that it is most valuable as a way for inventors to make money and support their operations for further inventiveness. It is, as she describes it, a great game of bluffing, blocking, and tackling.

Finally, I spoke with a patent attorney in the United Kingdom, whom I will call "Pavel."[16] Of all of the attorneys I interviewed, Pavel is the most knowledgeable about the arts and architecture, and he has worked on a number of issues related to arts IP, including architectural patents. Pavel claims that a patent is fundamentally not about the object it protects, but rather about the affordance the right offers. Here Pavel holds the same position as Carmen: that patents primarily serve neither inventors nor the public, but rather big business. Pavel also holds that big companies don't even really need patents and that their participation in the patent process is essentially a strategy to make the patent system useless to inventors with fewer resources. Pavel explained how patents for everyday objects (he used the example of a soup ladle) are essentially useless because patent lawyers are unable to reach the knockoff artists—many of whom will be manufacturing knockoffs in other countries—quickly enough and because thorough patent litigation can be expensive. Ten million US dollars is not an unusual figure for the cost of a patent case, which makes litigation for less lucrative (i.e., non-high-tech, non-pharmaceutical) inventions, which include many of the kind discussed in this book, not worth the effort.

The International Landscape

This array of patent scholars and attorneys provides excellent anecdotal evidence about the state of the commons, but it is too small to draw broad conclusions about patents as a feature of a truly global commons. To understand the international character of the commons and a patent's place in it, we must turn to historical and economic scholarship to understand, first, how patents were embedded in the colonial project.[17] The British patent system was by far the most consequential of those run by colonial powers. Prior to the reforms of 1852, the British patent system could, if inventors chose, be consonant with the British colonial footprint, and inventors were able to extend patent monopolies to important imperial sites of production, such as Jamaica and India, by paying a small fee to extend a patent's operation beyond England, Wales, and the town of Berwick-upon-Tweed to the entirety of the British empire.[18] About 15 percent of the patentees between 1716 and 1814 opted to do just this. Between 1846 and 1853 this number increased by nearly threefold.[19] Quantitative research on the British colonial patent system, such as the work of Aaron Graham, offers very valuable information about who patented what and how, but so far this research has failed to offer anything but bromides about technological progress, leaving a major lacuna to be filled in by postcolonial critique. Graham describes the colonial patent system as a "flexible instrument that inventors could use strategically to promote transnational technological innovation, in which people, ideas and skills moved back and forth between Britain and colonies such as Jamaica."[20] In reality, patents directed at the exploitation of colonial territories for goods like sugar or cotton only served to speed up the resource extraction and human exploitation that were integral to the colonial project. Any "transnational technological innovation" that occurred was benefiting the coffers of English industrialists first and foremost, and the English consumer class second. Colonial subjects gained nothing from the fiscal feedback loop of the patent system.

With the patent reforms of 1852, however, this so-called "imperial patent" was replaced by a single patent system that covered the British Isles but not any of the colonial territories of the Americas, Asia, Australasia, or Africa. Adrian Johns characterizes the 1852 reforms as they apply to empires as "an avowed spatial distinction when it came to the empire. It embraced a fissure between the home country and the colonies that was quite unlike what had existed in the previous century. The combination meant that the new, modernized patent system led to a radical debate that embraces international trade and politics."[21] Indeed, national patent systems in those colonies grew more autonomous, albeit without access to the financial and legal resources of London. Technically, these subimperial patent systems predated the patent reforms of 1852 but did not enjoy the same reach as an imperial patent, functioning more like a contractual arrangement between the local colonial government and the patentee. However, subimperial patents, both before and after the central reforms, were more flexible in their terms and tended to offer significantly broader monopolies for the patentee. What this shift effectively did was to transfer the primary form of patent power from white inventors living in England to white inventors living on the ground in the colonies, a fact that incentivized patent-minded industrialists to actually relocate from England to one of its colonies, thus speeding up settler colonialism.

The introduction of nails in New Zealand is an instructive example of the inextricability of patents, construction, and colonial rule.[22] Builders in New Zealand, like those in Australia, were reliant on certain British products for settlement in the far-flung

Illustration from Pierre Pomet, "Slave labour on a sugar plantation in the West Indies," from *A Compleat History of Drugs*, plate 35 (London: R. and J. Bonwicke and R. Wilkin, etc., 1725), Book 2, p. 54.

frontier. Barbed wire to mark property limits and to pen in sheep, corrugated iron roofing, and simple nails for timber construction were all vital elements of frontier settlement for which imperial patents ensured the monopolies of manufacturers in England while stymying local production, which would of course have been more cost-effective.[23] A popular patented nail, the British Ewbank Patent Wrought Nail, manufactured by J. J. Corde & Co.'s Dos Works in Monmouthshire, first surfaced in wood-framed buildings in New Zealand around 1847.[24] Records indicate that the Ewbank nails were at least occasionally distributed in bulk at auction: an advertisement appearing in an Auckland newspaper announces that 20 kegs of Ewbank's Patent Wrought nails, ranging from 1.5 to 3 inches in length, are to be sold at auction on July 12, 1848.[25] Similar auction announcement records exist for other types of patented nails as well as other types of building supplies. It would not have been unusual for New Zealand colonists to acquire nails after looking at guidebooks like *Brett's Colonists' Guide and Cyclopaedia* and the detailed cottage designs and construction instructions contained therein.[26] Each cottage required an average of 209 pounds of nails, and it was simply assumed that patented nails originating in England would be used for their construction.[27]

It was not until the 1892 census of manufacturing that New Zealand recorded its first local nail factory in Christchurch, which makes sense given several facts taken together: the earlier abolition of imperial patents in 1852, the expiration of earlier nail patents, and the growth of British settlement in New Zealand, which allowed for an economical local nail industry. The practice of patenting nails, each of which necessitated some kind of change that marked an inventive step, moved into New Zealand's national (subimperial) patent system, and between 1883 and 1909 the patent office received 41 applications for nail patents, filed predominantly by plumbers, engineers, and carpenters.[28] These included Alfred Robb and William Stoke's 1890 nails for roofing, which had a number of permutations for "dressing" to fit surface materials such as corrugated iron, lathe, and timber beams.[29]

In 1921, nearly 70 years after the abolition of the option for imperial-wide patents, a special committee of the Imperial Conference recommended reconstituting this option. Representatives from all of the Crown territories, including New Zealand, met to discuss the matter in London in June of 1922.[30] J. C. Lewis, the New Zealand Registrar of Patents, entered the conference seeing

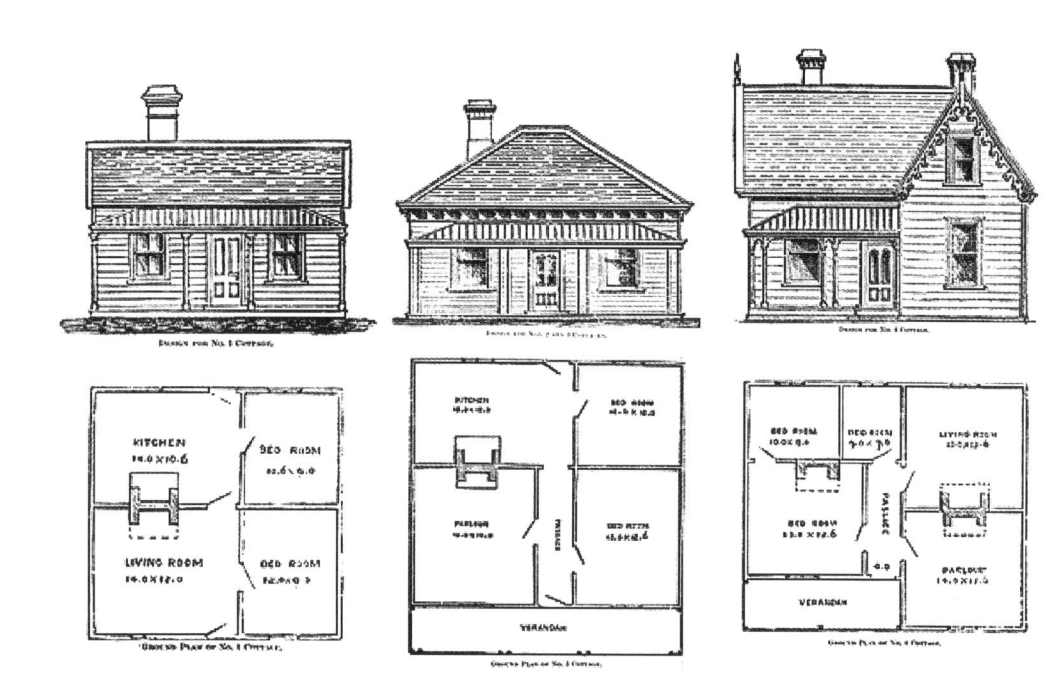

Illustration in Thomson W. Leys, *Brett's Colonists' Guide and Cyclopædia of Useful Knowledge* (Auckland: H. Brett, 1883).

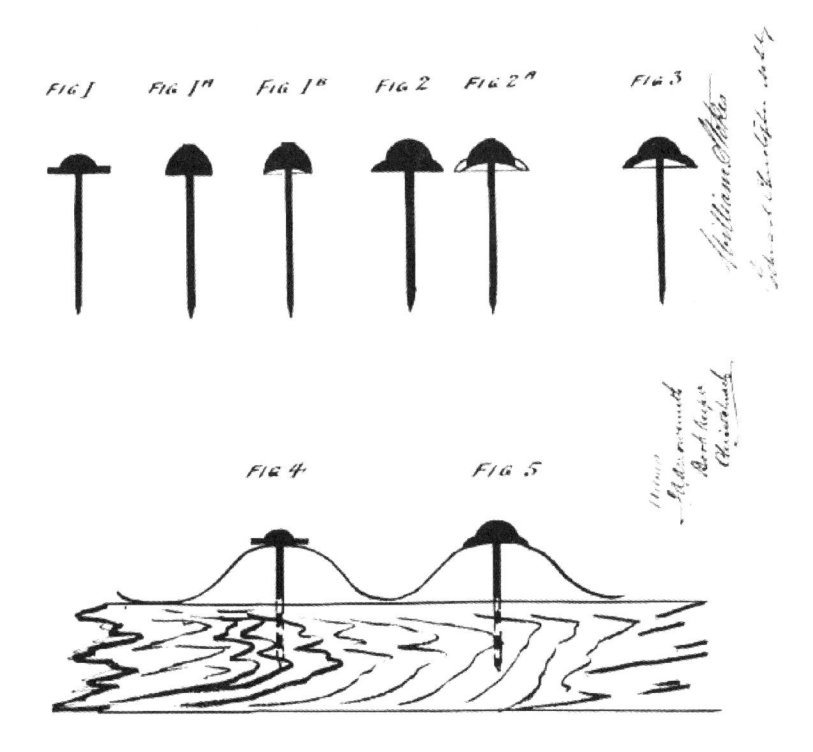

Alfred Robb and William Stoke's patent drawing of roofing nails, New Zealand, Patent 1,152, issued 1890. Nail types: (1) Venables; (1a and b) Parker; (2a and b, 3) Stoke & Robb "cup-headed self-adjusting roofing nail"; (4) Venables "without being dressed down"; (5) Stoke & Robb "requiring no dressing whatever."

some advantages to reconstituting the imperial patent: "In New Zealand, South Africa, and other parts of the Empire," he wrote, "adequate provision for examination has not yet been made; and while concurring that the proposals did not seem feasible, as outlined in the memorandum, I considered it possible that, rather than incur the heavy expense of establishing and maintaining a large examining staff, the Government of this country would prefer to limit its expenditure as far as practicable in this direction by depending to the fullest extent it reasonably could on the examination effected by the English."[31] The result of the conference, however, was not the reconstitution of the imperial patent but merely a retinkering of the various subimperial patent systems to make them relatively friendly to one another while still being independent. As Lewis put it, "something had been accomplished in the desired direction by bringing the laws of the various parts of the Empire into line with one another to a considerable extent, and according a certain limited priority for an applicant in one country to apply in others."[32] Patent law was hardly the vanguard of anticolonial activism, but the resolute result of the conference proved that the Crown's dominions and colonies were unwilling to facilitate foreign monopolies on local manufacturing. One can therefore see why New Zealand sought full independence, which it was granted eight years later.

Imperial patents, to be sure, were in their own right a kind of commons, and the monadization of patent systems that characterizes the decades leading up to independence is, in a sense, a kind of reactionary anti-commons. The important question about this change is whether a true intellectual commons was even possible within the imperial framework, where patent examination and administration acted as an extension of imperial power. Was it perhaps necessary for the anti-commons of the subimperial patent system to exist in order to subsequently instantiate the globally oriented (as opposed to imperially oriented) patent systems of the later twentieth century? One way to address this question is to counterpose it with the matter of indigenous rights and IP, a subject that continues to be meaningful even after the demise of the colonial project.

The concepts of novelty, utility, and nonobviousness that undergird Western definitions of IP are almost never transferable to the communal models of knowledge that characterize indigenous and preindustrial societies.[33] It was not until the late twentieth century that any substantive discussion began on extending IP rights to traditional forms of knowledge. Nor had there been any flicker of an idea as to a more radical way to protect indigenous knowledge through a new mode of protection, sui generis.[34] Defining the term "traditional" or "indigenous knowledge" had been tricky enough, but the task was somewhat simplified by the World Intellectual Property Organization when it put forth the following definition: "tradition-based literary, artistic, or scientific works, performances, inventions, scientific discoveries, designs, marks, names and symbols, undisclosed information, and all other tradition-based innovations and creations resulting from intellectual activity in the industrial, scientific, literary or artistic fields."[35] With these new concepts, "traditional" and "indigenous" knowledge, like patents, were not limited to tangible objects but also included systems and operations that came to encompass many forms of traditional medicine, weaving, and construction, among others. Some of this is akin to what UNESCO has described as "intangible heritage."

The Declaration on the Rights of Indigenous People, adopted in September 2007 by the United Nations, went one step further in protecting these forms of knowledge.[36] Article 11 is particularly germane to the topic of IP, noting that

> [i]ndigenous peoples have the right to practice and revitalize their cultural traditions and customs. This includes the right to maintain, protect and develop the past, present and future manifestations of their cultures, such as archaeological and historical sites, artefacts, designs, ceremonies, technologies and visual and performing arts and literature.... States shall provide redress through effective mechanisms, which may include restitution, developed in conjunction with indigenous people, with respect to their cultural, intellectual, religious and spiritual property taken without their free, prior and informed consent or in violation of their laws, traditions and customs.[37]

Here, too, definitions pose problems. Even the most progressive laws make distinctions between things like land forms, cultural monuments, the performing arts, and scientific knowledge, despite the fact that many indigenous communities do not make these distinctions. Competing claims over the design of Peruvian ceramics in the twenty-first century are just one of myriad examples.[38] By applying these distinctions, one can still undermine indigenous institutions.[39] This demonstrates well Siva Vaidhyanathan's argument that "[t]he fight over the global standardization of intellectual property has become one of the most important sites of tension in North-South global relations."[40]

It is also important to remember that flourishing technological innovation is not synonymous with a well-coordinated international system. The sewing machine is a case in point: its technology evolved very rapidly because patented innovations were picked up freely in countries where the laws of the preceding patent didn't apply.[41] While that may have benefited sewing technology, it also posed serious issues of piracy and unjust enrichment.[42] The United States, for example, did not recognize the rights of foreign IP holders until 1891.[43]

No single country looms larger in international patent politics in the twenty-first century than China. With the proclamation of China's first patent system, the Patent Law of the People's Republic of China, in 1984, its accession to the Paris Convention for the Protection of Industrial Property in 1985, and the status it acquired in 2001 as a member of TRIPS (Agreement on Trade-Related Aspects of Intellectual Property Rights, an international legal agreement between all member nations of the World Trade Organization, first effective in 1995) and of the World Trade Organization, China has become a full-fledged player in the international patent landscape. China's trade partners nonetheless continue to cite widespread IP infringement activity originating in that country.[44] China, like most of its east Asian neighbors (including Japan, South Korea, and Taiwan), has steadily moved away from the historical primacy of Confucianism, in which individual morality rather than the mechanisms of law acts as the central tenet for governance, instead making the importance of the individual (and any monopolies the individual might acquire) secondary to that of the community and thus preventing the public disagreements and "disruption of social harmony caused by civil litigation."[45] Where China differs from these neighbors is that its IP rules, including the 1984 patent system, were adopted somewhat reluctantly, largely due to economic pressure placed on the government as a result of trade agreements that mandated a certain level of IP protection. Unlike Japan, for example, where the government was able to test the IP law's impact on economic development over a long period, China implemented IP laws at breakneck speed at the beginning of the twenty-first century as part of an effort to become globally competitive.[46] China's new patent system, in which a patent now typically takes six years from application to grant, was part of a new economic order in which the privatization of the industrial sector and the introduction of Western technology and management techniques would be both steady and gradual.[47]

Foreign frustration with IP infringement in China is not without its own epistemic conditions. As Peter Ganea and Haijun Jin have noted, "Some commentators refer to an indigenous affinity of the Chinese towards copying which stems from the high regard in which the imitation of styles of old masters on the fields of poetry or calligraphy was held over millennia."[48] The absence of a stable legal environment does not help the Chinese patent system's ability to execute its own agenda either.[49] For several decades, foreign pressure on China to reform its IP laws has primarily focused on copyright, not on patents, leaving considerable space for a lack of oversight and increased infringement.[50] The United States has problematically divided counterfeit goods seized from China by US Customs into "harmless" (e.g., counterfeit fashion or household goods) and "harmful" (counterfeit pharmaceuticals).[51]

Like China, most nations in Africa have developed autonomous patent systems under the economic pressure that has come with global trade since the last decades of the twentieth century. But the similarities mostly stop there. Unlike China, African nations retain the memory of patent regimes imposed on them by colonial powers.[52] African nations have also done a much better job of coordinating their patent systems carefully with their neighbors, a fact that buoys the relatively small and underdeveloped systems of individual nations. Any discussion of patents in Africa would be incomplete without a discussion of the impact of TRIPS.[53] The various principles articulated by TRIPS have essentially acted as a road map for the creation of formal patent offices in Africa, which thus share many similarities. At the time of this writing, TRIPS member states include all nations on the continent of Africa except Algeria, Libya, South Sudan, Sudan, Ethiopia, and Somalia.[54] TRIPS, its proponents argue, acts as a bulwark against patent piracy, which has turned piracy itself into "a phenomenon of geopolitical thresholds" between nations that adhere to standard international IP laws and those that do not.[55]

View of the half-sized copy of the Eiffel Tower and Parisian-style architecture in Tianducheng, China.

Despite its intention to bring about development through an internationally coordinated IP system, TRIPS is not without its critics. Ikechi Mgbeoji expresses skepticism that the patent system can do anything for Africa, given that its priorities are so firmly steeped in Western industrial logic: "[F]or more than a century, African states have participated in IPR [intellectual property rights] regimes with little or nothing to show for it in terms of economic development and transfer of technology. Like a mirage, the wondrous promises of domestic innovation and technological development recede from grasp no matter how long African states tread on the hard path of strong IPR regimes."[56] Another critic, Andreas Rahmatian, articulates how some view TRIPS as having a problematic role in Africa: "An essential instrument in the process of neo-colonialism by economic means is the establishment of a legal framework of international trade which confers legally enforceable rights that support and safeguard economic penetration and control. This includes, in a similar way as in colonial times, the guarantee of protection of foreign property rights in dependent regions. Today, intellectual property rights fulfill this colonizing role to a large extent."[57]

Indeed, even as TRIPS penetrates the last corners of Earth that are without formal patent systems, the citizens of more economically advanced countries will continue to make greater use of their patent systems than those of more recent adopters, creating further imbalances in the revenues afforded by patents between the Global North and the Global South.[58] Rahmatian also contends that provisions in TRIPS that are ostensibly intended to protect indigenous knowledge and its products are, in fact, deleterious to it: "Besides the international expansion of Western-style intellectual property rights, there is another, seemingly contrasting and alternative nonproprietarian, legislative project, which nevertheless has neo-colonial effects: the protection of traditional cultural expressions in the context of 'traditional' arts."[59] It is not hard to see why several observers view the infiltration of TRIPS policy in Africa (and elsewhere) as merely a neocolonial mechanism to transfer wealth from the Global South to the Global North.[60] Caroline Ncube has shown that foreign and nongovernmental assistance in the formation of African patent law has also proved problematic in that the different actors offering guidance often have competing agendas. United Nations agencies, for example, have been guided by progressive development theories, but the same cannot be said for the World Intellectual Property Organization.[61]

Although in a less pronounced way, patents may also have helped to cultivate neocolonial logic in the patent systems of Latin America and Asia. After gaining independence, a number of Latin American republics opted for a blueprint for patent development modeled on the American system, as was the case in Mexico and Argentina, where Ferrari Hardoy had patented his butterfly chair.[62] Others opted for the more liberal, author-oriented French model, as in Chile and Peru.[63] Meanwhile, the uneven patent rules in Asia more or less chart each country's developmental history. Countries like Japan and Korea, which became industrial well before many of their neighbors, exhibit an approach to patents in which political stability is prized over foreign investment.[64] Patentable designs formulated by state employees in Indonesia, where the patent system began in 1991, are automatically transferred to government ownership, while those in private employment retain the rights to their inventions.[65] In Cambodia, whose IP system was founded only in 2001, patents have a 20-year period of validity and follow the common requisites of novelty and nonobviousness while also requiring an additional measure of "industrial applicability."[66] Despite their common heritage as colonies of the Crown, the patent system that emerged in India was quite different from that of New Zealand. Whereas the New Zealand patent has been relatively successful at stimulating new inventions within its relatively small economy, the Indian system has not.[67] Patent applications in India must include the location of origin for sources of biological materials integral to a patent, a requirement intended to root Indian patents in Indian materials but one that may in fact be holding them back, given the lack of availability of certain materials within Indian borders.[68]

Finally, patent skeptics and patent abolitionists will also be quick to point out that a handful of industrial or industrializing societies have produced globally significant inventions without a patent system at all, such as the invention of the diamond drill and of margarine in the Netherlands before it instituted a patent system.[69] A "differential" historical analysis of inventive societies without patents—such as the Netherlands and Switzerland in certain periods of their history—makes the excellent point that those countries with the best-developed patent systems—the United States, Germany, and Great Britain—were also the most formidable in terms of waging war and generating the horrors that come with it.[70] Is this the price of a robust patent system?

There are a number of important lessons to glean from patent regimes that have existed under socialism. In many ways, communism and socialism can be grouped with political philosophies like Confucianism where law was held in relatively low regard. In most socialist contexts, those claiming legal property rights tended to be those placing private interests over the interests of the community, thus representing something of an insurgent practice.[71] A special exception is East Germany, where there was robust patent activity, much of it developed under the auspices of the state. The Deutsche Bauakademie (DBA), an architecture school and research institute that combined the previous Institute for City Planning and Architecture and Institute of Building and Construction (originally directed by the architect Hans Scharoun), is a case in point.[72] The DBA had several specialized central research institutes where faculty were actively encouraged to patent new building technologies, most of which were concerned with the mass production of housing for the state. Faculty (though not their assistants or students who also developed inventions) would be credited as "inventors" in the patents, but the DBA was the assignee, effectively acting the same way corporations with research laboratories acted in the capitalist West.[73] Patent research at the DBA was at times so central to its mission that the development of new patents monopolized the curricula of many design studios, fostering two generations of architects who knew patenting inventions as a core mission of architectural design. Between 1972 and 1990, architects working under the auspices of the DBA filed an impressive 338 patents, or about one new patent every three weeks.[74] The patents, now part of a kind of socialist commons, were conceived as a way to pay tax dollars forward, making construction innovation a state operation and generating a modicum of revenue therefrom.

The DBA's patent venture was closely linked to a major state initiative alternatively known as *Neuererwesen* and the *Neuererbewegung*, or the "innovator movement," which was launched in 1948 and terminated a few months before reunification in 1990.[75] The purpose of the *Neuererbewegung* was to increase industrial productivity and bolster the flagging economy. It directly mirrored initiatives in the private sector of the Federal Republic of Germany.[76] A unique characteristic of the movement was its integration of the concerns of citizens, who could put forward written or oral

Two women working on patent development at the Werk für Fernsehelektronik (Factory for Television Technology) as part of the *Neuererbewegung*, East Berlin, March 1964.

suggestions, information, concerns, and complaints to economic management bodies, state-owned companies, and socialist cooperatives. A good number of the concerns and suggestions regarding building design, particularly housing, were passed on to faculty at the DBA, who would tackle these matters as studio projects and lab research that often featured patents as a demonstrable result of the "success" of the *Neuererbewegung*. Boosterist tracts extolling the virtues of the movement relay the goal of overtaking the West:

> With the further development of socialism in the DDR, there is an objective need to mobilize all existing creative potential to achieve the scientific and technical advance according to the principle of "overtaking without catching up" and to include this potential in the solution of our overall social task. For the institutes and facilities of the DBA, this means a high political and ideological responsibility to actively shape your activities in accordance with these requirements, which is an expression of the high awareness of the employees of the DBA and of the socialist community of people that is perpetually developing under the conditions of socialism.[77]

The deep enmeshment of East Germany's premier architecture school with a state economic doctrine was, on the face of it, beneficial to the public good in the way patents are supposed to be. Inventions like the *Raumzellenbauweise* (cellular room construction), developed through the DBA in 1973, modeled a start-to-finish prefabrication system that drastically cut down on construction costs for new housing, which would pass considerable savings on to the proletariat.[78] The system, which cited a considerable amount of prior art from Soviet fabrication systems (and freely distributed Soviet research), included a vehicular mechanism for transport and rapid delivery.[79] There were, however, also considerable downsides. By the 1970s, some of the DDR's "innovation managers" grew concerned that the movement was not sufficiently understood by state researchers such as the DBA faculty. These managers believed that the "innovators" were more interested in the academic and technological challenges of innovation than in its political content.[80] Another concern was that the degree to which patent activity increased was greater than the degree to which innovation was unfolding, demonstrating that the two were not one and the same.[81] Bureaucrats managing the *Neuererbewegung*

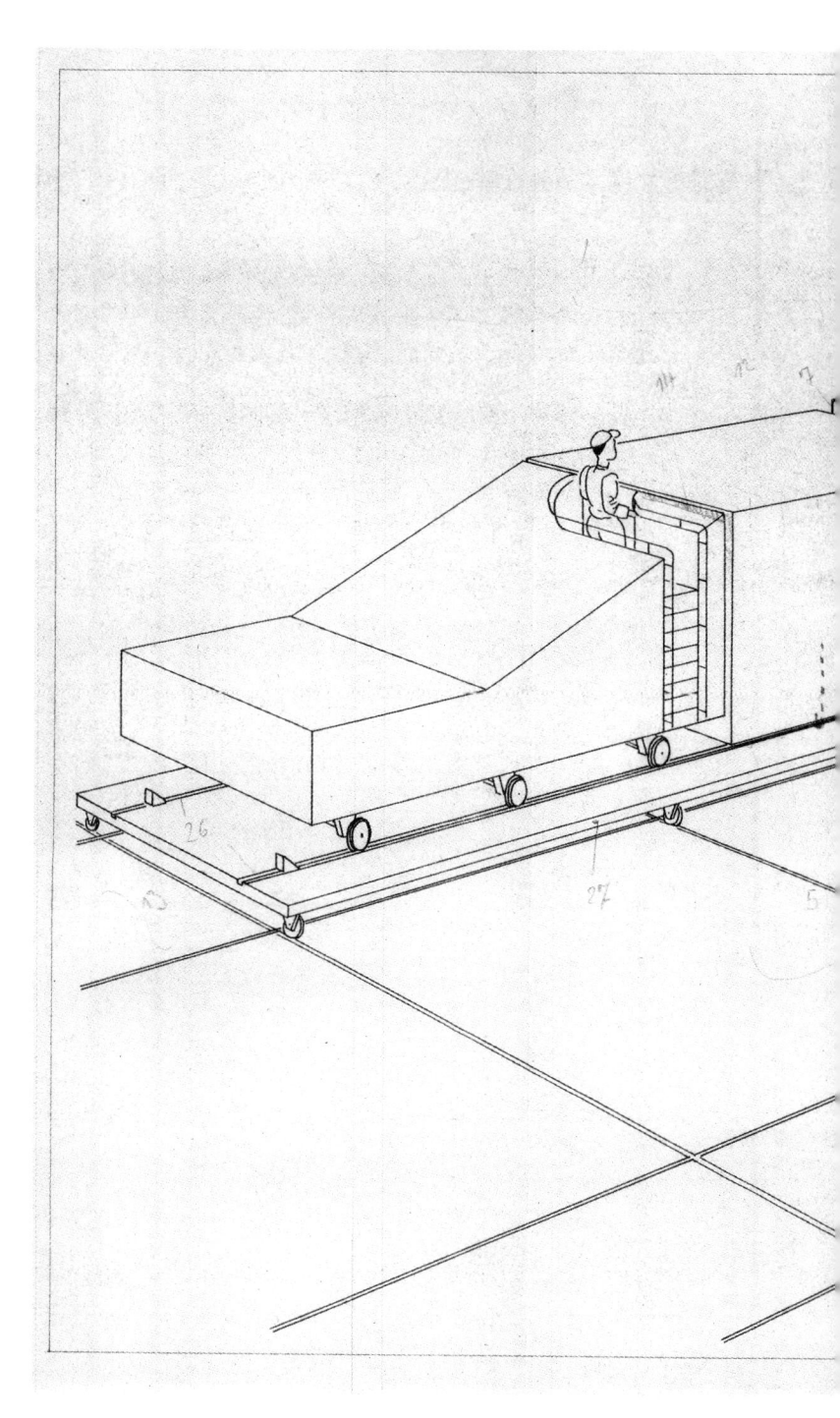

Illustration of the *Raumzellenbauweise* (cellular room construction) system produced by the Deutsche Bauakademie as a patent in 1973. Bundesarchiv Berlin Akte DH 2/5378.

Fig. 3

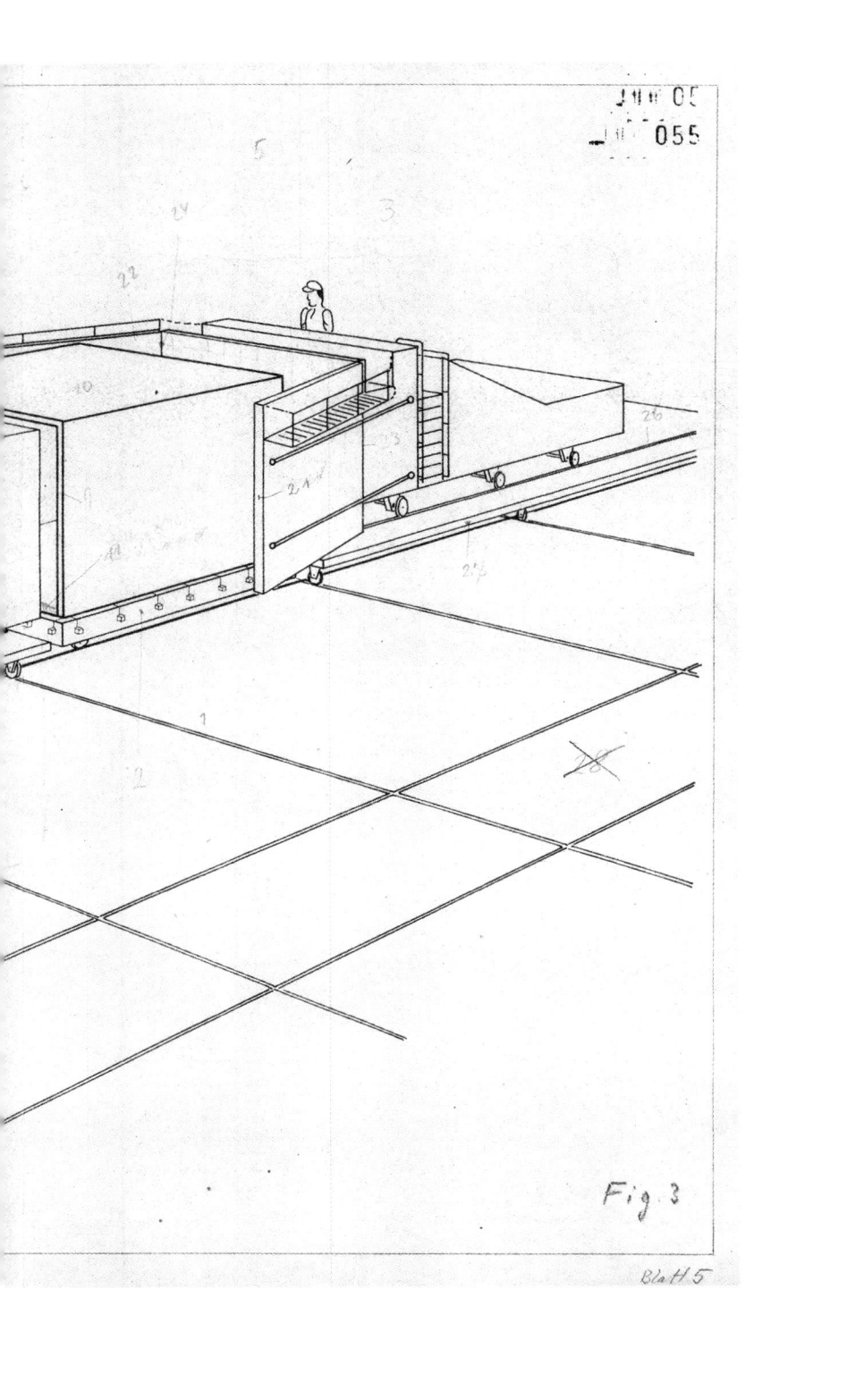

Blatt 5

had grown skeptical that patents were a pure measurement of collective innovation and suspected, even when they were held under the auspices of state institutions like the DBA, that they were at least in part vehicles for inventors to gain personal credentials that prioritized their research accomplishments over the needs of the populace. The problem with patents in this socialist milieu lay in the tension between patents as a form of inventive recognition given to an individual and patents as a mechanism for improving the condition of the masses.

Another problem with East German (and other socialist countries') patents was their inability to gain a commercial foothold in the West. Any truly successful patent under socialism would have to rely on adoption in the West to generate revenue scalable to state-sponsored research. DBA administrators and faculty regularly traveled to Munich to prosecute their patents at the West German patent office. By and large, their inventions were well researched, cited the correct international prior art, and met all the criteria for functionality, newness, and nonobviousness from a purely theoretical position. But the inventions were not suitable to the market tastes of consumers in the capitalist West, who often associated systems like those of the *Raumzellenbauweise* and mass prefabrication with socialism itself. This was also the case for a patent for a node connection for a half-timbered structure, developed by a DBA researcher by the name of Stollberg in 1966.[82] Stollberg and DBA colleagues traveled to Munich to prosecute the patent in March of 1967 and, after some back and forth, successfully obtained a patent. But the construction industry outside of East Germany had almost entirely abandoned industrial-scale half-timber construction, and the patent generated only negligible revenue in the ensuing years.

Philosophical Discourse

The history of legal battles, philosophical discourse, and the development of international patent law all reveal a similar central concern about the commons: the specter of its abuse. Copyism is a prominent type of that abuse. Yet, despite being one of the creative fields in which the flow of knowledge is least regulated, architecture remains one of the least problematic fields when it comes to unwanted copyism. This, of course, speaks to the argument made by Michael Graves to the United States Congress in 1990 that architecture has since time immemorial been a system of appropriation, citations, and borrowings that represent creativity and invention, not plagiarism. It may also speak to a kind of "gentleman's understanding" in the practice, as outlined in chapter 2, where copyright or patent infringement might be readily obvious to other practitioners.

Although they have been articulated less vocally than concerns about design plagiarism, general concerns about the ability to access architectural information and knowledge are also a recurrent theme in the entangled history of patents and architecture. Those defending the use of patents have argued that patents represent a bargain of sorts between the inventor and the general public, "such that the inventor [gets] protection for a limited period in return for not just making the invention but revealing it, and giving it to the public at the end of that period. In that light a patent [is] not an untrammeled private property held in defiance of the public . . . but actually *include*[s] a public interest."[83] Architecture, given the lengthiness of its gestation in comparison to other arts—buildings, of course, take years to design and build—may be particularly well suited to the delay of free use embedded in patent law. Concerns over the lengthy gestation of architectural projects may also mark one of architecture's unique contributions to the theoretical commitments of the "access to knowledge" (A2K), "open access," "voluntary information common," and "free culture impulse" movements in the early decades of the twenty-first century.[84]

The A2K movement's foremost objective is to challenge the ways in which IP, as a right, legitimates the production of artificial scarcity, which it does so purely through theory.[85] Proponents of A2K contend that IP is a "despotic dominion . . . that treats the privatization of information as the necessary condition for its efficient production and exploitation."[86] The elusive "commons" may be at the center of the A2K movement, but it remains to be seen whether the movement is propagating the freedom of information or the freedom of knowledge. If we return to the spatial analogy offered at the beginning of this chapter, we can put it another way: Is a purist idea of the commons a space that everyone must be able to access as built, or is it a space that everyone must build from scratch? More recent economic theory, like that of Fritz Machlup, contends that knowledge—or constructed information—is increasingly important to economic growth, which explains why the world's leading economies have become less agricultural and ever more knowledge-intensive.[87]

To be sure, patents are a form of knowledge, not mere information, as they are carefully constructed through the filters of novelty, utility, and nonobviousness and regulated through the screens of any number of international IP regimes. By extension, patents are propagative of the knowledge economy and the concentration of capital that comes with it in the era of neoliberalism that Machlup describes. If a commons constituted of knowledge—not information—is indeed irreversibly tethered to neoliberalism, is a commons constituted of information—unassembled knowledge—the recipe for an emancipation from neoliberalism? The IP scholar Amy Kapczynski has argued that information (as distinct from knowledge) acts as an artificial force of inflation in the commons as it is currently constituted: "Because the marginal cost of information is zero, the ideal price of information in a competitive market is also zero. As a result, IP rights create 'static' (short-run) inefficiencies. They tend to raise the prices of informational goods above their marginal cost of production, meaning that fewer people have access to these goods than should."[88]

It seems, then, as though patents have little to no role to play in a new utopian commons that is radically emancipatory. Are we thus in need of an entirely new analogy? Perhaps we can benefit from thinking alongside the sociologist Manuel Castells. The information economy that Kapczynski describes is consonant with what Castells describes as "informationalism."[89] Although information has always been vital to the economy and an industrial mode of development, informationalism, as Castells conceives it, is a phenomenon of only the last few decades. Informationalism comprises information that has been coalesced as knowledge that is in turn folded back onto itself to "act," creating a system where "the action of knowledge upon itself [is] the main source of productivity" in a society.[90] Knowledge acting on itself is plainly analogous to the structure of patents, constantly folding onto one another and converting that which is new into "prior art." Informationalism is thus a system where the human is a "direct productive force, not just a decisive element of the production system."[91] This is not to say that the making of the tangible world—as in agriculture, manufacturing, or construction—no longer occurs; it is merely to say that informationalism predetermines the terms of the materialization and productivity of these activities. Informationalism supports a commons not of information but of an ever-quickening feedback loop of technical knowledge actualized by human brainpower.

The architectural theorist Pier Vittorio Aureli has done the most work on extending the conceptual motif of the commons into architectural thought, and consideration of his characterization of the commons points to how informationalism is naturalized in the discipline.[92] For Aureli, the commons is unsynthesized architectural knowledge—technological, historical, and theoretical. The commons of architectural knowledge precedes individual instantiations and "exceeds its technical and commercial determination."[93] In other words, recombined architectural knowledge is greater than the sum of its parts. Urtzi Grau and Christina Goberna have argued that Aureli's knowledge-centered commons has two important implications for IP: "First, it differentiates between architectural knowledge and individual architects, negating the possibility of personal ownership. Similar to language, architectural knowledge has no authors because its existence precedes individual authorship. It is a shared culture. Second, disciplinary culture does not belong in the space that IP regulation provides for collective knowledge."[94] Grau and Goberna propose converting the term "commons"—with all of its analogical spatial potential—back into the adjectival form from which it originated, i.e., *common*, a term that "does not belong to the marketplace of ideas. Its disciplinary autonomy is not a disengagement from the world, but rather a political refusal to accept the prevalent models of production and consumption of culture."[95] Here the commons and the patent offer a roadmap to emancipation.

Coda

> Hail to the Patents! which enables Man
> To vend a folio . . . or a warming-pan
> This makes the windlass work with double force,
> And smoke-jacks whirl more rapid in their course.
> Hail to the patent! That at Irwin's shop
> Improves the flavour of a currant drop.

"The Patent, a Poem" by the British law clerk William Woty, penned in 1776, is a satirical takedown of the patent system.[1] Embedded in his verse is a familiar critique of the modern system of capital wherein an extrinsic construct forces humans into an endless cycle of labor and innovation to a seemingly insignificant end. Written on the eve of the advent of modern patents, Woty's poem captures a certain cynicism that all but disappeared with the Industrial Revolution, when patents became synonymous with a new democratic ideal of invention, entangled with the constituent components prized by modern society—creativity, novelty, property—as major industries were fundamentally transformed by their reach and power to govern at national and global levels.

Architecture, however, adopted the patent system only provisionally, an arm's-length relationship that has lasted nearly two and a half centuries. That distance remains a beguiling feature of the creation of our built environment, where patents are at once everywhere (in the discrete technological components of our buildings and cities) and nowhere (in the absence of IP as a major motivator of the architectural profession). This conundrum is important, because in a world in which the flows of global capital are shaped by who reaps the benefits of innovation and creative production and who dictates how science and art come into being, the architectural profession must ask itself whether it has—in the past as in the present—done everything it can to manifest its full potential as a transformational force for the human condition.[2] Moreover, the insights that patents offer into the nature of creativity, novelty, and property furnish an ideal framework for approaching a more rigorous questioning of the very nature of invention in architecture. This volume may not resolve such questions, but it is my hope that it has provided readers with some of the necessary tools to invent their own answers.

Acknowledgments

The intellectual kernel of this book rests in work I did as a junior curator in the Department of Architecture and Design at the Museum of Modern Art between 2006 and 2008 for the preparation of the two-part exhibition "Home Delivery: Fabricating the Modern Dwelling" with Barry Bergdoll. What I discovered in researching the history of prefabrication in architecture, the subject of our exhibition, were the intrinsic ties between the "dream of the factory-made house," as Konrad Wachsmann put it, and the culture of invention that the act of patenting promoted on a global scale. Here, a survey of patents used to constitute buildings helped form the distinction between tinkerers and architects, between inventing and creating. Inventors and industrialists like Thomas Edison and Carl Strandlund sat side by side with the likes of Jean Prouvé and Frank Lloyd Wright. The fascinating expository genre of patent drawings even adorned the endpapers of our exhibition catalogue. Still, the story remained underdeveloped, and for the time being I put aside my interest in patents and the nature of "invention" that it codifies in architecture.

Returning to the topic more than a decade later has been an incredibly rewarding experience. My research in the United States, the United Kingdom, Canada, France, Germany, Sweden, Italy, and Japan received generous support from a number of institutions, including fellowships and research grants from the John Solomon Guggenheim Foundation, the Institute for Advanced Study (IAS) in Princeton, the National Endowment for the Humanities, the Gerda Henkel Stiftung, the Paul Mellon Centre for British Art, and the dean's office at the University of Rochester. At IAS, my writing benefited tremendously from joining a brilliant group of historians of science in discussing our work, led by Myles Jackson.

Ripp Greatbatch assisted with archival preparation for patent research at the British Library. Ola Svenle helped me navigate the superb patent files of the Riksarkivet in Stockholm. Mirto Vitturi similarly helped me navigate the byzantine procedures of the Biblioteca Nazionale Centrale di Firenze. Back home, Stephanie Frontz helped me obtain rare volumes and images with her signature cheer at the University of Rochester's Art and Music Library, and Lisa Wright diligently produced publication-ready images from an array of ephemera and yellowing periodicals. Lorna Maier and Marty Collier, administrators in the university's Department of Art and Art History, processed countless invoices and reimbursements with diligence, allowing me to focus on research.

Thomas Weaver at the MIT Press has been a superlative collaborator: supportive, transparent, and visionary. Thanks also to the editorial and production team of Gabriela Bueno Gibbs, Matthew Abbate, Erin Hasley, Mary Reilly, and Jim Mitchell.

A number of people who supported me through the period of researching and writing this book deserve thanks for the invaluable gift of friendship: Nick Adderley, Kliment Alexiev, Tanya Bakhmetyeva, Natasha Case, Edward Elles, Freya Estrellar, Will Hamilton, Hunter Hargraves, Peter Hermann, Ben Homrighausen, Marguerite Humeau, Matt Jakob, Maureen Jeram, Mary Johnson, Diego Lourenço, Casey Miller, Reinhard Ostendorf, Miriam Peterson, Ann Reynolds, Nathan Rich, Eyal Rozmarin, Joan Saab, Ben Sales, Jenny Sedlis, and Stewart Weaver. At the end of this rainbow of good fortune is Ripp, my anchor and my best friend. Nothing I do would be possible without you.

Notes

Abbreviations
BaB: Bundesarchiv, Berlin
BaK: Bundesarchiv, Koblenz
BL: The British Library, London
CCA: Canadian Centre for Architecture, Montreal
CUL: Columbia University Libraries, New York
DM: Deutsches Museum, Munich
FLC: Fondation Le Corbusier, Paris
HFM: Henry Ford Museum of American Innovation, Dearborn, MI
HM: Hagley Museum and Library, Wilmington, DE
HUL: Harvard University Libraries, Loeb Library, Rare Books and Special Collections, Cambridge, MA
LMA: London Metropolitan Archives, London
LOC: Library of Congress, Washington, DC
NA: National Archives of the United Kingdom, London
NARA: National Archives and Records Administration, College Park, MD
Ra: Riksarkivet, National Archives of Sweden, Stockholm
RWWA: Rheinisch-Westfälische Wirtschaftsarchiv, Cologne
SUL: Stanford University Libraries, Palo Alto, CA
TEP: Thomas A. Edison Paper, Rutgers University, New Brunswick, NJ
UAH: University of Alabama–Huntsville, Archives and Special Collections, Huntsville, AL
UBA: University of Bristol Archives, Bristol
UBM: Ufficio Brevetti e Marchi, Ministero dello Sviluppo Economico, Rome.
UGAS: University of Glasgow Archive Services, Glasgow
VPUL: Virginia Polytechnic University Libraries, Blacksburg, VA
WAVP: International Archive of Women in Architecture, Virginia Polytechnic Institute and State University
YUL: Yale University Libraries, Manuscripts and Rare Books, New Haven, CT

Introduction
1. See, for example, Biagioli, "Patent Specification"; Griset, *European Patent*; MacLeod, *Inventing the Industrial Revolution*; Pila and Torremans, *Intellectual Property Law*; van Dulken, *British Patents*.
2. Kostylo, "From Gunpowder to Print," 34.
3. The phrase comes from Hobsbawm, *Age of Extremes*.
4. Pila and Torremans, *Intellectual Property Law*, 69.
5. MacLeod, *Inventing the Industrial Revolution*, 3.
6. Garcia, "Architectural Patents," 94; Griset, *European Patent*, 12–14; MacLeod, *Inventing the Industrial Revolution*, 11; Pila and Torremans, *Intellectual Property Law*, 15; Vaidhyanathan, *Intellectual Property*, 42.
7. Biagioli, "Patent Specification." On "negative rights" see van Dulken, *British Patents*, 1.
8. Biagioli, "Patent Specification," 35.
9. Pila and Torremans, *Intellectual Property Law*, 9.
10. Morgan, *Vitruvius*, 197–198. See also Long, "Invention, Authorship, 'Intellectual Property,'" 855.
11. Ganea and Jin, "China," 23.
12. Ibid., 17. China did not have any kind of rigorous intellectual property policy until 1978.
13. Pila and Torremans, *Intellectual Property Law*, 14–15.
14. MacLeod, *Inventing the Industrial Revolution*, 18.
15. See Long, "Invention, Authorship, 'Intellectual Property,'" 860.
16. MacLeod, *Inventing the Industrial Revolution*, 15.
17. A popular account of the process of construction can be found in King, *Brunelleschi's Dome*.
18. Long, "Invention, Authorship, 'Intellectual Property,'" 878.
19. Cited by Pila and Torremans, *Intellectual Property Law*, 15. The Venetian system was also crucial to the development of copyright. See Pila and Torremans, *Intellectual Property Law*, 11.
20. Long, "Invention, Authorship, 'Intellectual Property,'" 881.
21. Van Dulken, *British Patents*, 2.
22. Ibid. See also Foullon, *Usage et description*.
23. Biagioli, "Patent Specification," 26.
24. MacLeod, *Inventing the Industrial Revolution*, 29.
25. Griset, *European Patent*, 23; MacLeod, *Inventing the Industrial Revolution*, 34.
26. Hurx, "Most Expert in Europe."

323

27. Ibid., 12–13; MacLeod, *Inventing the Industrial Revolution*, 88. On Jefferson see Jasanoff, *Ethics of Invention*, 183–185.

28. Johns, *Piracy*, 251.

29. MacLeod, *Inventing the Industrial Revolution*, 50.

30. Schiff, *Industrialization without National Patents*, 21.

31. Johns, *Piracy*, 289; Schiff, *Industrialization without National Patents*.

32. Johns, *Piracy*, 247.

33. Petroski, *Invention by Design*, 44–45.

34. Johns, *Piracy*, 274; Otto and Klippel, *Geschichte*, 51.

35. Johns, *Piracy*, 274.

36. MacLeod, *Inventing the Industrial Revolution*, 78–79.

37. MacLeod notes that for an earlier period there were two main incentives behind patented inventions: saving labor and saving capital. While this certainly continued into the nineteenth and twentieth centuries, a number of inventions were also patented that would more aptly be described as theoretical, speculative, and defensive, none of which fit into these two neat premodern categories. See MacLeod, *Inventing the Industrial Revolution*, 158–181.

38. Griset, *European Patent*, 12–14.

39. Grothe, *Das Patentgesetz*, 14–15.

40. Otto and Klippel, *Geschichte*, 2, 49; Seckelmann, *Industrialisierung*, 2.

41. World Intellectual Property Organization, "Paris Convention."

42. Griset, *European Patent*, 13; Pila and Torremans, *Intellectual Property Law*, 17.

43. Grabicki, *Breaking New Ground*, 32.

44. Otto and Klippel, *Geschichte*, 152–153.

45. Nazi intellectual property law adamantly emphasized the autonomy of the (non-Jewish) individual. See Otto and Klippel, *Geschichte*, 129.

46. Otto and Klippel, *Geschichte*, 13–14.

47. Ibid., 80.

48. Vinsel and Russell, *Innovation Delusion*, 24.

49. See Grau and Goberna, "Should We Patent Architectural Knowledge?," 103, for an analysis.

50. Vaidhyanathan, *Intellectual Property*, 12, 20.

51. Kevin Emerson Collins has come to be recognized as the foremost authority on issues regarding patent law, copyright, and trade dress as these apply to architecture. See Collins, "Patent Law's Authorship Screen" and "Architectural Patents."

52. Collins, "Architectural Patents," 189; MacLeod, *Inventing the Industrial Revolution*, 5.

53. Collins, "Patent Law's Authorship Screen," 1646–1647.

54. Collins, "Architectural Patents," 208.

55. Ibid., 186.

56. Ibid.

57. Osman, "Managerial Aesthetics of Concrete," 69.

58. See Giedion, *Bauen in Frankreich*.

59. Ibid., 150. See also Osman, "Managerial Aesthetics of Concrete," 69.

60. Garcia, "Architectural Patents," 97.

61. Reckwitz, *Invention of Creativity*, 4, 202.

62. Ibid., 29.

63. Ibid., 2.

64. Ibid.

65. Johns, *Piracy*, 248.

66. Reckwitz, *Invention of Creativity*, vii.

67. Ibid., 6.

68. Ibid., 27.

69. Long, "Invention, Authorship, 'Intellectual Property,'" 883.

70. Johns, *Piracy*, 269.

71. As cited in ibid.

72. Merges, *Justifying Intellectual Property*, ix.

73. Ibid., 17.

74. Kivenson, *Art and Science of Inventing*, 169.

75. Ibid., 169–172.

76. Reckwitz, *Invention of Creativity*, 23.

77. Biagioli, "Patent Specification," 30.

78. Reckwitz, *Invention of Creativity*, 36.

79. Usselman, "Patents Purloined," 1059.

80. Colman, "Design and Deviance, Part 1," 44.

81. Ibid.

82. Merges, *Justifying Intellectual Property Law*, 13.

83. Ibid., x.

84. Ibid, 5, 71. See also Pila and Torremans, *Intellectual Property Law*, 69.

85. Merges, *Justifying Intellectual Property*, 75, 80.

86. Ibid., 78.

87. Ibid., 72.

88. Pila and Torremans, *Intellectual Property Law*, 69.

89. Nozick, *Anarchy, State, and Utopia*, 174–175. See also Merges, *Justifying Intellectual Property*, 14–15.

90. See Pila and Torremans, *Intellectual Property Law*, 79.

91. Hegel, *Philosophy of Right*, 5. See also Pila and Torremans, *Intellectual Property Law*, 79.

Chapter 1

1. James Eugene Wierzbicki offers an original interpretation of Nadine's silent drape runners through the lens of sound in *Music, Sound and Filmmakers*, 179. Holly Rogers reflects on the wider meaning of curtains in *Twin Peaks* and Lynch's work in "The Audiovisual Eerie," esp. 253.

2. On the origins of the lazy Susan see Gross, "Lazy Susan."

3. In his evocation of this term, Negroponte was not specifically referring to gender or even domesticity but instead to the empowerment of a much broader swath of the population through access to tools that make research and experimentation easier. This is most famously evident in his project to invent a laptop that was affordable and easy enough to produce that it could be deployed to every child on Earth, a project known as the "One Laptop Per Child" (OLPC) project. See Das, "Frugal Innovation."

4. On women inventors and the US Patent Office see Khan, *Inventing Ideas*, esp. 255–286. On women inventors in Germany and Austria see Vare and Ptacek, *Patente Frauen*. On women inventors in France see Chanteux, "Les inventives." See also Lai, *Patent Law and Women*; Mozans, *Woman in Science*, esp. 334–355; Stanley, *Mothers and Daughters*.

5. Merritt, "Hypatia," 237.
6. Ibid., 289, 298.
7. Voltaire, *Dictionnaire philosophique*, 98. Also quoted in Merritt, "Hypatia," 235.
8. Merritt, "Hypatia," 245.

9. Mott: US Patent 202,115; Benjamin: US Patent 386,289; Shufelt: US Patent 435,461; Merlien: Brevet n. 371,915; Jagdmann: Reichspatentamt Patentschrift Nr. 331,064; Strowich: Reichspatentamt Patentschrift Nr. 331,065; Sackur: Deutsche Patentamtschrift Nr. 443,726C; Zeisel: US Patent 2,574,367. Secondary literature exists only for Benjamin and Zeisel. On Benjamin see Sluby, *Inventive Spirit*, esp. 153. On Zeisel see Kirkham, *Eva Zeisel*.

10. MacLeod, *Inventing the Industrial Revolution*, 77. Bristol probably held this position due to its prominence as a port in the Triangle of Trade and the associated inventions that came with it. Bristol was also a very wealthy city with considerable resources for experimentation in the realm of invention.

11. Great Britain Patent 3,405.
12. Elton, "Sarah Guppy."
13. Ibid.
14. *The Monthly Magazine* (May 1, 1811). UBA DM1140/1.
15. *Felix Fairley's Bristol Journal* (July 27, 1811). Cited by Elton, "Sarah Guppy."
16. *Felix Fairley's Bristol Journal* (August 3, 1811). Cited by Elton, "Sarah Guppy."
17. *The Repertory of Arts, Manufactures and Agriculture* XIX (1811): 242. Cited by Elton, "Sarah Guppy."
18. Ibid.
19. Elton, "Sarah Guppy."
20. Bentham and Hallett, *Project Boast*, 1.
21. Great Britain Patent 3,549.
22. Elton, "Sarah Guppy."
23. Ibid.
24. On Keichline see Allaback, *First Women Architects*; Lewis, *Women of Steel and Stone*; Perkins, "Women Designers," 120–125. Limited files on Keichline are also housed at WAVP.

25. US Patents 1,047,072; D46,679; 1,612,730; 1,736,653; 1,647,733; 1,653,771; 1,838,839.

26. US Patent 1,653,771. In *Cheap, Quick, and Easy*, 166, Pamela Simpson traces a genealogy for the concrete block that does not include Keichline. She cites patents given to Thomas Grant (1632), John Francis Watt (1814), William Ranger (1832), Joseph Gibbs (1850), L. D. Owen (1856), C. S. Hutchenson (1866), Frederick Ransome (1866), T. B. Rhodes (1874), and Harmon S. Palmer (1900), among others. Simpson's genealogy seems focused on larger blocks with at least four closed edges. Keichline's invention was smaller, and its shape afforded it more versatile types of hollow wall construction, including walls with irregular shapes and those that were particularly thin. However, Simpson also qualifies this genealogy with a hint about how Keichline's invention may have fit into later innovation, even when she is not cited directly:

> This fact has led earlier writers on this subject to the mistaken conclusion that the concrete block industry grew gradually over the last thirty years of the nineteenth century and the first few decades of the twentieth. The real story is much more dramatic. None of the nineteenth-century patents led to any widespread production of concrete block. They were all isolated experiments that

resulted in a few buildings and a push for the idea of block, but not in its practical mass production. The mass manufacture of block started only with Palmer's invention of a cast-iron machine with removable core and adjustable sides. He had experimented with it for ten years and had built several structures, including six houses in Chicago in 1897, in order to tests its strength. It was his durable, practical design that spelled the beginning of the modern industry. (Simpson, *Cheap, Quick, and Easy*, 109)

27. Keichline, "Modern Wall Construction," located in WAVP Ms1989-016.

28. Reichspatentamt Patentschrift 660,453.

29. Facebook Messenger conversation between the author and Ted Szego, February 6–7, 2022. Ted Szego is Stefan and Ilse Szegö's grandson and lives in North Carolina. He provided several details from his family files and oral history of Ilse and Stefan Szegö.

30. United States Holocaust Museum, "Antisemitic Legislation."

31. Despite her maiden name of Siegel, Ilse Szegö's grandson is certain that she was not Jewish, a fact that seems supported by her ability to file the patent in the first place.

32. *Federal Register* (August 30, 1944): 10595–10596.

33. US Patent 2,352,296.

34. *Federal Register* (August 30, 1944): 10596.

35. Facebook Messenger conversation between the author and Ted Szego.

36. Hoover maintained a deep-seated skepticism regarding African-American activists throughout his career and directed the FBI to conduct extensive surveillance of Martin Luther King Jr., Elijah Muhammad, and Malcolm X, among others. See Moore, "Strategies of Repression," 10–16.

37. Duberman, *Paul Robeson*, 365.

38. See Goodman, *Paul Robeson*, 124–125.

39. Ibid.

40. Ibid.

41. Walker, "Everything You Need." See also Fox, "Frances Gabe."

42. Video interviews with various news outlets reveal that the only room Gabe ever completed in toto was the kitchen, and that even here the sprinkler system was riddled with problems. Hilary Sample has written a brief thought piece imagining the acoustics of Gabe's invention in "Ambient Maintenance."

43. HM accession no. 2005.21.7a,c.

44. McMurran, "Frances Gabe's Self-Cleaning House."

45. Cited in ibid.

46. Brown, "Son of Carwash." See also Walker, "Everything You Need."

47. US Patent 4,428,085.

48. US Patents 3,381,312; 3,713,176; 3,720,961.

49. On this tradition see Brekke, Nordhagen, and Lexau, *Norsk arkitekturhistorie*.

50. On prefabrication and settler colonialism see Katz, "Mobile Colonial Architecture."

51. Ra Kommerskollegium, Huvudarkivet, DVII:7 1881 E. Ericsson, Metod att uppföra boningshus och andra byggnader af trä utan timmer och utan så kallat kors- eller resvirke: 21/12 1880 No. 376.

52. Ra Kommerskollegium, Huvudarkivet, DVII:7 1882 P. J. Ekman, Nytt sätt att sammansätta och bygga boningshus och andra byggnader af trä, såväl fasta som flyttbara: 15/2 1882 No. 50; Ra Kommerskollegium, Huvudarkivet, DVII:7 1883, K. Wildhagen, Taktäckning af tjär-cement 17/9 1883 No. 351.

53. Great Britain Patent 2151.

54. "Ninety-Year-Old Precast System," 28–29.

55. Ibid. See Shaw and Adams, *Sketches for Cottages*. See also Saint, "Thoughts," 3–12, esp. 5.

56. See "Ninety-Year-Old Precast System," 28; Saint, "Thoughts," 5.

57. "Ninety-Year-Old Precast System," 28.

58. Ibid.

59. Ibid.

60. Ibid.

61. Ibid.

62. US Patent 1,242,872. See also Collins, "Patent Law's Authorship Screen," 1650.

63. US Patent 1,242,872.

64. Ibid.

65. See Freeman, *Clarence Saunders*; Petroski, "Shopping by Design," 491–495.

66. Duboy, "L'Esprit Nouveau," 77.

67. Ibid.

68. Ford, *Raymond Roussel*, 170–171.

69. Ibid., 170.

70. As quoted by Ford, *Raymond Roussel*, 170.

71. Jones, "Architect's Lazy Susan Home."

72. It is also important to note the large number of patents for the rapid

manufacture of building parts that arose in this same period of machines, particularly those parts used en masse, such as nails and bricks. On nail machines see Phillips, "Mechanic Geniuses," 47–49. On brickmaking machines see Broeksmit and Sullivan, "Dry-Press Brick," 48. On Gotthard Wählstrom's patent, see Ra Kommerskollegium, Huvudarkivet, DVII:7 1883, K. Wildhagen, Taktäckning af tjär-cement: 17/9 1883 No. 351.

73. Simpson, *Cheap, Quick, and Easy*.

74. As quoted by Simpson, *Cheap, Quick, and Easy*, 1.

75. Ra Kommerskollegium, Huvudarkivet, DVII:7 1868 Gust. M. Dahlström, Sätt att tillverka artificiel sten: 27/3 1868 No. 44.

76. "Floorcloth and Linoleum Industry." See also Simpson, *Cheap, Quick, and Easy*, 20.

77. Gross, "Evolution of Linoleum."

78. From an advertisement for Nairn's Art Linoleum, ca. 1878–1880, National Library of New Zealand, reference Eph-A-DECOR-1878-01-06.

79. Rhude, "Structural Glued Laminated Timber." See also Otto Hetzer AG Weimar, *Neue Holzbauweisen*; Rug, "100 Jahre Hetzer-Patent."

80. Rhude, "Structural Glued Laminated Timber."

81. Rusak, "Wooden Churches."

82. Giedion, *Space, Time and Architecture*, 191.

83. Ibid.

84. See Lambert, *Building Seagram*.

85. See Christensen, "Precious Metal."

86. Ra Kommerskollegium, Huvudarkivet, DVII:7 1874 Nicolai Ivanowitsch Poutiloff, Nytt sätt att af jernvägsskenor bygga hus: 2/10 1874 No. 159.

87. See Grant, *Big Business in Russia*.

88. Ra Kommerskollegium, Huvudarkivet, DVII:7 1874 Nicolai Ivanowitsch Poutiloff, Nytt sätt att af jernvägsskenor bygga hus: 2/10 1874 No. 159. I thank Ola Svenle for his generous help in navigating the Riksarkivet and providing me with translations of several key patents, including those mentioned in this chapter.

89. US Patent 3,643,910. See also Heifetz, *Description*; Heifetz, "Progrès réalisés"; Kromoser and Huber, "Pneumatic Formwork Systems."

90. I thank Neta Feniger for sharing her research on Heifetz with me. Neta Feniger, email to the author, December 1, 2021.

91. Collins, "Patent Law's Authorship Screen," 1644–1650.

92. US Patent 4,015,385.

93. Ibid.

94. Ibid.

95. Ibid.

Chapter 2

1. On patents and copyrights in the arts see Bertoni and Montagnani, "Public Art and Copyright Law"; Collins, "Architectural Patents"; Hick, *Artistic License*; Long, *Openness, Secrecy, Authorship*; Weizman, "Architecture and Copyright."

2. The American Institute of Architects, for example, has outlined some ways to recognize the difference between plagiarism and inspiration. Such analyses tend to act as unofficial guidelines and probably keep a significant amount of copyright (and patent) infringement at bay. See Bowser, "Understanding the Scope."

3. See, for example, UK Patents 1,147, 2,001, 6,294, 9,936, 24,278, 107,673, 155,857, and 157,471.

4. For predesigned dwelling plans within patents see, for example, UK Patents 3,804, 9,580, 14,579, 120,746, 162,845, 174,067, and 216,977.

5. Michael Graves: US Patent D320,206; Norman Foster: US Patent D556,466.

6. Ochsendorf and Freeman, *Guastavino Vaulting*.

7. Ibid., 198.

8. Ibid.

9. In a special issue in 1999, the *APT Bulletin* republished all 24 of the Guastavino patents in their entirety. See Waite and Gioia, "Patents Held." On the early history of the firm see Austin, "Rafael Guastavino's Construction Business." Janet Parks has published a useful primer on researching the Guastavinos; see Parks, "Documenting the Work." See also Ochsendorf, "Los Guastavino."

10. Kiln: US Patent 670,777; Vault: US Patent 383,050; Structure of Masonry and Steel: US Patent 915,026.

11. The architectural records of the Guastavino Fireproof Construction Company are held at CUL, 12/31/99 T. I thank Janet Parks who, despite being retired, helped me navigate this trove of files.

12. UK Patent 10,777.

13. Ochsendorf and Freeman, *Guastavino Vaulting*, 165, 167.

14. In an email to the author, John Ochsendorf noted that when he presented a lecture on the Guastavino patents to a

Spanish audience, there was a sense that the Guastavinos were patenting "something as common as bread or wine, and they thought it was bizarre." He also notes that "with nearly 20 years of reflection, I would say that most of their patents did not have significant innovation. The main exception is the acoustical patents which had significant new innovations." John Ochsendorf, email to the author, March 27, 2022.

15. US Patent 323,930.

16. Fuller's patents and his reflections on them were compiled in a single volume in 1983, shortly after he filed the last of his 28 patents. See Fuller, *Inventions*.

17. 4D House: US Patent 1,633,702 (Fuller, *Inventions*, 11); Dymaxion Bathroom: US Patent 2,220,482 (Fuller, *Inventions*, 41); Dymaxion Deployment Unit: US Patent 2,343,764 (Fuller, *Inventions*, 56–57); Dymaxion House (Wichita): US Patent 2,351,419 (Fuller, *Inventions*, 95); Geodesic Dome: US Patent 2,682,235 (Fuller, *Inventions*, 127, 129); Octet Truss: US Patent 2,986,241 (Fuller, *Inventions*, 167).

18. SUL Manuscripts Division, Collection M1090.

19. Fuller, *Inventions*. Even the introductory text for this book does not reveal anything that could be described as a philosophical position on the practice of patenting.

20. Fuller, *Inventions*, 1. The Stockade patent is US Patent 1,633,702.

21. On the Biosphere see Massey, "Buckminster Fuller's Reflexive Modernism."

22. Fuller, *Inventions*, 127.

23. US Patent 3,063,521.

24. Fuller, *Inventions*, 179.

25. US Patent 2,682,235.

26. Minnesota Mining and Manufacturing Company, *Our Story*, 101–102.

27. Ibid.

28. Fuller did inspire many students and was famous for his student following. Günter Günschel, a German architect, while not having studied under Fuller, emulated his work, including his practice of patenting. Records at the Canadian Centre for Architecture reveal Günschel's c. 1951 design of a "shell-shaped structure patent model" that stands alone in one photograph and acts as a chair in another, echoing the scalelessness of Fuller's design. See CCA AP198.S1.1949.PR01.001.

29. John Dixon to Ronald Whiteley, February 25, 1954, SUL Manuscripts Division, M1090 Box 80 Series 2 Folder 7.

30. See Kikutake, *From Tradition to Utopia*; Nyilas, "On the Formal Characteristics"; Oshima, *Kiyonori Kikutake*; Pernice, "Japanese Urban Artificial Islands"; Wagenknecht, *Kiyonori Kikutake*.

31. Method of Building Floating Structure: Japan Patent 5,275,840; Floating Structure: Japan Patent 5,275,841; Artificial Ground: Japan Patent 5,373,845; Aquatic Construction: Japan Patent 5,310,639; Structure on Water I: Japan Patent 5,410,534; Structure on Water II: Japan Patent 5,434,530; Floating Structure for Watersurface City: Japan Patent 5,416,001; Structure for Artificial Multilayer Land: Japan Patent 5,526.307; Floating Structure for Building: Japan Patent 5,526,368.

32. See Kumagai, *History of Japanese Industrial Property*.

33. See Ching, Jarzombek, and Prakash, *Global History*, 738. On the Metabolist manifesto see Gardner, *Metabolist Imaginations*, esp. 32–39.

34. See Kurokawa, *Philosophy of Symbiosis*.

35. Hoberman's files through 2003 are held by CCA, Chuck Hoberman fonds, AP165.S1–7.

36. CCA Chuck Hoberman fonds, 165-001-14.

37. US Patent 4,780,344. See also Lynn and Hoberman, "Expanding Geometries."

38. US Patent 4,981,732.

39. CCA fonds Chuck Hoberman, 165-001-14.

40. Ibid.

41. On Alvar Aalto see Aalto, *Alvar Aalto*, 3 vols.; Pelkonen, *Alvar Aalto*; Reed and Frampton, *Alvar Aalto*. On Jean Prouvé see Burkhalter and Sumi, *Konrad Wachsmann*; Bollinger and Medicus, *Stressing Wachsmann*; Sulzer, *Jean Prouvé: Œuvre complète*, 4 vols. On Konrad Wachsmann see Grüning and Wachsmann, *Der Architekt*. On Renzo Piano see Buchanan, *Renzo Piano Building Workshop*, 5 vols.

42. Heikenheimo, "Alvar Aalto's Patents," 10.

43. Ibid., 9. These include: Finland Patent 18,256 (A Method of Making Furniture and Other Objects of that Nature, and Chairs and Other Items of Furniture Using the Method, 1938, also patented in Sweden [1938], Denmark [1936] and Great Britain [1935]); Finland Patent 18,666 (A Bending Method for Wood and the Articles Produced by This Method,

1940, also patented in Belgium [1935], Great Britain [1935], Italy [1935], France [1935], the Netherlands [1936], Austria [1936], Switzerland [1936], Denmark [1936], USA [1936], and Czechoslovakia [1937]); Finland Patent 23,421 (A Method of Bending Pieces of Wood and the Bent Wood Products Made with It, 1949); and Finland Patent 28,191 (Combining Pieces of Bent Wood, 1956, also patented in Sweden [1956]).

44. Ibid. These include Finland Patent 16,222 (Metal Leg Chair, 1934, also patented in Germany [1935]); Finland Patent 19,798 (Stair Tread, 1943, also patented in Sweden [1945]); and Finland Patent 35,162 (An Anti-glare Shade for Light Fittings, 1965, also patented in Sweden [1965] and Denmark [1965]).

45. Heikenheimo, "Alvar Aalto's Patents," 11.

46. Ibid.

47. Ibid., 12.

48. Aalto was extremely attentive to construction details like lamination. See Dietziker and Gruntz, *Aalto in Detail*.

49. Heikenheimo, "Alvar Aalto's Patents," 12.

50. Ibid., 15.

51. Ibid.

52. Picchi and Piano, "Prouvé inventore."

53. See Barré-Despond, *UAM*.

54. Ibid.

55. Picchi and Piano, "Prouvé inventore."

56. https://www.centrepompidou.fr/en/program/calendar/event/caGRyk (accessed March 31, 2022). On Prouvé and the discourse around the poetics of technical objects see von Vegesack, *Jean Prouvé*.

57. Improvements in the Construction of Soundproof Panels: France Patent 717,866; Double-Hung Window Whose Sashes Open by Rotating around a Vertical Axis: France Patent 853,226; Improvements to Buildings with Box-Girder Structures and Walls in Double-Wall Panels: France Patent 921,344; Sun-Louvre Stretchers: France Patent 1,411,236. See also Picchi and Piano, "Prouvé inventore."

58. Picchi and Piano, "Prouvé inventore."

59. Metal Construction System: France Patent 922,364; Construction System for Single-Story Buildings: France Patent 943,352; Elements for Erecting Buildings and Their Installation: France Patent 1,138,751, extensions 69,072, 69,173, 71,655; Structural Building Method and Systems Applying Said Method: France Patent 1,389,572. See also Picchi and Piano, "Prouvé inventore."

60. See Rubin and Cinqualbre, *Jean Prouvé*.

61. The terms here are qualified with the word "ostensible" because Prouvé's system was not deployed in France alone but also in French colonies in Africa, where they were used to quickly establish or enforce colonial rule, which is of course patently undemocratic. On the recombinant nature of Prouvé's design systems, see Huber, *Jean Prouvé*.

62. See Bergdoll and Christensen, *Home Delivery*, 13. It was in this position that Wachsmann designed a summer house for Albert Einstein, a longtime friend, in Caputh, Brandenburg.

63. See Herbert, *Dream of the Factory-Made House*, esp. 200–212.

64. Ibid., 243.

65. Ibid., 246.

66. Ibid., 248.

67. Ibid., 249.

68. Ibid., 254. US Patent 2,355,192 (Prefabricated Building, 1944, cross-listed in France, Great Britain and Canada).

69. It was quite common for the official patent title to have a different name in the everyday parlance of its designers and interlocutors. In this case, "Prefabricated Building" is expressive of the technological terminology that would appear to give the invention the widest possible patent protection, while the name "Packaged House" was marketing-friendly and touched on visceral connections for a consumer public.

70. US Patent 2,355,192.

71. The site was also known as the Dugway Proving Ground. See Cohen, *Architecture in Uniform*, 231–239; Davis, "Berlin's Skeletons in Utah's Closet."

72. Herbert, *Dream of the Factory-Made House*, 258–259. In this connection, one might also consider Wachsmann's *later* articulations (in both writing and design) on mass timber construction, which seem to have given him some independence from Gropius. See Wachsmann, Grüning, and Reuss, *Technik und Gestaltung*.

73. Herbert, *Dream of the Factory-Made House*, 266–298.

74. US Patent 2,421,305 (Building Structure, 1947, cross-listed in France, Great Britain, and Canada); US

Patent 2,426,802 (Sectional Wall Structure System, 1947); US Patent 2,391,882 (Building Construction, 1949); US Patent 2,559,741 (Building Structure, 1951).

75. Herbert, *Dream of the Factory-Made House*, 302.

76. Ibid., 302.

77. US Patent 3,175,657.

78. US Patent D219,165.

79. I am grateful to Lorenzo Ciccarelli for generously sharing his work on Piano with me. See Ciccarelli, "Architecture as Construction."

80. Piano, "Modulazione e coordinamento modulare."

81. Ciccarelli, "Architecture as Construction," 448.

82. Ibid., 447. The five patent dossiers are housed at the UBM.

83. See Dal Co, *Centre Pompidou*.

84. On Frank Lloyd Wright see Levine, *Architecture of Frank Lloyd Wright*; Loper and Schreiber, *This American House*; Neumann, "'Century's Triumph in Lighting.'" On Sauvage see Minnaert, *Henri Sauvage*; Minnaert, "Henri Sauvage." On I. M. Pei see Collins, "Architectural Patents," 201–205; Jodidio and Strong, *I. M. Pei*, esp. 22. On Skidmore, Owings & Merrill see Adams, *Gordon Bunshaft and SOM*; Adams, *Skidmore, Owings & Merrill*. On Santiago Calatrava see Tzonis, *Santiago Calatrava*.

85. On Edison see Israel, *Edison*; Nettles, *Thomas A. Edison's Advancements*. Special thanks are due to Paul Israel who met with me not far from Menlo Park, New Jersey, to discuss his thoughts on the nature of Edison's approach to patenting. There has simply never been anyone else as knowledgeable about Thomas Edison as he.

86. Wachhorst, *Thomas Alva Edison*, 5.

87. Ibid.

88. Israel, "'Claim the Earth,'" esp. 19, 37–38.

89. Ibid., 38.

90. Ibid. See also Schallenberg, *Bottled Energy*, 359.

91. Israel, "'Claim the Earth,'" 38–39. See also Choate, *Hot Property*, 67–76; Lamoreaux, Raff, and Temin, *Learning by Doing*.

92. Israel, "'Claim the Earth,'" 19.

93. Testimony of Thomas Alva Edison, p. 50, *Boehm v. Edison*. Pat. Int. 7943, TEP, W100DED032. Also see Israel, "'Claim the Earth,'" 26.

94. See Baas, "Concrete in the Steel City"; Bergdoll and Christensen, *Home Delivery*, 42; Lee, "The Second Nature"; Slaton, *Reinforced Concrete*; Wermiel, "California Concrete."

95. US Patent 1,219,272.

96. Extant Edison concrete houses in New Jersey can be located on an interactive map published by TEP, Rutgers University: https://uploads.knightlab.com/storymapjs/a12cc1cc54c4c8e0842oa3954abf6183/thomas-edisons-new-jersey-1/index.html (accessed March 31, 2022).

97. "Building of Concrete," *Boston Daily Globe*, September 24, 1908. See Baas, "Concrete in the Steel City," 253n24.

98. "Seven Great Wonders of Science and Industry." See Baas, "Concrete in the Steel City," 253n25.

99. David, Frohne, and Wright, "The Advent of the Fireproofed Dwelling," 309. See also Baas, "Concrete in the Steel City," 253n26.

100. Baas, "Concrete in the Steel City," 255.

101. US Patents D27,977–28,015, D28,017, D28,020. Only D27,977 was realized. See Neumann, "'Century's Triumph in Lighting,'" 28–29.

102. Neumann, "'Century's Triumph in Lighting,'" 28–29. Wright also conducted other work for the Luxfer company, including the design of a skyscraper in 1898. Wright's Luxfer designs were widely used in India. See Neumann, "'Century's Triumph in Lighting,'" 31, 33.

103. US Patents D27,977–28,015, D28,017, D28,020.

104. Ibid.

105. http://www.steinerag.com/flw/Artifact%20Pages/AmerSystBltHomes.htm (accessed March 31, 2022).

106. https://franklloydwright.org/site/american-system-built-homes/ (accessed March 31, 2022).

107. Wright, "To the Young Man."

108. US Patent D114,204. Wright received two other design patents in the interim: D108,473 (Design for a Chair, 1938) and D114,203 (Design for a Desk, 1939).

109. US Patent D114,204.

110. See "Ennis House."

111. Donald Barker to Frank Lloyd Wright, January 26, 1933, CUL, Frank Lloyd Wright Foundation Archives, B019E02. See also B019E03.

112. US Patent 1,535,030.

113. US Patent 1,351,086.

114. Minnaert, "Henri Sauvage," 41. Sauvage was also very much influenced by, and perhaps to some degree in competition with, the architect Auguste Perret, whose

own reinforced concrete construction company combined architectural design, technical studies, and implementation maximization.

115. Ibid., 42.
116. Ibid., 43
117. Ibid.
118. Loyer, *Henri Sauvage*.
119. France Patent 439,292.
120. Minnaert, "Henri Sauvage," 41.
121. Ibid.
122. Ibid., 42, 45. See also Rabinow, *French Modern*, 214–217.
123. France Patent 678,632. Minnaert, "Henri Sauvage," 43.
124. France Patent 684,089.
125. US Patent 2,698,973. Collins, "Architectural Patents," 201–205.
126. US Patent 2,670,859. Zeckendorf and McCreary, *Autobiography of William Zeckendorf*, 97–99.
127. Switzerland Patents 638,581 (Self-Supporting, Collapsible, 1983), 662,602 (Roof-Supporting Structure, 1987), 664,769 (Gate with Up-and-Over Folding Sections, 1988), 675,002 (Supporting Structure for Cupola for Large Building, 1990); German Patent 3,538,236 (Tipping Door, 1987); Austria Patent 349,583 (Shadow Covering Device, 2007); European Patents 1,027,934 (Mill, Especially Tube Mill, 2003) and 5,010,736 (Device with Adjustable Elements, 2003), US Patents D479,657 (Chair, 2003), D607,133 (Lamp, 2009), D581,706 (Table, 2009), D587,557 (Handles, 2009).
128. US Patents D554,271 (Building), D615,211 (Bridge and Gondola System).
129. Melis Abdula, email to author, December 1, 2021.

130. "Syntax: Contribution of the Curtain Wall," 308.
131. Insights into SOM's research and innovation practices were provided to the author by Chas Peppers, the firm's director of innovation. Chas Peppers, telephone interview with author, January 12, 2022.
132. See US Patent Application 2021/0087810A1, filed March 25, 2021 (Precast Wall Panels and Method of Erecting a High-Rise Building Using the Panels).
133. Chas Peppers, telephone interview with author, January 12, 2022.
134. On the themes of housing, mass production, and prefabrication in the work of Le Corbusier see Fondation Le Corbusier, *Une petite maison*; Le Corbusier, *City of Tomorrow*; Le Corbusier, *Vers une architecture*.
135. France Patents 492,385, 492,386, 492,387, 492,388, 492,389, 492,390.
136. France Patents 496,913, 21,671E, 619,254, 996,664, 1,012,589, 60,067E, 1,029,744.
137. Canada Patent 470,356.
138. On the saga of the prospective Modulor patent see Matteoni, "16 brevetti," 78–79.
139. Canada Patent 470,356.
140. Tim Benton has done this with aplomb. See Benton, "Domino."
141. Ibid., 30.
142. Ibid.
143. Ibid., 31, 36.
144. Boesiger, *Le Corbusier*, vol. 7, 240.
145. Tim Benton, email to author, March 31, 2022.
146. Ibid. See also Benton, "Villa de Mandrot."

147. Benton, "Domino," 30.
148. Ibid., 24.
149. Ibid.
150. Matteoni, "16 brevetti," 70.
151. Ibid. See also Janson et al., *Le Corbusier*.
152. Matteoni, "16 Brevetti," 70.
153. Ibid.
154. US Patent 3,337,999. See also CCA fonds Victor Prus, ARCH267964 3-22-15T; Gaudette et al., "D'un stade mal-aimé"; Prus, *Essays in Architecture*.
155. US Patent 3,337,999.
156. Shepard, "Victor Prus."
157. US Patent 2008/0143696 (Display System, 2008, also filed with the World Intellectual Property Organization, Japan, Australia, and Great Britain). See also Goulthorpe, *Possibility of (an) Architecture*.
158. Mark Goulthorpe, email to author, November 30, 2021.
159. Ibid.

Chapter 3

1. Abelshauser et al., *German Industry*, 36. See also Reinhardt and Travis, *Heinrich Caro*.
2. On BASF see Abelshauser, *Die BASF*; Becker, Grim, and Krameritsch, *Zum Beispiel*; Räuschel, *Die BASF*.
3. Hounshell and Smith argue that Caro's innovations at BASF were at least partly derivative from those taking hold at Bayer, although my assessment is that the developments were concurrent and that Caro's efforts were more forceful. See Hounshell and Smith, *Science and Corporate Strategy*, 4–5.

4. Seckelmann, *Industrialisierung*, 402.
5. Merges, *Justifying Intellectual Property*, 22.
6. Ibid., 23.
7. Hounshell and Smith, *Science and Corporate Strategy*, 6. On the Sherman Antitrust Act, see NARA, Act of July 2, 1890 (Sherman Anti-Trust Act) and Enrolled Acts and Resolutions of Congress, 1789–1992; General Records of the United States Government, Record Group 11; *Encyclopaedia Britannica Online*, s.v. "Sherman Antitrust Act," accessed May 31, 2002, https://www.britannica.com/event/Sherman-Antitrust-Act.
8. Seckelmann, *Industrialisierung*, 163.
9. Hounshell and Smith, *Science and Corporate Strategy*, 4; Otto and Klippel, *Geschichte*.
10. Hounshell and Smith, *Science and Corporate Strategy*, 7.
11. Seckelmann, *Industrialisierung*, 19.
12. Carlson, *Innovation*, 241.
13. MacLeod, *Inventing the Industrial Revolution*, 77.
14. Van Dulken, *British Patents*, 21.
15. Thomson, *Selections*, 351.
16. Carlson, *Innovation*, 6.
17. Ibid., 268.
18. Abelshauser et al., *German Industry*, 36–37.
19. Ibid.
20. Ibid., 84–87.
21. Grabicki, *Breaking New Ground*, 18, 37.
22. Abelshauser et al., *German Industry*, 85.
23. Ibid.
24. On the Swiss patent system see Schiff, *Industrialization without National Patents*. See also Abelshauser et al., *German Industry*, 140–141, 146.

25. Abelshauser et al., *German Industry*, 79, 86–87.
26. Ibid., 159.
27. Ibid., 150.
28. Ibid.
29. Ibid., 195.
30. A. C. Horn Company, "Keramik: A Color Penetrant for Concrete Surfaces," promotional brochure, 1925, CUL.
31. Hounshell and Smith, *Science and Corporate Strategy*, 5.
32. Ibid., 525–526.
33. G. B. Malone, "E. I. DuPont de Nemours and Company Central Research Department: Patent Policies and Procedures" (internally published handbook, July 1, 1964), HM Accession No. 2380, Box 1, ii.
34. Ibid.
35. Ibid.
36. "The DuPont Experimental Station, Issued on the Occasion of the Dedication of the New Research Laboratories, Wilmington, Delaware, May 10, 1951" (commemorative brochure), HM Accession No. 2380, Box 1, Folder 9, 27.
37. Malone, "E. I. DuPont," iii.
38. Ibid., 1–34.
39. Ibid., 14.
40. Hounshell and Smith, *Science and Corporate Strategy*, 8.
41. "DuPont Experimental Station," 6.
42. Minnesota Mining and Manufacturing Company, *Our Story*, 114.
43. Ibid., 22.
44. Ibid., 96, 104. In 2001, 3M had 60 patent attorneys on its staff.
45. Hounshell and Smith, *Science and Corporate Strategy*, 369.

46. Minnesota Mining and Manufacturing Company, *Our Story*, 31.
47. Ibid., 34–35, 141.
48. Stalder, "Turning Architecture Inside Out."
49. Ibid., 70.
50. Ibid., 71.
51. On DYWIDAG see Dyckerhoff und Widmann A. G., *Öffentliche Gebäude*; Dyckerhoff und Widmann A. G., *Dywidag*; Böhm, Jungwirth, and Neunert, *Festschrift*.
52. Christensen, *Precious Metal*, 133–137; Ilkosz, *Die Jahrhunderthalle*.
53. "DYWIDAG: Bauunternehmung und Betonwerke," DM FA010/164.
54. Osman, "Managerial Aesthetics of Concrete," 71.
55. "Betonphalgründungen Patent Strauß," DM FA010/224.
56. Ibid.
57. "Betonpfähle Patent Strauß. Eine neueres Gründungsverfahren. Vortrag gehalten im Sächsischen Ingenieur- und Architekten-Verein zu Dresden am 16. März 1908 und ergänzt nach den Ausführungen der neuesten Zeit von W. Gehler, Regierungsbaumeister a. D., Oberingenieur der Firma Dyckerhoff & Widmann, A. G.," DM FA010/225.
58. Ibid., 3.
59. Minnesota Mining and Manufacturing Company, *Our Story*, 21, 35, 98, 123.
60. Ibid., 60.
61. Ibid., 58, 65.
62. Ibid., 60–61.
63. On Junkers see Lorenz, *Kennzeichen Junkers*; Wagner, *Hugo Junkers*; Weber, "National Character," 386–395.
64. On Messerschmitt see Ebert, Kaiser, and Peters, *Willy Messerschmitt*; Pabst, *Willy*

Messerschmitt; van Ishoven, *Willy Messerschmitt*.

65. On Junkers and the Bauhaus see Droste and Bauhaus-Archiv, *Bauhaus 1919–1933*, 216–218; Guillén, *Taylorized Beauty*, 58–61; Welter, "Limits of Community," 63–80.

66. DM FA Junkers 766.

67. Ibid.

68. Hans Drescher to Junkerswerke, November 9, 1934, DM FA Junkers 766.

69. Hans Drescher to Junkerswerke, November 20, 1934, DM FA Junkers 766.

70. Bergdoll and Christensen, *Home Delivery*, 62–64; Menzel, *Das Messingwerk*; von Borries and Fischer, *Heimatcontainer*.

71. Hans Drescher to Junkerswerke, November 20, 1934, DM FA Junkers 766. See also Herbert, *Dream of the Factory-Made House*, 179.

72. Hans Drescher to Junkerswerke, November 20, 1934, DM FA Junkers 766.

73. DM FA Junkers 502.

74. "Zur Frage wirtschaftlich wertvollen Patentschutzes auf dem Gebiet des 'Hausbaues,'" 2, DM FA Junkers 721.

75. See Musciano, *Messerschmitt Aces*, 168.

76. DM FA003/0618; DM FA003/619; DM FA003/1200. Messerschmitt's patent activities appear to have been aided greatly by a man named Rudolf Busselmeier, who is alternately referred to as "patent engineer" and "patent lawyer" in the firm's records.

77. "Germany: Into Plowshares."

78. DM FA 003/1200; DM FA 003/1250.

79. DM FA 003/619; DM FA 003/1251; DM FA 003/1252.

80. DM FA 003/619, 1.

81. Ibid, 2–3.

82. DM FA 003/619.

83. "New Markets in Building" (unidentified newspaper clipping), HM Accession 2380, Box 4. Patent information in HM Accession 2632-II, Box 1, File 54. See also Memorandum from N. A. Higgins to J. R. McCartney, December 31, 195, HM Accession 2632-II, Box 1, File 54.

84. "Summary Report of Flash-Spun Polyethylene," May 5, 1969, HM Accession 2632-II, Box 1, File 54.

85. Memorandum from J. M. Griffin to F. R. Milhiser, October 23, 1969, HM Accession 2632-II, Box 1, File 54.

86. Memorandum from J. M. Griffin to J. P. Sinclair, April 14, 1969, HM Accession 2632-II, Box 1, File 54.

87. Memorandum from N. A. Higgins to J. R. McCartney, December 31, 1956, HM Accession 2632-II, Box 1, File 54.

88. Patent credits are discussed in an internal management memorandum. G. M. Rothrock and L. B. Chandler to J. M. Griffin, April 1, 1969, HM Accession 2632-II, Box 1, File 54.

89. "FDA Accepts DuPont Test Protocols for Transition of Tyvek to Upgraded Manufacturing Process" (press release), December 1995, HM Accession 2632-II, Box 7, File 20.

90. Hounshell and Smith, *Science and Corporate Strategy*, 436–437.

91. On Peterson see ibid., 524.

92. "Thirty Years of Technology Proves DuPont Tyvek is a Leader in Nonwoven Industry" (press release), January 1996, HM Accession 2631, Box 4.

93. Ibid.

94. Compendium of Memoranda, "Explorative Idea for Growing Shelters Synthetically," June 22, 1964–September 1, 1964, HM Accession 2122, Box 1, Domenico Mortellito papers.

95. Domenico Mortellito was featured prominently in a series of three issues of *American Artist* in 1942, covering his experimental techniques with plastic rock, monocork, and leatherwall marine veneer (April 1942: 32–34), plastic on glass, lacquers on glass, glass cloth, and plastic compounds (February 1942: 26–28), and lucite and battleship linoleum (January 1942: 10–11, 27). Lacquered linoleum was explored in an earlier issue (December 1937: 4–9).

96. Mortellito's family, under the auspices of the Domenico Mortellito Trust, maintains an informative website on the artist with detailed information about his biography, including his education. See http://www.domenicomortellito.com (accessed March 5, 2022).

97. On TRAP see Grieve, *Federal Art Project*, 88. On Mortellito and TRAP see United States Treasury Department Procurement Division, *Treasury Department Art Projects*, 5–6.

98. http://www.domenicomortellito.com (accessed March 5, 2022).

99. Ibid.

100. Ibid.

101. Compendium of Memoranda, "Explorative Idea for Growing Shelters Synthetically," June 22, 1964–September 1, 1964. HM

Notes

Accession 2122, Box 1, Domenico Mortellito papers.
102. Ibid.
103. HM Accession No 2122, Box 1, Domenico Mortellito papers. For example, see "Two Step Foam Home," 61. He also had a clipping on the topic in the following year: "Designer's Forecast," 51.
104. Domenico Mortellio to Samuel Lehner, June 23, 1964, HM Accession 2122, Box 1, Domenico Mortellito papers.

Chapter 4

1. Çelik Alexander, "Managing Iteration."
2. Post, *American Enterprise*; Dood, *Patent Models*; Ferguson and Baer, *Little Machines*; Janssen, *Icons of Invention*; Ray and Ray, *Art of Invention*; Rothschild and Rothschild, *Inventing a Better Mousetrap*. Alexander Tolhausen indicated that some patent models existed in European, South and Southeast Asian, Australian, and North American repositories around 1857, even though they were not required for patent applications, with the following locations: Austria: the Ministry of Commerce and Trades in Vienna; Bavaria: the Ministry of Commerce and Public Works in Munich; Belgium: the Museum of Industry in Brussels; Canada: the Record Office of the Ministry of Agriculture in Ottawa; France: the Conservatoire des Arts et Métiers in Paris; Great Britain and Ireland: the Museum of the Commissioners of Patents in London; Hannover: the Ministry of the Interior in Hannover; India: the Office of the Secretary of Government of India in Delhi; Parma, Piacenza, and Guastalla in Parma: the Ministry of Finances and the Interior; Prussia: the Ministry of Commerce and Public Works in Berlin; Roman states: the Treasury at the Vatican; Sardinia: the Museum of Industry at the Royal Polytechnic Institute in Cagliari; Saxony: the Ministry of the Interior in Dresden; Spain: the Royal Conservatory of Arts in Madrid; Sweden and Norway: the Records of the Chamber of Commerce in Stockholm; Victoria (Australia): the Office of the Registrar General in Melbourne; Dutch West Indies: the Colonial Ministry at The Hague; Württemberg: the Ministry of the Interior in Stuttgart. See Tolhausen, *A Synopsis*. See also Biagioli, "Patent Specification," 31.
3. Frances Gabe's invention (see chapter 1) provides one such example.
4. See Brückner, Isenstadt, and Wasserman, *Modelwork*; Mindrup, *Architectural Model*; Oswald, *Meister der Miniaturen*; Schilling, *Architektur und Modellbau*; Smith, *Architectural Model as Machine*; Wells et al., *Alphabet of Architectural Models*.
5. Emmons, "Preface," in Mindrup, *Architectural Model*, xviii.
6. Pottage and Sherman, *Figures of Invention*, 13. See also Gochberg, "Useful Arts."
7. Gochberg, "Useful Arts," 129.
8. Ray and Ray, *Art of Invention*, 1–3.
9. Ibid., 4.
10. Ibid.
11. Ibid.
12. Ibid.
13. Ibid., 5.
14. Ibid.
15. Dood, "Why Models?"
16. Patent Act of July 4, 1836, quoted in Ray and Ray, *Art of Invention*, 1.
17. Gochberg, "Useful Arts," 128.
18. On Mills see Bryan, *Robert Mills*. The carpenter Henry S. Davis, who was commissioned to build the vitrines for displaying the patent models, had a number of disputes with Robert Mills regarding their design as well as his compensation. See NARA Record Group 48, Entry No. A1 269 (Records Relating to the Construction of the Patent Office Building 1849–1875), Box 3. During the American Civil War, the Confederacy established its own Patent Commission in the Mechanics Building in Richmond, Virginia. Rufus B. Rhodes was the Commission's first and only commissioner of patents.
19. Commissioner of Patents, *Annual Report*, January 17, 1838, H. Doc. 112. Henry Ellsworth, the Commissioner of Patents, announced the opening of the Patent Office's National Gallery in letters to various newspapers. See letters from Ellsworth in the *National Intelligencer*, November 21, 1840, and December 18, 1840, in the Columbia Historical Society's Machen Collection. See also Evelyn, "Exhibition America," 36n12.
20. NARA Record Group 48, Entry No. A1 269 (Records Relating to the Construction of the Patent Office Building 1849–1875), Box 3.
21. Ray and Ray, *Art of Invention*, 14–16.
22. Ibid., 14.
23. Ibid.
24. Ibid.
25. Ibid.

26. Ibid., 15.
27. Evelyn, "The Patent Models on Display," 17.
28. Ray and Ray, *Art of Invention*.
29. The model is now held by the Hagley Museum, object no. 61.47.254, as part of the E. Tunnicliff Fox Collection.
30. US Patent 175,765.
31. See Haas, *Alexandria in Late Antiquity*.
32. See Lemoine, "L'entreprise Eiffel," 255; Lyonnet du Moutier, *L'aventure de la tour Eiffel*, 202. See also Langmead and Garnaut, *Encyclopedia*, 106.
33. US Patents 199,076, 202,460, 212,941, 224,491 276,673.
34. Ray and Ray, *Art of Invention*, 1–12.
35. Ibid., 8, 10.
36. This sum was probably grossly insufficient for the inventors tasked with replacing their models, considering that one hour of a modelmaker's labor alone cost 50 cents.
37. Ray and Ray, *Art of Invention*, 10.
38. Ray and Ray, *Art of Invention*.
39. Ibid., 10. The former patent building now houses the National Portrait Gallery, and the courtyard is enclosed by a dome by the British architecture firm Foster + Partners.
40. Ibid.
41. Ray and Ray, *Art of Invention*. The patent collections of the Henry Ford Museum include an exhaustive collection of barbed wire patents and samples thereof.
42. "Antique Gadgets."
43. Museum of Modern Art (MoMA) Object No. 443, 1956. Although there is no evidence of plans for an exhibition of patent models in Drexler's files at MoMA, Christopher Mount, a curator who worked with Drexler, recalls seeing files for research on patent models and a potential patent model exhibition at the museum; Christopher Mount, message to the author via Facebook Messenger, January 31, 2022. I thank Michelle Elligott, Chief of Archives, Library and Research Collections at the Museum of Modern Art, for confirming the absence of patent research in Drexler's files while the museum was closed to researchers. Nevertheless, MoMA has celebrated a number of patented inventions through small monographic exhibitions, including Buckminster Fuller's Octet Truss Space Frame in 1959, organized by Arthur Drexler, and Chuck Hoberman's Retractable Iris Dome in 1994, organized by Matilda McQuaid. In the pamphlet for the latter, Hoberman directly referenced Fuller's patents and the practice of patenting more generally. I suspect that Drexler may have been discouraged from organizing further exhibitions when he learned that the Cooper-Hewitt Museum was organizing its own exhibition of 514 patent models in 1984.
44. See Kriegel, *Labor, Empire, and the Museum*; Physick, *Victoria and Albert Museum*; Wright, "National Museum."
45. Science Museum, "Brief History."
46. As quoted by H. T. Wood, "Letter to the Editor."
47. Ibid.
48. CCA DR1989:0026:048–049, 32.
49. CCA DR1989:0026:050–051, 10–11.
50. MacLeod, *Inventing the Industrial Revolution*, 53.
51. Rankin, "'Person Skilled in the Art,'" 58.
52. Ray and Ray, *Art of Invention*, 12.
53. Rankin, "'Person Skilled in the Art,'" 58.
54. Ibid. In the United States, ten photolithographic copies of the drawings were required for each drawing claim from 1936 onward.
55. Ibid., 60.
56. Ibid., 59–60.
57. Ibid., 64.
58. Canada Patent 29,533. See Polk, "Sire of the Skyscraper." Polk relays a description of a so-called "cloudscraper" written by Buffington: "Imagine, then, if you please, a building framed completely of latticed riveted girders of iron and covered with an envelope of light, cheap and absolutely non-conducting substance, which admits of forms of great beauty and corrects entirely the only trouble that might arise in a structure of this nature. To speak more particularly, the system of building which is recommended herein may be described as a series of laminated riveted iron posts diminishing in size as they ascend from the foundation and braced diagonally after the manner of lattice girders, as well as by the horizontal bracing of the iron beams of each floor which form an integral structure and are no small factor in its stability. The exterior is formed of stone, brick or terra cotta, which further precludes any expansion or contraction of the iron." (Polk, "Sire of

the Skyscraper," 15.) See also Christison, "How Buffington Staked His Claim."

59. Germany Patent 2,719,953.

60. The Graphos technical pen, introduced in 1934, and pens like the Rapidograph line of technical pens with internal ink cartridges significantly improved the quality, consistency, and precision of the technical ink drawings typically produced for patent applications.

61. The Plant Patent Act was enacted on June 17, 1930. Design patents are allowed to employ photodocumentation in certain historical and geographic contexts.

62. See Kimmerer, *Venerable Trees*, 192.

63. The original drawings are held by HFM. See Object IDs 71.1.1421, 00.4.7142.1, 71.1.1422, 00.4.7142.2, 71.1.1420.

64. Abelshauser et al., *German Industry*, 200.

65. Ibid.

66. Otto and Klippel, *Geschichte*, 294–295. See also Johns, *Piracy*, 1.

67. Van Dulken, *British Patents*, 25. See also Great Britain Patent Office, *List of Abandoned and Void Applications*.

68. Crim, *Our Germans*. See also Farquharson, "Governed or Exploited?"; Jacobsen, *Operation Paperclip*; O'Reagan, *Taking Nazi Technology*, 8. When these men were gradually released from their work for federal agencies, they often sought work in the private sector. Certain companies, DuPont foremost among them, actively discouraged the hiring of German émigrés irrespective of their participation in Operation Paperclip.

See Hounshell and Smith, *Science and Corporate Strategy*, 92.

69. The Hilten archives are held at UAH, identifier MC-96.

70. See LOC, HAER ALA, 45-HUVI.V, 7F-6.

71. See Central Intelligence Agency, "Summary of Operation Ossavakim"; Christopher, *Race for Hitler's X-Planes*; Hall, "'Other End of a Trajectory.'"

72. Otto and Klippel, *Geschichte*, 295.

73. O'Reagan, *Taking Nazi Technology*, 8.

74. DM, G.P B.I.O.S. 538, 1.

75. O'Reagan, *Taking Nazi Technology*, 69.

76. Ibid., 2.

77. Ibid., 69–70.

78. Ibid. Ironically, the French established a patent office in their own occupation zone in 1946 that served as a destination for many German inventors eager to establish international patent priority since being barred from the system.

79. Ibid., 7. Albert Einstein, although somewhat insulated from industry as a theoretical scientist, is something of an emblem of this generation of German Jewish scientists. While he found a certain insulation from US military research as a professor at the Institute for Advanced Study in Princeton, New Jersey, most of his colleagues were able to make their living in the US through corporate and governmental research work. The best biographical text on Einstein with many elements of his cultural experience as a refugee remains Holton and Elkana, *Albert Einstein*.

80. O'Reagan, *Taking Nazi Technology*, 22. Also see Moser,

Voena, and Waldinger, "German Jewish Émigrés."

81. NARA, Record Group 131.

82. "Seized German Millions."

83. Gross, "The U.S. Confiscated."

84. Coates, "The Secret Life of Statutes." See also O'Reagan, *Taking Nazi Technology*.

85. NA, Reference Group METL.

86. DM, G. P B.I.O.S. 538, p. 1.

87. On I. G. Farben's dissolution see Hounshell and Smith, *Science and Corporate Strategy*, 204.

88. Ibid., 12. On Krupp and the Nazis see the latter chapters of Manchester, *Arms of Krupp*.

89. Hounshell and Smith, *Science and Corporate Strategy*, 204.

90. Ibid., 3.

91. O'Reagan, *Taking Nazi Technology*, 67.

92. Ibid.

93. See Harries, *Nikolaus Pevsner*, 364. Pevsner's colleagues on the study trip included Enid Marx, Allan Walton, Hendrik Albert Nieboer, John Beresford-Evans, Edward Dickey, Lorna Hubbard, Margaret Lambert, and another German immigrant, Margaret Leischner. Pevsner and Leischner became British citizens on the occasion of the BIOS trip to Germany—Pevsner on July 20, 1946, and Leischner in the first week of September. See Sudrow, *Geheimreport*, 111.

94. Harries, *Nikolaus Pevsner*, 365. BIOS had been set up prior to the end of World War II with the objective of collecting German technical

Notes

knowledge as it emerged through Allied victories in Europe.

95. Pevsner, *Enquiry*.
96. As quoted in Sudrow, *Geheimreport*, 109.
97. Pevsner, *Enquiry*, 7–8.
98. Sudrow, *Geheimreport*, 147.
99. Ibid., 107.
100. Ibid., 113.
101. Sudrow, *Geheimreport*, 113.
102. Ibid., 116.
103. Ibid.
104. Ibid., 141.
105. Ibid., 140.
106. Ibid., 155.
107. Ibid., 148.
108. Ibid., 149.
109. Ibid., 150.
110. Ibid., 159. See also a discussion of the *Musterschutz* on p. 157.
111. Sudrow, *Geheimreport*, 159.
112. Ibid., 176–177.
113. Ibid. On the Normen-Ausschuss see Sudrow, *Geheimreport*, 142.

Chapter 5

1. Dickens, "Poor Man's Tale." On patent fees as a method to exclude the lower classes from the patent system in the nineteenth century see Johns, *Piracy*, 261.
2. On Dickens and copyright see Fielding, "Dickens and International Copyright"; Houtchens, "Charles Dickens and International Copyright"; Hudon, "Literary Piracy."
3. This is the famous quip of US Supreme Court Justice Potter Stewart when he was asked to identify a concrete set of rules or a test that defines materials as "obscene" under the First Amendment. In lieu of a legal answer explaining what constituted "hard core" pornography in *Jacobellis vs. Ohio* (1964), he simply said, "I know it when I see it."
4. Collins, "Patent Law's Authorship Screen," 1637.
5. On Jefferson and patents see Matsuura, *Jefferson vs. The Patent Trolls*.
6. Collins, "Patent Law's Authorship Screen," 1639.
7. Ibid., 1639–1640.
8. The major exception here is for inventions in the chemical arts, which do not require an aspect of utility, either proven or projected. Collins, "Patent Law's Authorship Screen," 1632–1633.
9. Ibid. On the perils of ambiguity in the textual descriptions of patents see Johns, *Piracy*, 259.
10. Collins, "Patent Law's Authorship Screen."
11. Ibid., 1616–1617.
12. Ibid., 1633–1635; the cited quotation is on page 1633.
13. Ibid., 1634.
14. Ibid.
15. See Kivenson, *Art and Science of Inventing*, 52–53; Sibley, *Eureka Myth*, 22.
16. Collins, "Patent Law's Authorship Screen," 1639–1640.
17. Garcia, "Architectural Patents," 95–96.
18. Biagioli, "Patent Specification," 27.
19. Martinez, "Patent Pending," 38.
20. Collins, "Patent Law's Authorship Screen," 1612.
21. Ibid.
22. Petroski, *Invention by Design*, 60. The paper clip is part of the permanent design collection of the Museum of Modern Art, object no. 150.2005.
23. The Patent Office as such would be created in 1802, relieving the office of the Secretary of State of patent examination duties. See Matsuura, *Jefferson vs. the Patent Trolls*.
24. Harper, "Life and Work of Clara Barton." See also Khan, *Democratization of Invention*, 136.
25. "Woman Clerk," 5. LOC, Clara Barton papers, MSS11973, Box 84, Microfilm Reel 66.
26. See Galison, *Einstein's Clocks*; Stix, "Patent Clerk's Legacy."
27. Paid, *Subtle Is the Lord*, 48.
28. https://history.aip.org/exhibits/einstein/ae10.htm (accessed May 9, 2022).
29. See Swanson, "Emergence of the Professional Patent Practitioner."
30. Johns, *Piracy*, 251.
31. Otto and Klippel, *Geschichte*, 60.
32. "Raul," email to author, May 8, 2022.
33. Collins, "Patent Law's Authorship Screen," 1637.
34. "Raul," telephone interview by author, January 12, 2022.
35. Van Dulken, *British Patents*, 28.
36. R. B. Fuller typescript, March 24, 1954, SUL Department of Special Collections, Collection M1090, Box 80, Series 2, Folder 6, Ser. No. 261,168.
37. Ibid.
38. Ibid.
39. Donald W. Robertson to R. Buckminster Fuller, February 24, 1954, SUL Department of Special Collections, Collection M1090, Box 80, Series 2, Folder 6.
40. Ibid.

41. Otto and Klippel, *Geschichte*, 304–305.
42. See Hart, *Skelettbauten*.
43. Ibid., 304.
44. Bosconi et al., *Brevetti del design italiano*, 53–58.
45. Ibid., 55.
46. Ibid.
47. Grothe, *Das Patentgesetz*, 19, 30.
48. Courtney, "Building Research Establishment," 285–291; Lea, *Science and Building*.
49. See Siporex, *Dach + Decke + Wand*.
50. NA DSIR 4/2502.
51. NA DSIR 4/2512.
52. R. Desbrière to Building Research Station, February 19, 1946, NA DSIR 4/2495.
53. France Patent 495,487.
54. G. H. Whiting for the Building Research Station to R. Desbrière, May 30, 1946. NA DSIR 4/2495.
55. Comité de propagande colonial de la Charente-Inférieure, *Notes et souvenirs*.
56. See European Patent Office, *E. P. O. + 10*; European Patent Office, *Wheel of Invention*; Griset, *European Patent*.
57. Scheuchzer, *Nouveauté et activité inventive*, 24–26.
58. Council of Europe Committee of Experts on Patents, Classification Working Party, *Convention sur l'unification*. The treaty was ratified by 13 European nations: Belgium, Denmark, France, Germany, Ireland, Italy, Liechtenstein, Luxembourg, the Republic of Macedonia, the Netherlands, Sweden, Switzerland, and the United Kingdom.
59. See van Empel, *Granting of European Patents*.
60. Griset, *European Patent*, 125. See also Vaidhyanathan, *Intellectual Property*, 65. The countries included Belgium, West Germany, France, Luxembourg, the Netherlands, Sweden, Switzerland, and the United Kingdom.
61. Jasanoff, *Ethics of Invention*, 194.
62. See Seckelmann, *Industrialisierung*, 313.
63. Johns, *Piracy*, 263.
64. Griset, *European Patent*, 89.
65. Ibid.
66. See Hell, *München '72*; Hennecke, Keller, and Schneegans, *Demokratisches Grün*; Schulze, "Der Park als Spiellandschaft."
67. Kurt Haertel to Hermann Maassen, July 7, 1791. BaK B/141/47802.
68. von Gerkan, *Von Gerkan, Marg und Partner*, vol. 2. The complex was listed as a historic monument by the central German preservation agency in 2020. Second prize went to Brunnert, More, Osterwalder and Vielmo. See BaK B/141/47802.
69. Griset, *European Patent*, 112.
70. Althammer, *Das deutsche Patentamt*, 44. In the United Kingdom, the patent library (known as the Science Reference and Information Service) remains located outside of the patent office itself, as part of the British Library since 1973. See van Dulken, *British Patents*, 4.
71. Griset, *European Patent*, 141.
72. Ibid., 108.
73. Ibid., 123.
74. BaK B/141/47802.
75. See Luginbühl and Luginbühl, *Bernhardluginbühlstiftung*; Monteil, "Bernhard Luginbühl"; Wilhelm, Siegmantel, and Kunz, "Bernhard Luginbühl."

76. European Patent Office, "Bernard Luginbühl," last modified February 12, 2016, https://www.epo.org/about-us/office/social-responsibility/art/workplace/artists/luginbuehl_de.html.
77. See Violand, "Jean Tinguely's Kinetic Art."
78. On IBM and architecture see Harwood, *The Interface*.
79. Griset, *European Patent*, 169, 172–173.
80. Ibid.
81. See Hofer and Bommersbasch, *Gabriel von Seidl*.
82. Davey, "Offices, Munich," 342.

Chapter 6
1. Robert W. Fiddler, esq. to Professor Seymour Melman, August 29, 1958, CUL, Rare Books and Manuscripts, MS 0863.
2. Subcommittee on Patents, Trademarks, and Copyrights of the Committee on the Judiciary, United States Senate, Eighty-Fifth Congress, Second Session, "Impact of the Patent System," iii.
3. Robert W. Fiddler, esq. to Professor Seymour Melman, August 29, 1958, CUL, Rare Books and Manuscripts, MS 0863.
4. See Khan, *Democratization of Invention*, 1–15.
5. US Constitution, Article I, Section 8, Clause 8.
6. Hermann Grothe later characterized the American system from the perspective of Germany as follows: "The legal view of invention in the American legal codes and their commentaries is such that the invention is regularly regarded as labor-based, so that the patent is

not a privilege but a reward; an equivalent, an intellectual service rendered to society as a whole work, is apprehended." Grothe, *Das Patentgesetz*, 2–3.

7. Griset, *European Patent*, 23.

8. Ibid.

9. Colman, "Design and Deviance, Part 1"; Colman, "Design and Deviance, Part 2." See also Lai, "The Role of Patents."

10. Colman, "Design and Deviance, Part 2," 22.

11. Ibid., 1.

12. Ibid.

13. Franklin Knitting Mills vs. Gropper Knitting Mills, 15 F.2d 375, 375 (2d Cir.). Quoted in Colman, "Design and Deviance, Part 2," 19–20.

14. Mallinson vs. Ryan, 242 F. 951, 952 (S.D.N.Y. 1917). Quoted in Colman, "Design and Deviance, Part 2," 24.

15. Kirkham and Ogata, "Europe 1830–1900," 429. See also Colman, "Design and Deviance, Part 2," 4.

16. Colman, "Design and Deviance, Part 2," 20.

17. Ibid., 9.

18. *Encyclopaedia Britannica Online*, s.v. "James Renwick," accessed May 31, 2022, https://www.britannica.com/biography/James-Renwick-Jr.

19. CUL RBML MS#1063, James Renwick correspondence with Edward Sabine Renwick.

20. Renwick, *Patentable Invention*, iii.

21. Ganea and Nagaoka, "Japan," 153.

22. See Fisk, *Working Knowledge*, 1; Seckelmann, *Industrialisierung*, 46, 372.

23. Fisk, *Working Knowledge*, 1.

24. Collins, "Patent Law's Authorship Screen," 1644–1645.

25. Ibid., 1650–1651. See also Wong, "Ambience as Property."

26. Merges, *Justifying Intellectual Property*, 24.

27. Van Dulken, *British Patents*, 27.

28. Bottomley, *British Patent System*, 65.

29. Ibid., 53.

30. Vaidhyanathan, *Intellectual Property*, 65

31. Johns, *Piracy*, 249. See also Janis, "Patent Abolitionism."

32. See Bastable, *Arms and the State*; Grove, *Correlation of Physical Forces*; MacFie, *Recent Discussions*. The MacFie family files are held at UGAS, GB 248 DC 120.

33. MacFie, *Recent Discussions*. See also Machlup and Penrose, "The Patent Controversy."

34. MacFie, *Recent Discussions*, 237.

35. Ibid.

36. Ibid.

37. Johns, *Piracy*, 249.

38. Abelshauser et al., *German Industry*, 80.

39. See Aso, Rademacher, and Dobinson, *History of Design and Design Law*.

40. Mott, "Standard of Ornamentality," 549.

41. Ibid., 549–551.

42. Rowe vs. Blodgett & Clapp Co., 112 Fed. 61, 62 (2d Circ., 1901). Cited by Mott, "Standard of Ornamentality," 550.

43. Forestek Plating and Mfg. Co. v. Knapp-Monarch Co., 106 F.2d 554 (6th Circ. 1939).

44. Ibid. See also Mott, "Standard of Ornamentality," 551–552.

45. Mott, "Standard of Ornamentality," 550.

46. Ibid.

47. See Yabumoto, "History of Design Protection," 23.

48. Ganea and Nagaoka, "Japan," 139.

49. Ibid., 136–137.

50. See Sagar, "Patent Policy."

51. Garde, "India," 59.

52. See Hirschman, "Legalizing Architecture."

53. http://www.govtrack.us/congress/bills/101/s198. Kastenmeier was the chair of the Senate subcommittee administering the hearing, the Judiciary Subcommittee on Courts, Intellectual Property, and the Administration of Justice. He represented a primarily rural district in southwestern Wisconsin. Boehlert's district was centered in Utica, New York. Procedural strategy wound up dictating that H.R. 3990 be incorporated as a component of the Computer Software Rental Amendments Act of 1989 (S. 198, 101st Congress), sponsored by Republican Senator Orrin Hatch of Utah and passed on September 27, 1990. The bill was passed by voice vote, indicating that it was uncontroversial.

54. Copyright Act of 1909, Pub.L. 60-349, 35 Stat. 1075, enacted March 4, 1909.

55. H.R. Rep. No. 1476, at. 55 (176).

56. Robert R. Jones Assocs. vs. Niro Homes, 858 F.2d 274, 279 (6th Cir. 1988); Acorn Structures, Inc. vs. Swants, 657 F. Supp. 70, 75 (W.D.Va. 1987).

57. See Broussard, "Intellectual Property Food Fight."

58. World Intellectual Property Organization,

"Berne Convention for the Protection of Literary and Artistic Works," accessed May 31, 2022, https://www.wipo.int/treaties/en/ip/berne/.

59. Bowser, "Understanding the Scope."

60. United States Congress, *Architectural Design Protection*, 1.

61. Ibid., 1–2.

62. Ibid., 9–10.

63. Hirschman, "Legalizing Architecture," 21–24.

64. United States Congress, *Architectural Design Protection*, 185.

65. Ibid., 186.

66. Hirschman, "Legalizing Architecture," 23–24.

67. United States Congress, *Architectural Design Protection*, 111.

68. Ibid., 97

69. Ibid.

70. Ibid., 76

71. Ibid., 42, 44.

72. Hirschmann, "Legalizing Architecture," 25.

73. United States Congress, *Architectural Design Protection*, 25–29.

74. Hirschman, "Legalizing Architecture," 25.

75. United States Congress, *Architectural Design Protection*, 50–51; Graves, "A Case for Figurative Architecture," 86–90.

76. Colman, "Design and Deviance, Part 2," 27.

77. Ibid., 7, 17.

78. J. Ferrari Hardoy to Modern Color Inc., May 15, 1951, HUL Des-199-0001-007823498, Folder E110.

79. Ibid.

80. Dorothy Schindele to J. Ferrari Hardoy, June 5, 1951, HUL Des-199-0001-007823498, Folder E110.

81. HUL Des-199-0001-007823498 Folder D082b.

82. Ganea and Nagaoka, "Japan," 145.

83. Ibid., 135–136.

84. Ibid., 145. In the early 1990s several courts in the United States also demanded that Japanese corporations pay backdated fines for infringement of US patents.

85. Among these are Demetriades vs. Kaufman, 690 F. Supp. 289 (S.D.N.Y. 1988). Another example involved the Glasgow architect Jim Cuthbertson. In 1989 he was in the process of renovating a pharmacy conversion in Glasgow, and the bids for the project came in higher than the client had expected, which placed the project on hold. The client already had Cuthbertson's drawings, a fact that became critical when Cuthbertson passed the property some time later and noticed that it was being renovated by a contractor using his drawings. Cuthbertson was found to have a rightful claim to stolen intellectual property. In yet another example, the Italian architect Grazia Repetto claimed that Renzo Piano had plagiarized Repetto in Piano's design for Japan's Kansai Airport, citing a competition entry that Repetto had submitted for the Genoa Airport in 1970. The judge in this case asked both parties to create models that would allow the court to compare the designs and ultimately found them to be too different to merit a claim of plagiarism. See Richardson, "Lost Property," 12–13. Most prominent was the case of *Shine vs. Childs*, wherein Thomas Shine, a Yale University architecture student, claimed that David Childs, a partner at Skidmore, Owings & Merrill, had copied his design for a skyscraper after seeing it in a design review and applied it to SOM's design for the reconstruction of the World Trade Center in New York City. The similarities indeed struck many in the architecture community, but it was ultimately deemed that Shine had no standing for the claim because the design had been produced while he was a student and that the design was therefore actually the property of Yale. See Thomas Shine vs. David M. Childs 382 F. Supp. 2d 602 (Dist ct. S.D.N.Y. 2005). In England, the June 2000 buckling of piles for a dock in the city of Southampton, which killed one and injured four, prompted an inquiry into whether faulty designs that were granted patents could warrant some form of culpability on the part of the British patent office. The piles in question were part of the Hennebique pile system, which had been invented in France and patented in the United Kingdom in the early twentieth century. The case did not wind up designating any responsibility for the UK patent office. See LMA LCC AR / BA / 03 / 001. For additional commentary on architectural patent disputes see Woltron, "Patent Architektur."

Chapter 7

1. Johns, *Piracy*, 273.

2. See Standing, *Plunder of the Commons*.

3. Ibid., ii.

4. Collins, "Architectural Patents"; Hindle, "Prototyping the Mississippi Delta"; Wong, "Ambience as

Property." With regard to the protection of "ambience" through the legal mechanism of trade dress, Wong contends in "Ambience as Property" that there is a breakdown in the function-ornament dichotomy: "It . . . implies a need to redraw the existing parameters for understanding the designed commodity, while forgoing the opportunity to articulate an alternate set of distinctions. The adjustment in the court's position [on trade dress] from 1993 to 2000 over the nature of ambience as commodity points to the contradictions arising out of a wider understanding that commodification had grown to encompass not just products and things, or even spaces, but also the individual and idiosyncratic experiences of consumption construed as common ones" (96). In "Architectural Patents," Collins has contended that the primary goal behind the rise of patents for "disposition of space," representing an "awkward zone of copyright-patent overlap" in the American patent system, is programmatic affordances, not technical objects per se: "To recognize the programmatic affordance of a disposition of space," he says, "only a weaker and less controversial iteration of the modernist 'form follows function' credo, stripped of its determinist and aesthetic connotations, is needed: spaces facilitate behavior. It is in this eminently practical sense that dispositions of space are functional, patentable inventions, provided that they are sufficiently different from prior dispositions of space"

(193). A good example of such a patent is United States Patent no. 4,852, 313, Building Arrangement Maximizing Views.

5. Merges, *Justifying Intellectual Property*, 36.

6. Locke, *Two Treatises*, second treatise, chapter V, paragraph 27. See also Merges, *Justifying Intellectual Property*, 26, 61, 64.

7. Author, email to Robert Merges, November 16, 2021.

8. Robert Merges, email to author, November 27, 2021.

9. Author, email to Robert Merges, November 16, 2021.

10. Robert Merges, email to author, November 27, 2021.

11. Author, telephone interview with "Carmen," January 27, 2022.

12. Author telephone interview with "Dave," Monday, January 24, 2022.

13. Crouch, "Tracing the Quote."

14. The doctrine of equivalents is a legal concept embedded in most of the world's patent systems that permits courts to penalize a party for patent infringement even when the infringement is not tied directly to a claim of the patent infringed upon but is equivalent. The goal, as Judge Learned Hand put it, is "to temper unsparing logic and prevent an infringer from stealing the benefit of the invention." Royal Typewriter Co. v. Remington Rand, Inc., 168 F.2d 691, 692 (2d Cir. 1948).

15. Author telephone interview with "Hilde," January 24, 2022.

16. Author telephone interview with "Pavel," January 26, 2022.

17. On patents and the colonial project see Barton, "Issues Posed"; Bently, "'Extraordinary Multiplicity'"; Hrdy, "State Patent Laws"; Robinson, "James Watt"; Wadlow, "British Empire Patent"; Zahedieh, "Colonies, Copper, and the Market."

18. Graham, "Patents and Invention," 944.

19. Ibid.

20. Ibid.

21. Johns, *Piracy*, 264.

22. Isaacs, "Nails in New Zealand," 83.

23. Ibid. Isaacs cites the *North Otago Times* (July 1, 1882), 2.

24. Isaacs, "Nails in New Zealand," 84.

25. Ibid.

26. Ibid., 88n35.

27. Ibid., 88.

28. Ibid., 89.

29. New Zealand Patent 1,152. See also Isaacs, "Nails in New Zealand," 91.

30. Lewis, "Report on the British Empire Patent Conference," 1–3. See also van Dulken, *British Patents*, 5.

31. Ibid., 1.

32. Ibid.

33. Jasanoff, *Ethics of Invention*, 203.

34. Correa, "Access to Knowledge," 239.

35. Ibid.

36. United Nations, "United Nations Declaration on the Rights of Indigenous People," accessed August 20, 2022, https://www.un.org/development/desa/indigenouspeoples/wp-content/uploads/sites/19/2018/11/UNDRIP_E_web.pdf.

37. Ibid.

38. Goldgel-Carballo and Poblete, *Piracy and Intellectual Property*, 18.

39. Barsh, "How Do You Patent?," 14.

40. Vaidhyanathan, *Intellectual Property*, 14. See also Schiff, *Industrialization without National Patents*, 13.

41. Vaidhyanathan, *Intellectual Property*, 56.

42. Jasanoff, *Ethics of Invention*, 202–203.

43. Goldgel-Carballo and Poblete, *Piracy and Intellectual Property*, 8.

44. Clark, *Patent Litigation in China*; Ganea and Pattloch, *Intellectual Property Law*.

45. Goldstein and Kraus, *Intellectual Property in Asia*, viii, 35.

46. Ganea and Jin, "China," 17–18. On Japan and the adoption of a patent system in the Meiji Restoration see Ganea and Nagaoka, "Japan," 129–131.

47. Ganea and Jin, "China," 21, 43.

48. Ibid., 34. See also Goldgel-Carballo and Poblete, *Piracy and Intellectual Property*, 4; Wong, *Van Gogh on Demand*.

49. Ganea and Jin, "China," 18, 37.

50. Ibid., 32.

51. Ibid., 34.

52. Ncube, *Intellectual Property Policy*, 3–4.

53. See Pila and Torremans, *European Intellectual Property Law*, 30.

54. World Intellectual Property Organization, "Member States," accessed June 9, 2022, https://www.wipo.int/members/en/.

55. Johns, *Piracy*, 13. Lawrence Lessig famously said that the United States was "born a pirate nation" in *Free Culture*, 63. See also Goldgel-Carballo and Poblete, *Piracy and Intellectual Property*, 8.

56. Mgbeoji, "Comprador Complex," 1.

57. Rahmatian, "Neo-Colonial Aspects," 40.

58. Goldstein and Kraus, "Introduction," vii.

59. Rahmatian, "Neo-Colonial Aspects," 40.

60. Kapczynski, "Access to Knowledge," 25–26.

61. Ncube, *Intellectual Property Policy*, 4.

62. On the superimposition of Anglo-American patent norms on Latin America see Zimmer, "Between Abundance and Appropriation," 179.

63. Goldgel-Carballo and Poblete, *Piracy and Intellectual Property*, 9.

64. Park, "South Korea," 270. Japan's patent system does have a handful of heterogenous features, including universities holding the patents of their professoriate and the attractiveness of utility models as an alternative to patents. See Ganea and Nagaoka, "Japan," 141, 144.

65. Antons, "Indonesia," 90, 95.

66. Ganea, "Cambodia," 1–2.

67. Garde, "India," 59.

68. Ibid., 78.

69. Schiff, *Industrialization without National Patents*, 28, 53–54.

70. Ibid., 11, 14.

71. Ganea and Jin, "China," 26.

72. Bauakademie der DDR, *25 Jahre Bauforschung*; Bauakademie der DDR, *Städtebau und Architektur*.

73. For the various institutes see BaB DH/1/24947, Anlage VD 367/74, January 18, 1974, p. 1.

74. BaB DH/2/5237, printout of patents filed 1972–1990.

75. See Hemmerling, *Neuererbewegung*; Eilemann et al., *Neuererbewegung und Investitionen*. DDR patent law was enshrined in 1950. See also Otto and Klippel, *Geschichte*, 253.

76. Otto and Klippel, *Geschichte*, 240–244. Otto and Klippel note that patent statistics were a valuable metric for economists in measuring the health of national economies, and this is likely vital for understanding why East Germany, with its particular history, was one of the few socialist countries to place a premium on patents.

77. BaB DH/2/5654, "Grundsätze und Verfahrensregelung für die Behandlung von Neuerervorschlägen und Neuerervereinbarungen in der Deutshcen Bauakademie," October 29, 1970, p. 1. This kind of language was also enshrined in Neuererbewegung legal policy. See BaB DH/1/35242, *Neuererrecht* (Berlin: Staatsverlag, 1983), p. 10.

78. BaB DH/2/5378.

79. On the employment of Soviet precedents see, for example, BaB DH/1/25733, "Analyse der Entwicklung der Arbeit mit Schutzrechten im Bauwesen im Jahre 1974 und im 1. Halbjahr 1975," January 18, 1974. The Soviet patent system was nominal. Inventors typically received certificates of appreciation from the state along with a cash reward. The state, meanwhile, retained ownership rights over inventions. Certain areas thrived under Soviet support, especially physics, mathematics, and cybernetics, and some major inventions, such as the AK-47, came to pass under the Soviet system in its

74-year history. But the magnitude and clip of innovation was not comparable to that of Western patent powers. See Vaidhyanathan, *Intellectual Property*, 64.

80. BaB DH/1/24947, Amt für Erfindungs- und Patentwesen der DDR, "Information über die Entwicklung der Neuererbewegung im 1. Halbjahr 1974 sowie über die Ergebnisse einer Untersuchung zu Problemen der Leitung und Planung der Neuererbewegung," September 10, 1974, pp. 5–6.

81. BaB DH/1/24947, Anlage VD 367/74, January 18, 1974, p. 1; BaB DH/1/25733, "Analyse der Entwicklung der Arbeit mit Schutzrechten im Bauwesen im Jahre 1974 und im 1. Halbjahr 1975," January 18, 1974, p. 2.

82. BaB DH/2/5420, "Bericht über die Dienstreise nach Westdeutschland vom 30.8 bis 1.9.1970," September 24, 1970, pp. 1–2.

83. Johns, *Piracy*, 273.

84. Kapczynski, "Access to Knowledge," 48; Merges, *Justifying Intellectual Property*, 86; Zimmer, "Between Abundance and Appropriation," 175.

85. On artificial scarcity and patents see Vaidhyanathan, *Intellectual Property*, 11.

86. Kapczynski, "Access to Knowledge," 48.

87. Kapczynski, "Access to Knowledge," 18–19. See also Machlup, *Production and Distribution of Knowledge*, 9.

88. Kapczysnki, "Access to Knowledge," 28.

89. Castells, *Rise of the Network Society*, 13–21, 30–31, 101. See also Kapczynski, "Access to Knowledge," 19n11, 19n12, 19n13.

90. Castells, *Rise of the Network Society*, 225–226. See also Kapczynski "Access to Knowledge," 19.

91. Ibid.

92. Aureli, "The Common." See also Grau and Goberna, "Should We Patent Architectural Knowledge?," 103.

93. Aureli, "The Common," 25. See also Grau and Goberna, "Should We Patent Architectural Knowledge?," 103.

94. Grau and Goberna "Should We Patent Architectural Knowledge?," 103.

95. Ibid.

Coda

1. See MacLeod, *Inventing the Industrial Revolution*, 87.

2. See Sibley, *Eureka Myth*, 5.

Bibliography

Journals and Periodicals
The journals and periodicals listed below were consulted for all years in circulation between 1800 and 2000.

Architecture d'Aujourd'hui
Der Civilingenieur
Deutschen Industriebeamten-Zeitung
European Intellectual Property Review
Le Journal officiel de l'OEB
Journal of Intellectual Property Rights
Journal of World Intellectual Property Law
Newsletter of the European Patent Office
The New York Times
Progressive Architecture
Scientific American
Die Technik
Zeitschrift für gewerblichen Rechtsschutz / Gewerblicher Rechtschutz und Urheberrecht
Zeitschrift für Industrierecht

Books and Articles
Aalto, Alvar. *Alvar Aalto: Das Gesamtwerk.* Vol. 1, *1922–1962*. Ed. Karl Fleig. Basel: Birkhäuser, 1999.
Aalto, Alvar. *Alvar Aalto: Das Gesamtwerk.* Vol. 2, *1963–1970*. Ed. Karl Fleig. Basel: Birkhäuser, 1999.
Aalto, Alvar. *Alvar Aalto: Das Gesamtwerk.* Vol. 3, *Projekte und letzte Bauten*. Ed. Elissa Aalto. Basel: Birkhäuser, 1999.
Abelshauser, Werner. *Die BASF: Eine Unternehmensgeschichte.* Munich: C. H. Beck, 2002.
Abelshauser, Werner, Wolfgang von Hippel, Jeffrey Allan Johnson, and Raymond G. Stokes. *German Industry and Global Enterprise: BASF: The History of a Company.* New York: Cambridge University Press, 2003.
Adams, Nicholas. *Gordon Bunshaft and SOM: Building Corporate Modernism.* New Haven: Yale University Press, 2019.
Adams, Nicholas. *Skidmore, Owings & Merrill: SOM since 1936.* Milan: Electa, 2006.
Allaback, Sarah. *The First Women Architects.* Urbana: University of Illinois Press, 2008.
Althammer, Werner. *Das deutsche Patentamt: Aufgaben, Organisation und Arbeitsweise.* Cologne: Heymann, 1970.

"Antique Gadgets Put on Exhibition: Items from Animal Traps to Gatling Gun Will Go on Sale Monday at Gimbels." *New York Times*, June 3, 1950.

Antons, Christoph. "Indonesia." In Goldstein and Kraus, *Intellectual Property in Asia*, 87–128.

Arwade, Sanjay R., Liakos Ariston, and Thomas Lydigsen. "Structural Systems of the Bollman Truss Bridge at Savage, Maryland." *APT Bulletin: The Journal of Preservation Technology* 37, no. 1 (2006): 27–35.

Aso, Tsukasa, Christoph Rademacher, and Jonathan Dobinson, eds. *History of Design and Design Law: An International and Interdisciplinary Perspective*. Singapore: Springer Nature, 2022.

Aureli, Pier Vittorio. "The Common and the Production of Architecture." In *Common Ground: A Critical Reader*, edited by David Chipperfield, Kieran Long, and Shumi Bose, 24–27. Venice: La Biennale di Venezia, 2012.

Austin, Peter. "Rafael Guastavino's Construction Business in the United States: Beginnings and Development." *APT Bulletin: The Journal of Preservation Technology* 30, no. 4 (1999): 15–19.

Baas, Christopher. "Concrete in the Steel City: Constructing Thomas Edison's House for the Working Man." *Indiana Magazine of History* 108, no. 3 (2012): 245–273.

Banham, Reyner. "Ransome at Bayonne." *Journal of the Society of Architectural Historians* 42, no. 4 (December 1983): 383–387.

Barré-Despond, Arlette. *UAM: Union des Artistes Modernes*. Paris: Editions du Regard, 1982.

Barsh, Russel Lawrence. "How Do You Patent a Landscape? The Perils of Dichotomizing Cultural and Intellectual Property." *International Journal of Cultural Property* 8, no. 1 (1999): 14–47.

Barton, John. "Issues Posed by a World Patent System." *Journal of International Economic Law* 7 (2004): 341–357.

Bastable, Marshall J. *Arms and the State: Sir William Armstrong and the Remaking of British Naval Power*. Farnham, UK: Ashgate, 2004.

Bauakademie der DDR. *25 Jahre Bauforschung und neue Aufgaben der Bauakademie der DDR zur Beschleunigung des wissenschaftlich-technischen Fortschritts im Bauwesen in Verwirklichung der Beschlüsse des IX. Parteitages der SED 35: Außerordentliche Planartagung der Bauakademie der DDR, Berlin, 16 und 17 Dezember 1976*. Berlin: Bauinformation DDR, 1977.

Bauakademie der DDR, Institut für Städtebau und Architektur. *Städtebau und Architektur in der DDR: Historische Übersicht*. Berlin: Bauinformation DDR, 1989.

Becker, Britta, Maren Grim, and Jakob Krameritsch. *Zum Beispiel: BASF; Über Konzernmacht und Menschenrechte*. Vienna: Mandelbaum, 2018.

"Beijing Patent Office Building, Fengtai, Beijing, 2013–2017: China Architecture Design & Research Group." *Jian Zhu Xue Bao* 9 (September 2018): 78–82.

Beltran, Alain, Sophie Chauveau, and Gabriel Galvez-Behar. *Des brevets et des marques: Une histoire de la propriété industrielle*. Paris: Fayard, 2001.

Bentham, Rachel, and Alyson Hallett, eds. *Project Boast*. Chatmouth, UK: Triarchy, 2018.

Bently, Lionel. "The 'Extraordinary Multiplicity' of Intellectual Property Laws in the British Colonies in the Nineteenth Century." *Theoretical Inquiries in Law* 12 (2011): 161–200.

Benton, Tim. "Domino and the Phantom *Pilotis*." *AA Files* 69 (2014): 13–37.

Benton, Tim. "The Villa de Mandrot and the Place of the Imagination." In *Massilia 2011: Annuaire d'études corbuséennes*, edited by Michel Richard, 92–105. Marseille: Editions Imbernon, 2011.

Bergdoll, Barry, and Peter Christensen, eds. *Home Delivery: Fabricating the Modern Dwelling*. New York: Museum of Modern Art, 2008.

Bertoni, Aura, and Maria Lillà Montagnani. "Public Art and Copyright Law: How the Public Nature of Architecture Changes Copyright." *Future Anterior: Journal of Historic Preservation, History, Theory, and Criticism* 12, no. 1 (2015): 47–55.

Besen, Stanley M., and Leo J. Raskind. "Introduction to the Law and Economics of Intellectual Property." *Journal of Economic Perspectives* 5, no. 1 (1991): 3–27.

Biagioli, Mario. "Patent Specification and Political Representation: How Patents Became Rights." In *Making and Unmaking Intellectual Property*, edited by Mario Biagioli, Peter Jaszi, and Martha Woodmansee, 25–40. Chicago: University of Chicago Press, 2011.

Boeckl, Matthias. "The Modern Design between Norms and Creativity 1918–1938." *Rassegna* 13, no. 46 (1991): 62–69.

Boesiger, Willy. *Le Corbusier: Œuvre Complète*. Vol. 7, *1957–1965*. Basel: Birkhäuser, 2015.

Böhm, Christian, Dieter Jungwirth, and Berthold Neunert, eds. *Festschrift Ulrich Finsterwalder: 50 Jahre für Dwyidag*. Karlsruhe, Germany: Dyckerhoff & Widmann, 1973.

Bollinger, Klaus, and Florian Medicus, eds. *Stressing Wachsmann: Structures for a Future / Strukturen für eine Zukunft*. Basel: Birkhäuser, 2020.

Bosconi, Giampiero, Francesca Picchi, Marco Strina, and Nicola Zanardi. *Brevetti del design italiano: 1946–1965*. Milan: Electa, 2000.

Bottomley, Sean. *The British Patent System During the Industrial Revolution 1700–1852: From Privilege to Property*. Cambridge: Cambridge University Press, 2016.

Bowser, David H. "Understanding the Scope of Architectural Copyright Protection." American Institute of Architects website, accessed May 31, 2022, https://www.aia.org/articles/26591-understanding-the-scope-of-architectural-cop.

Brandt, E. N. *Growth Company: Dow Chemical's First Century*. Lansing: Michigan State University Press, 1997.

Brekke, Nils Georg, Per Jonas Nordhagen, and Siri Skjold Lexau. *Norsk arkitekturhistorie: frå steinalder og bronsealder til det 21. Hundreåret*. Oslo: Oslo Samlaget, 2008.

Broeksmit, Susan Begley, and Anne Sullivan. "Dry-Press Brick: A Nineteenth-Century Innovation in Building Technology." *APT Bulletin: The Journal of Preservation Technology* 37, no. 1 (2006): 45–52.

Broussard, J. Austin. "An Intellectual Property Food Fight: Why Copyright Law Should Embrace Culinary Innovation." *Vanderbilt Journal of Entertainment and Technology Law* 10, no. 3 (2008): 691–728.

Brown, Patricia Leigh. "Son of Carwash, The Self-Cleaning House." *New York Times*, January 17, 2002.

Brückner, Martin, Sandy Isenstadt, and Sarah Wasserman, eds. *Modelwork: The Material Culture of Making and Knowing*. Minneapolis: University of Minnesota Press, 2021.

Bryan, John M. *Robert Mills: America's First Architect*. New York: Princeton Architectural Press, 2001.

Buchanan, Peter. *Renzo Piano Building Workshop: Complete Works*. 5 vols. London: Phaidon, 1993.

Bugbee, Bruce. *The Genesis of American Patent and Copyright Law*. Washington, DC: Public Affairs Press, 1967.

"Building of Concrete." *Boston Daily Globe*, September 24, 1908.

Burkhalter, Marianne, and Christian Sumi, eds. *Konrad Wachsmann and the Grapevine Structure*. Zurich: Park Books, 2018.

Carlson, W. Bernard. *Innovation as a Social Process: Elihu Thomson and the Rise of General Electric*. Cambridge: Cambridge University Press, 1991.

Casciato, Maristella. "Construction 'objective' et préfabrication: Le brevet no 14521 de Duiker et Wiebenga." *Cahiers de la Recherche Architecturale* 40 (1997): 67–78, 134, 137.

Castells, Manuel. *The Rise of the Network Society*. Chichester, MA: Wiley-Blackwell, 1996.

Çelik Alexander, Zeynep. "Managing Iteration: The Modularity of the Kew Herbarium." In *Iteration: Episodes in the Mediation of Art and Architecture*, edited by Robin Schuldenfrei, 1–24. London: Routledge, 2020.

Central Intelligence Agency. "Summary of Operation Ossavakim." January 13, 1947, declassified document published online,

accessed April 17, 2022, https://www.cia.gov/readingroom/document/cia-rdp82-00457r000200570007-4.
"Centre européen des brevets: La Hage." *Architecture d'Aujourd'hui* (June 1991): 74–75.
Chanteux, Anne. "Les inventives: Femmes, inventions et brevets en France à la fin du XIXe siècle." *Documents pour l'histoire des techniques* 17 (2009): 90–97.
Ching, Francis D. K., Mark M. Jarzombek, and Vikramaditya Prakash. *A Global History of Architecture*. New York: Wiley, 2010.
Choate, Pat. *Hot Property: The Stealing of Ideas in the Age of Globalization*. New York: Alfred A. Knopf, 2005.
Christensen, Peter H. *Precious Metal: German Steel, Modernity, Ecology*. University Park: Penn State University Press, 2022.
Christensen, Peter H. "Precious Metal: The I-Beam in the Late Ottoman Empire." In *Making Modernity in the Islamic Mediterranean*, edited by Margaret S. Graves and Alex Dika Seggerman, 214–233. Bloomington: Indiana University Press, 2022.
Christison, Muriel B. "How Buffington Staked His Claim: An Analysis of His Memories and Skyscraper Drawings." *Art Bulletin* 26, no. 1 (1944): 13–24.
Christopher, John. *The Race for Hitler's X-Planes: Britain's 1945 Mission to Capture Secret Luftwaffe Technology*. Cheltenham, UK: History Press, 2013.
Ciccarelli, Lorenzo. "Architecture as Construction in the Beginnings of Renzo Piano: Five Patents for Construction Systems and 'Pieces' of Buildings." In *Proceedings of the Fifth International Congress on Construction History*, vol. 1, edited by Brian Bowen, Donald Friedman, Thomas Leslie, and John Ochsendorf, 447–454. Woodstock, IL: Construction History Society of America, 2015.
Clark, Douglas. *Patent Litigation in China*. 2nd ed. Oxford: Oxford University Press, 2015.
Coates, Benjamin A. "The Secret Life of Statutes: A Century of the Trading with the Enemy Act." *Modern American History* 1, no. 2 (2018): 151–172.
Cohen, Jean-Louis. *Architecture in Uniform: Designing and Building for the Second World War*. Montréal: Canadian Centre for Architecture / Paris: Hazan / New Haven: Yale University Press, 2011.
Collins, Kevin Emerson. "Architectural Patents beyond Bucky Fuller's Quadrant." In Lawrence and Miljački, *Terms of Appropriation*, 186–212.
Collins, Kevin Emerson. "Patent Law's Authorship Screen." *University of Chicago Law Review* 84, no. 4 (2017): 1603–1673.
Colman, Charles E. "Design and Deviance: Patent as Symbol, Rhetoric as Metric, Part 1." *Jurimetrics* 55 (Summer 2015): 419–462.

Colman, Charles E. "Design and Deviance: Patent as Symbol, Rhetoric as Metric, Part 2." *Jurimetrics* 56 (Fall 2015): 1–45.

Comité de propagande colonial de la Charente-Inférieure. *Notes et souvenirs sur l'exposition colonial de La Rochelle juillet-août 1927*. La Rochelle: Impr. de Massons fils, 1928.

Correa, Carlos M. "Access to Knowledge: The Case of Indigenous and Traditional Knowledge." In Krikorian and Kapczynski, *Access to Knowledge in the Age of Intellectual Property*, 237–252.

Council of Europe Committee of Experts on Patents, Classification Working Party. *Convention sur l'unification de certains éléments du droit des brevets d'invention*. Strasbourg: Council of Europe, 1963.

Courtney, Roger. "Building Research Establishment: Past, Present, and Future." *Building Research and Information* 25, no. 5 (1997): 285–291.

"Cranstons' Patent Greenhouses & Co." *American Architect and Builders' Monthly* 1 (March 1870): 7–8.

Crim, Brian E. *Our Germans: Project Paperclip and the National Security State*. Baltimore: Johns Hopkins University Press, 2018.

Crosbie, Michael J. "Selected Detail: Fabric Formwork for Concrete." *Progressive Architecture* 75, no. 7 (1994): 114.

Crouch, Dennis. "Tracing the Quote: Everything That Can be Invented Has Been Invented." *Patently-O*, January 6, 2011. https://patentlyo.com/patent/2011/01/tracing-the-quote-everything-that-can-be-invented-has-been-invented.html.

Csikszentmihalyi, Mihaly. *Creativity: The Psychology of Discovery and Invention*. New York: Harper Perennial, 2013.

Dal Co, Francesco. *Centre Pompidou: Renzo Piano, Richard Rogers, and the Making of a Modern Monument*. New Haven: Yale University Press, 2016.

Das, Saswato R. "Frugal Innovation: India Plans to Distribute Low-Cost Handheld Computers to Students." *Scientific American*, September 28, 2010. https://www.scientificamerican.com/article/india-35-dollar-tablet/.

Davey, Peter. "Offices, Munich; Architects: Von Gerkan, Marg and Partners." *Architectural Review* 169 (June 1981): 342–345.

David, A. C., H. W. Frohne, and P. B. Wright. "The Advent of the Fireproofed Dwelling: Some Structural Aspects of the Fire-Proofed Dwelling; Some Fire-Resisting Country Houses." *Architectural Record* 25, no. 5 (1909): 309–314.

Davis, Mike. "Berlin's Skeletons in Utah's Closet." In *Dead Cities: And Other Tales*, 64–83. New York: New Press, 2002.

"Designer's Forecast, a Spray-Foam Home." *New York Times*, October 12, 1965.

Dewan, Shaila. "Bring on the Lawyers: Will Patents Turn Architecture into a Litigious Dogfight?" *Architecture* 88, no. 12 (1999): 128–130.

Dickens, Charles. "Poor Man's Tale of a Patent." *Household Words* 2, no. 30 (October 19, 1850): 73–75.

Dietziker, Céline, and Lukas Gruntz. *Aalto in Detail: A Catalogue of Components*. Basel: Birkhäuser, 2022.

"Dispute of Patent: Leroy S. Buffington vs. W. L. B. Jenney." *Western Architect* 3 (October 1904): 11.

Dood, Kendall J. *Patent Models and the Patent Law: 1790–1880*. Alexandria, VA: Patent Office Society, 1983.

Dood, Kendall J. "Why Models?" In Post, *American Enterprise*, 14–15.

Droste, Magdalena, and Bauhaus-Archiv. *Bauhaus 1919–1933*. Cologne: Taschen, 1990.

Duberman, Martin. *The Martin Duberman Reader: The Essential Historical, Biographical, and Autobiographical Writings*. New York: New Press, 2015.

Duberman, Martin. *Paul Robeson: A Biography*. New York: New Press, 1965.

Duboy, Philippe. "L'Esprit Nouveau: La 'Villa Nomade' de Raymond Roussel." *Architecture d'Aujourd'hui* 328 (June 2000): 72–77.

Dutfield, Graham. *The Highest Design of Purest Gold: A Critical History of the Pharmaceutical Industry, 1880–2020*. Singapore: World Scientific, 2020.

Dutton, H. I. *The Patent System and Inventive Activity During the Industrial Revolution, 1750–1852*. Manchester: University of Manchester Press, 1984.

Dyckerhoff und Widmann A. G. (München). *Dwywidag*. Munich: Bruckmann, 1962.

Dyckerhoff und Widmann A. G. (München). *Öffentliche Gebäude, Geschäfts- und Wohnhäuser*. Biebrich am Rhein: Dywidag, 1920.

Dyson, James. "Text Message: James Dyson Answers a Few Questions on Engineering, Hands-On Work, and Protecting Your Ideas." *Metropolis* 29, no. 8 (2010): 136.

Ebert, Hans J., Johann B. Kaiser, and Klaus Peters. *Willy Messerschmitt: Pionier der Luftfahrt und des Leichtbaues*. Bonn: Bernard & Graefe, 2008.

Eilemann, Dietrich, Gerhard Herrmann, Herbert Sobanski, and Amt für Erfindungs- und Patentwesen. *Neuererbewegung und Investitionen: Die Rolle der Neuererbewegung bei der Erhöhung des ökonomischen Nutzeffekts der Investitionen*. Berlin: Staatsverlag der DDR, 1967.

Elton, Julia. "Sarah Guppy and Her Bridge Patent No. 3405, 1811." *Women Engineers' History*, November 16, 2019, https://womenengineerssite.wordpress.com/2019/11/16/sarah-guppy-and-her-bridge-patent-no-3405-1811-guest-article-by-julia-elton/.

"Ennis House." Historic American Building Survey, Library of Congress, accessed March 31, 2022, https://www.loc.gov/pictures/item/ca0227/.

Epstein, R. "Industrial Inventions: Heroic or Systematic?" *Quarterly Journal of Economics* 40 (1926): 232–272.

"Europäisches Patentamt in München." *Detail* (November 1981): 815–822.

European Patent Office. *E. P. O. + 10: Regional Patent and Trademark Protection in Europe on the Tenth Anniversary of the European Patent Convention.* Munich: Ladas & Parry, 1989.

European Patent Office. *The Wheel of Invention from Ideas to Patent: An Exhibition of the European Patent Office.* Munich: European Patent Office, 2003.

Evelyn, Douglas E. "Exhibition America: The Patent Office as Cultural Artifact." *Smithsonian Studies in American Art* 3, no. 3 (1989): 24–37.

Evelyn, Douglas E. "The Patent Models on Display." In Post, *American Enterprise*, 17–19.

Farquharson, John. "Governed or Exploited? The British Acquisition of German Technology, 1945–48." *Journal of Contemporary History* 32, no. 1 (1997): 23–42.

Ferguson, Eugene S., and Christopher T. Baer. *Little Machines: Patent Models in the Nineteenth Century.* Wilmington, DE: Hagley Museum, 1979.

Fielding, K. J. "Dickens and International Copyright." *Bulletin of the British Association for American Studies* 4 (August 1962): 29–35.

Fisk, Catherine L. *Working Knowledge: Employee Innovation and the Rise of Corporate Intellectual Property, 1800–1930.* Chapel Hill: University of North Carolina Press, 2009.

Flint, Richard W. "Prosperity through Patents: The Furniture of George Hunzinger & Son." *Nineteenth Century* 8, no. 3-4 (1982): 115–130.

"The Floorcloth and Linoleum Industry." *Scottish Bankers Magazine* (July 1913), 5.

Fondation Le Corbusier. *Une petite maison.* Basel: Birkhäuser, 2020.

Ford, Mark. *Raymond Roussel and the Republic of Dreams.* Ithaca: Cornell University Press, 2019.

Foullon, Abel. *Usage et description de l'holomètre.* Paris: P. Béguin, 1567.

Fox, Margalit. "Frances Gabe, Creator of the Only Self-Cleaning Home, Dies at 101." *New York Times*, July 18, 2017.

Freeman, Mike. *Clarence Saunders and the Founding of Piggly Wiggly: The Rise and Fall of a Memphis Maverick.* Charleston, SC: History Press, 2011.

Fuller, R. Buckminster. *Inventions: The Patented Works of R. Buckminster Fuller.* New York: St. Martin's Press, 1983.

Galison, Peter. *Einstein's Clocks, Poincaré's Maps.* New York: W. W. Norton, 2003.

Galvez-Behar, Gabriel. *La république des inventeurs: Propriété et organisation de l'innovation en France, 1791–1922*. Rennes: Presses universitaires de Rennes, 2008.

Ganea, Peter. "Cambodia." In Goldstein and Kraus, *Intellectual Property in Asia*, 1–16.

Ganea, Peter, and Haijun Jin. "China." In Goldstein and Kraus, *Intellectual Property in Asia*, 17–54.

Ganea, Peter, and Sadao Nagaoka. "Japan." In Goldstein and Kraus, *Intellectual Property in Asia*, 129–154.

Ganea, Peter, and Heath Pattloch, eds. *Intellectual Property Law in China*. The Hague: Kluwer Law International, 2005.

Garcia, Mark. "Architectural Patents and Open-Source Architectures: The Globalisation of Spatial Design Innovations (or Learning from 'E99')." *Architectural Design* (September 2016): 95–96.

García Carbonero, Marta. "La arcadia de hormigón: Innovaciones Patentes / The Concrete Arcadia: Patent Innovations." *AV monografías / AV Monographs* 101 (May 2003): 60–61.

Garde, Tanuja. "India." In Goldstein and Kraus, *Intellectual Property in Asia*, 55–86.

Gardner, William O. *The Metabolist Imaginations: Visions of the City in Postwar Japanese Architecture and Science Fiction*. Minneapolis: University of Minnesota Press, 2020.

Gaudette, Marilyne, Romain Roult, Mohamed Reda Khomsi, and Sylvain Lefebvre. "D'un stade mal-aimé à un stade oublié: l'Autostade de Montréal." *Urban History Review / Revue d'histoire urbaine* 46, no. 1 (2017): 7–22.

Georgiadis, Sokratis. "Sigfried Giedion: Patents in Historical Investigation." *Rassegna* 13, no. 46 (1991): 54–61.

"Germany: Into Plowshares." *Time*, June 6, 1949, http://content.time.com/time/subscriber/article/0,33009,801903,00.html.

Giedion, Sigfried. *Bauen in Frankreich, Bauen in Eisen, Bauen in Eisenbeton*. Leipzig: Klinkhardt & Biermann, 1928.

Giedion, Sigfried. *Space, Time and Architecture*. Cambridge, MA: Harvard University Press, 1941.

Giele, Marieke. "Gesloten instituut met open uitstraling: European Patent Office in Rijswijk door Ateliers Jean Nouvel en Dam & Partners Architecten." *Architect* (Netherlands) 49, no. 2 (2018): 101–106.

Gieryn, Thomas. "What Buildings Do." *Theory and Society* 32 (2002): 35–74.

Gilfillan, S. Colum. *The Sociology of Invention*. Cambridge, MA: Follett, 1935.

Gochberg, Reed. "The Useful Arts of Nineteenth-Century Patent Models." In Brückner, Isenstadt, and Wasserman, *Modelwork*, 125–142.

Goldgel-Carballo, Víctor, and Juan Poblete, eds. *Piracy and Intellectual Property in Latin America: Rethinking Creativity and the Common Good.* London: Routledge, 2020.

Goldstein, Paul, and Joseph Kraus, eds. *Intellectual Property in Asia.* Singapore: Springer Nature, 2009.

Goldstein, Paul, and Joseph Kraus. "Introduction." In Goldstein and Kraus, *Intellectual Property in Asia*, v–xii.

Goodman, Jordan. *Paul Robeson: A Watched Man.* London: Verso, 2013.

Goulthorpe, Mark. *The Possibility of (an) Architecture: Collected Essays by Mark Goulthorpe, dECOi Architects.* London: Routledge, 2008.

Grabicki, Michael. *Breaking New Ground: The History of BASF in China from 1885 to Today.* Hamburg: Hoffmann und Campe, 2015.

Graham, Aaron. "Patents and Invention in Jamaica and the British Atlantic before 1857." *Economic History Review* 73, no. 4 (2020): 940–963.

Grant, Jonathan A. *Big Business in Russia: The Putilov Company in Late Imperial Russia, 1868–1917.* Pittsburgh: University of Pittsburgh Press, 1999.

Grau, Urtzi, and Cristina Goberna. "Should We Patent Architectural Knowledge?" *Volume* 38 (2013): 100–105.

Graves, Michael. "A Case for Figurative Architecture." In *Theorizing a New Agenda for Architecture: An Anthology of Architectural Theory, 1965–1995*, edited by Kate Nesbitt, 86–90. New York: Princeton Architectural Press, 1996.

Gray, Lee. *A History of the Passenger Elevator in the 19th Century.* Mobile, AL: Elevator World, 2002.

Great Britain Patent Office. *List of Abandoned and Void Applications for Patents by or on Behalf of (a) German Nationals and Companies, and (b) Japanese Nationals and Companies.* London: Patent Office, 1946.

Grieve, Victoria. *The Federal Art Project and the Creation of Middlebrow Culture.* Champaign: University of Illinois Press, 2009.

Griset, Pascal. *The European Patent: A European Success Story for Innovation.* Munich: European Patent Office, 2014.

Gross, Daniel A. "Lazy Susan, the Classic Centerpiece of Chinese Restaurants, Is Neither Classic nor Chinese: How the Rotating Tool Became the Circular Table That Circled the Globe." *Smithsonian Magazine*, February 21, 2014, https://www.smithsonianmag.com/arts-culture/lazy-susan-classic-centerpiece-chinese-restaurants-neither-classic-nor-chinese-180949844/.

Gross, Daniel A. "The U.S. Confiscated Half a Billion Dollars in Private Property During WWI." *Smithsonian Magazine*, July 18, 2014, https://www.smithsonianmag.com/history/us-confiscated-half-billion-dollars-private-property-during-wwi-180952144/.

Gross, Linda. "The Evolution of Linoleum." Hagley Museum and Library website, accessed March 25, 2022, https://www.hagley.org/librarynews/evolution-linoleum.

Grothe, Hermann. *Das Patentgesetz für das Deutsche Reich mit Erläuterungen zum praktischen Gebrauch für Patentnehmer, Ingenieure, Gewerbetreibende, Fabrikanten, mit Angaben über Auslandspatente, internationales Patentrecht.* Berlin: De Gruyter, 1877.

Grove, William Robert. *The Correlation of Physical Forces.* 6th ed. London: Longmans, Green, Reader, and Dyer, 1874.

Grüning, Michael, and Konrad Wachsmann. *Der Architekt Konrad Wachsmann: Erinnerungen und Selbstauskünfte.* Vienna: Löcker, 1986.

Guillén, Mauro F. *The Taylorized Beauty of the Mechanical: Scientific Management and the Rise of Modernist Architecture.* Princeton: Princeton University Press, 2020.

Haas, Christopher. *Alexandria in Late Antiquity: Topography and Social Conflict.* Baltimore: Johns Hopkins University Press, 1997.

"Habitations type Antilles." *Techniques et architecture* 9, no. 3–4 (1950): 22.

Hall, Charlie. "'The Other End of a Trajectory': Operation Backfire and the German Origins of Britain's Ballistic Missile Programme." *International History Review* 42, no. 6 (2019): 1118–1136.

Harper, Ida Husted. "The Life and Work of Clara Barton." *North American Review* 195, no. 678 (1912): 701–712.

Harries, Susie. *Nikolaus Pevsner: The Life.* London: Random House, 2013.

Hart, Franz. *Skelettbauten.* Munich: Callwey, 1956.

Harwood, John. *The Interface: IBM and the Transformation of Corporate Design, 1945–1976.* Minneapolis: University of Minnesota Press, 2016.

Hayes, Peter. "Carl Bosch and Carl Krauch: Chemistry and the Political Economy of Germany, 1925–1945." *Journal of Economic History* 47, no. 2 (1987): 353–363.

Hegel, Georg Wilhelm Friedrich. *Philosophy of Right.* Translated by S. W. Dyde. New York: Dover, 2002 (orig. 1821).

Heifetz, Haim. *Description of the Domecrete Building System.* Haifa: Domecrete Ltd., 1970 [?].

Heifetz, Haim. "Progrès réalisés dans le domaine des coffrages gonflables." *Cahiers Centre Scientifique et Technique du Bâtiment* 3, no. 1–2 (1970): 27–39.

Heikenheimo, Marianna. "Alvar Aalto's Patents." *ptah* 2 (2004): 9–16.

Heilmeyer, Florian. "Europäisches Patentamt: Zweigstelle Den Haag, Rijswijk." *Bauwelt* 96, no. 5 (2005): 6–7.

Hell, Matthias. *München '72: Olympia-Architektur damals und Heute; Gespräche mit prominenten Zeitzeugen und Akteuren.* Munich: MünchenVerlag, 2012.

Hemmerling, Joachim. *Neuererbewegung*. Berlin: Staatsverlag der DDR, 1970.

Hennecke, Stefanie, Regine Keller, and Juliane Schneegans, eds. *Demokratisches Grün: Olympiapark München*. Berlin: JOVIS, 2013.

Herbert, Gilbert. *The Dream of the Factory-Made House: Walter Gropius and Konrad Wachsmann*. Cambridge, MA: MIT Press, 1986.

Hick, Darren Hudson. *Artistic License: The Philosophical Problems of Copyright and Appropriation*. Chicago: University of Chicago Press, 2017.

Hilaire-Pérez, Liliane. *Inventions et inventeurs en France et en Angleterre au XVIIIe siècle*. Lille: Université de Lille, 1994.

Hilaire-Pérez, Liliane. *L'invention technique au siècle des Lumières*. Paris: Albin Michel, 2000.

Hindle, Richard L. "Prototyping the Mississippi Delta: Patents, Alternative Futures, and the Design of Complex Environmental Systems." *Journal of Landscape Architecture* 2 (2017): 32–47.

Hintz, Eric. *American Independent Inventors in an Era of Corporate R&D* (Cambridge, MA: MIT Press, 2021).

Hirschman, Sarah M. "Legalizing Architecture: How Congress Defined the Discipline." *Future Anterior: Journal of Historic Preservation, History, Theory and Criticism* 12, no. 1 (2015): 17–30.

Hobsbawm, Eric. *The Age of Extremes: The Short Twentieth Century, 1914–1991*. New York: Vintage, 1994.

Hofer, Veronika, and Irmgard Bommersbasch. *Gabriel von Seidl: Architekt und Naturschützer*. Kreuzlingen, Switzerland: Hugendubel, 2002.

Holton, Gerald, and Yehuda Elkana, eds. *Albert Einstein: Historical and Cultural Perspectives; The Centennial Symposium in Jerusalem*. Princeton: Princeton University Press, 1982.

Hounshell, David A., and John Kenly Smith. *Science and Corporate Strategy: Du Pont R&D, 1902–1980*. Cambridge: Cambridge University Press.

Houtchens, Lawrence H. "Charles Dickens and International Copyright." *American Literature* 13, no. 1 (1941): 18–28.

"How Seized German Millions Fight Germany: First Authoritative Account of the Many-Sided Activities of the Alien Property Custodian; Enemy Money Is Put into Liberty Bonds." *New York Times*, January 27, 1918.

Hrdy, Camilla Alexandra. "State Patent Laws in the Age of Laissez Faire." *Berkeley Technology Law Journal* 28, no. 1 (2013): 45–113.

Huber, Benedikt. *Jean Prouvé: Prefabrication; Structures and Elements*. New York: Praeger, 1971.

Hudon, Edward G. "Literary Piracy, Charles Dickens and the American Copyright Law." *American Bar Association Journal* 50, no. 12 (1964): 1157–1160.

Hurx, Merlijn. "The Most Expert in Europe: Patents and Innovation in the Building Trades in the Early Dutch Republic (1580–1650)." *Architectural Histories* 7, no. 1 (2019): 1–15, accessed online at https://doi.org/10.5334/ah.337.

Ilkosz, Jerzy. *Die Jahrhunderthalle und das Ausstellungsgelände in Breslaus Werk Max Bergs*. Munich: R. Oldenbourg, 2006.

Isaacs, Nigel. "Hollow Concrete Blocks in New Zealand 1904–10." *Construction History* 30, no. 1 (2015): 93–108.

Isaacs, Nigel. "Nails in New Zealand 1770–1910." *Construction History* 24 (2009): 83–101.

Israel, Paul. "'Claim the Earth': Protecting Edison's Inventions at Home and Abroad." In *Knowledge Management and Intellectual Property: Concepts, Actors and Practices from the Past to the Present*, edited by Stathis Araposthatis and Graham Dutfield, 19–38. Cheltenham, UK: Edward Elgar, 2013.

Israel, Paul. *Edison: A Life of Invention*. New York: John Wiley & Sons, 2000.

Jacoberger, Nicole A. "Sugar Rush: Sugar and Science in the British Caribbean." *Britain and the World* 9, no. 14 (2021): 128–150.

Jacobsen, Annie. *Operation Paperclip: The Secret Intelligence Program That Brought Nazi Scientists to America*. Boston: Little, Brown, 2014.

Janis, Mark D. "Patent Abolitionism." *Berkeley Technology Law Journal* 17, no. 2 (2002): 899–952.

Janson, Alban, Carsten Krohn, Anja Grunwald, and Axel Menges. *Le Corbusier: Unité d'habitation, Marseille*. Stuttgart: Axel Menges, 2007.

Janssen, Barbara Suit. *Icons of Invention: American Patent Models*. Washington, DC: National Museum of American History, 1990.

Jasanoff, Sheila. *The Ethics of Invention: Technology and the Human Future*. New York: W. W. Norton, 2016.

Jeremy, David J. *Transatlantic Industrial Revolution: The Diffusion of Textile Technologies between Britain and America, 1790–1830s*. Cambridge, MA: MIT Press, 1981.

Jodidio, Philip, and Janet Adams Strong. *I. M. Pei: Complete Works*. New York: Rizzoli, 2008.

Johns, Adrian. *Piracy: The Intellectual Property Wars from Gutenberg to Gates*. Chicago: University of Chicago Press, 2010.

Jones, Stacy V. "Architect's Lazy Susan Home Built about Revolving Hallway." *New York Times*, October 6, 1956.

Kapczynski, Amy. "Access to Knowledge: A Conceptual Genealogy." In Krikorian and Kapczynski, *Access to Knowledge in the Age of Intellectual Property*, 17–56.

Katz, Irit. "Mobile Colonial Architecture: Facilitating Settler Colonialism's Expansions, Expulsions, Resistance, and

Decolonisation." *Mobilities* 17, no. 2, 213–237. Published ahead of print, January 11, 2022. https://doi.org/10.1080/17450101.2021.2000838.

Keichline, Anna Wagner. "Modern Wall Construction." *Clay-Worker*, June 1, 1932.

Khan, B. Zorina. *The Democratization of Invention: Patents and Copyrights in American Economic Developments, 1790–1920*. Cambridge: Cambridge University Press, 2009.

Khan, B. Zorina. *Inventing Ideas: Patents, Prizes, and the Knowledge Economy*. Oxford: Oxford University Press, 2020.

Khan, B. Zorina. "Married Women's Property Right Laws and Female Commercial Activity." *Journal of Economic History* 56, no. 2 (1996): 356–388.

Khan, B. Zorina. "'Not for Ornament': Patenting Activity by Women Inventors." *Journal of Interdisciplinary History* 33, no. 2 (2000): 159–195.

Khan, B. Zorina, and Kenneth L. Sokoloff. "Entrepreneurship and Technological Change in Historical Perspective." *Advances in the Study of Entrepreneurship, Innovation, and Economic Growth* 6 (1993): 37–66.

Kikutake, Kiyonori. *From Tradition to Utopia*. Milan: L'Arca Edizioni, 1997.

Kimmerer, Tom. *Venerable Trees: History, Biology, and Conservation in the Bluegrass*. Louisville: University Press of Kentucky, 2015.

King, Ross. *Brunelleschi's Dome: How a Renaissance Genius Reinvented Architecture*. New York: Bloomsbury, 2000.

Kirkham, Pat. *Eva Zeisel: Life, Design, and Beauty*. San Francisco: Chronicle Books, 2013.

Kirkham, Pat, and Amy F. Ogata. "Europe 1830–1900." In *History of Design: Decorative Arts and Material Culture, 1400–2000*, edited by Pat Kirkham and Susan Weber, 394–435. New Haven: Yale University Press, 2013.

Kivenson, Gilbert. *The Art and Science of Inventing*. New York: Van Nostrand Reinhold, 1982.

Kostylo, Joanna. "From Gunpowder to Print: The Common Origins of Copyright and Patent." In *Privilege and Property: Essays on the History of Copyright*, edited by Ronan Deazley, Martin Kretschmer, and Lionel Bently, 21–50. Cambridge: Open Book Publishers, 2010.

Kriegel, Lara. *Labor, Empire, and the Museum in Victorian Culture*. Durham, NC: Duke University Press, 2007.

Krikorian, Gaëlle, and Amy Kapczynski, eds. *Access to Knowledge in the Age of Intellectual Property*. Brooklyn: Zone Books, 2010.

Krinsky, Carol Herselle. *Gordon Bunshaft of Skidmore, Owings & Merrill*. Cambridge, MA: MIT Press, 1988.

Kromoser, Benjamin, and Patrick Huber. "Pneumatic Formwork Systems in Structural Engineering." *Advances in Materials Science and Engineering* (2016): article ID 472403. https://doi.org/10.1155/2016/4724036.

Kumagai, Ken'ichi. *History of Japanese Industrial Property System*. Tokyo: Japan Patent Office, Asia-Pacific Industrial Property Center, 1999.

Kurokawa, Kisho. *The Philosophy of Symbiosis*. London: Academy Group, 1994.

Kurz, Peter. *Weltgeschichte des Erfindungsschutzes: Erfinder und Patente im Spiegel der Zeiten*. Cologne: Heymanns, 2000.

Lai, Jessica C. *Patent Law and Women: Tackling Gender Bias in Knowledge Governance*. London: Routledge, 2021.

Lai, Jessica C. "The Role of Patents as a Gendered Chameleon." *Social and Legal Studies* 30, no. 2 (2021): 203–229.

Lambert, Phyllis. *Building Seagram*. New Haven: Yale University Press, 2013.

Lamoreaux, Naomi R., Daniel M. G. Raff, and Peter Temin, eds. *Learning by Doing in Firms, Markets, and Countries*. Chicago: University of Chicago Press, 1999.

Lamoreaux, Naomi R., and Kenneth L. Sokoloff. "Long-Term Change in the Organization of Inventive Activity." *Science, Technology and the Economy* 93 (1996): 1286–1292.

Landes, David S. *Unbound Prometheus: Technological Change and Industrial Development in Western Europe from 1750 to the Present*. Cambridge: Cambridge University Press, 1969.

Langmead, Donald, and Christine Garnaut. *Encyclopedia of Architectural and Engineering Feats*. Santa Barbara, CA: ABC-CLIO, 2006.

Lauber, John. "And It Never Needs Painting: The Development of Residential Aluminum Siding." *APT Bulletin: The Journal of Preservation Technology* 31, no. 2–3 (2000): 17–24.

Lawrence, Amanda Resser, and Ana Miljački, eds. *Terms of Appropriation: Modern Architecture and Global Exchange*. London: Routledge, 2018.

Lea, F. M. *Science and Building: A History of the Building Research Station*. London: HMSO, 1971.

Le Corbusier. *The City of Tomorrow and Its Planning*. Translated by Frederick Etchells. London: John Rocker, 1929.

Le Corbusier. *Vers une architecture*. Paris: Crès, 1924.

Lee, Gabriel F. "The Second Nature of American Progressivism." PhD diss., Stanford University, 2019.

Lemoine, Bertrand. "L'entreprise Eiffel." In *Structural Iron and Steel, 1850–1900*, edited by Robert Thorne, 247–260. London: Routledge, 2000.

Lessig, Lawrence. *Free Culture: The Nature and Future of Creativity*. New York: Penguin, 2005.

Lévêque, François, and Yann Ménière. *Économie de la propriété intellectuelle*. Paris: La Découverte, 2003.

Levine, Neil. *The Architecture of Frank Lloyd Wright*. Princeton: Princeton University Press, 1997.

Lewis, Anna M. *Women of Steel and Stone: 22 Inspirational Architects, Engineers, and Landscape Designers*. Chicago: Chicago Review Press, 2014.

Lewis, J. C. "Report on the British Empire Patent Conference, 1922." *Appendix to the Journals of the House of Representatives* 1923 Session I-II, H-10a. The National Library of New Zealand, accessed June 9, 2002, https://atojs.natlib.govt.nz/cgi-bin/atojs?a=d&d=AJHR1923-I-II.2.2.5.12&e=-------10--1------0hZz-11--.

Locke, John. *Two Treatises of Government: In the Former, the False Principles, and Foundation of Sir Robert Filmer, and his Followers, are Detected and Overthrown; The Latter is an Essay Concerning the True Original, Extent, and End of Civil Government*. London: Awnsham Churchill, 1689.

Long, Pamela O. "Invention, Authorship, 'Intellectual Property,' and the Origin of Patents: Notes toward a Conceptual History." *Technology and Culture* 32, no. 4 (1991): 846–884.

Long, Pamela O. *Openness, Secrecy, Authorship: Technical Arts and the Culture of Knowledge from Antiquity to the Renaissance*. Baltimore: Johns Hopkins University Press, 2001.

Loper, Jason, and Michael Schreiber. *This American House: Frank Lloyd Wright's Meier House and the American System-Built Homes*. Richmond, VA: Pomegranate Press, 2015.

Lorenz, Holger. *Kennzeichen Junkers: Ingenieure zwischen Faust-Anspruch und Gretchenfrage; Technische Entwicklung und politische Wandlung in den Junkerswerken 1931 bis 1961*. Marienberg: Druck- und Verlagsgesellschaft, 2005.

Loyer, François. *Henri Sauvage: Les immeubles à gradins*. Wavre, Belgium: Mardaga, 1987.

Luginbühl, Ursi, and Bernhard Luginbühl. *Bernhardluginbühlstiftung*. Mötschwil, Switzerland: Bernhardluginbühlstiftung, 2001.

Lynn, Greg, and Chuck Hoberman. "Expanding Geometries." *Log* 36 (2016): 87–98.

Lyonnet du Moutier, Michel. *L'aventure de la tour Eiffel: Réalisation et financement*. Paris: Publications de la Sorbonne, 2009.

MacFie, Robert A. *Recent Discussions on the Abolition of Patents for Inventions in the United Kingdom, France, Germany, and the Netherlands: Evidence, Speeches, and Papers in Its Favour*. London: Longmans, Green, Reader, and Dyer, 1869.

Machlup, Fritz. *The Production and Distribution of Knowledge in the United States*. Princeton: Princeton University Press, 1962.

Machlup, Fritz, and Edith Penrose. "The Patent Controversy in the Nineteenth Century." *Journal of Economic History* 10, no. 1 (1950): 1–29.

Macht, William P. "Solution File: Patented Parking." *Urban Land* 59, no. 6 (2000): 24–25.

Mächtel, Florian. *Das Patentrecht im Krieg*. Tübingen: Mohr Siebeck, 2009.

MacLeod, Christine. *Inventing the Industrial Revolution: The English Patent System, 1660–1800*. Cambridge: Cambridge University Press, 1988.

Manchester, William Raymond. *The Arms of Krupp*. Boston: Little, Brown, 1968.

Marcos, Ignacio. "Early Concrete Structures: Patented Systems and Construction Features." *International Journal of Architectural Heritage* 12, no. 3 (2017): 310–319.

Martinez, Frank. "Patent Pending." *Interiors* 159, no. 8 (2000): 38.

Massey, Jonathan. "Buckminster Fuller's Reflexive Modernism." *Design and Culture* 4, no. 3 (2012): 325–344.

Matsuura, Jeffrey H. *Jefferson vs. the Patent Trolls: A Populist Vision of Intellectual Property Rights*. Charlottesville: University of Virginia Press, 2008.

Matteoni, Dario. "16 brevetti di le Corbusier 1918–1961." *Rassegna* 46 (1991): 70–79.

Mayer, Martin. *Olympia Triumphans: Skulptur, Architektur, Landschaft*. Munich: Callwey, 1992.

McMurran, Kristin. "Frances Gabe's Self-Cleaning House Could Mean New Rights of Spring for Housewives." *People*, March 29, 1982. https://people.com/archive/frances-gabes-self-cleaning-house-could-mean-new-rights-of-spring-for-housewives-vol-17-no-12/.

Meister, Chris. "Albert Kahn's Partners in Industrial Architecture." *Journal of the Society of Architectural Historians* 72, no. 1 (2013): 78–85.

Menzel, Christian. *Das Messingwerk und alte Hüttenamt*. Eberswalde, Germany: Denkmalschutzbehörde des Landkreises Barnim, 2007.

Merges, Robert P. *Justifying Intellectual Property*. Cambridge, MA: Harvard University Press, 2011.

Merritt, Deborah J. "Hypatia in the Patent Office: Women Inventors and the Law, 1865–1900." *American Journal of Legal History* 35, no. 3 (1991): 235–306.

Meza, Edwin Gonzalez. "The Triangle Grid, the Evolution of Layered Shells Since the Beginning of the 19th Century," *Curved and Layered Structures* 8, no. 1 (2021): 337–353.

Mgbeoji, Ikechi. "The Comprador Complex: Africa's IPRS Elite, Neo-colonialism and the Enduring Control of African IPRS Agenda by External Interests." *Osgoode Hall Law School Legal Studies Research Paper Series* 10, no. 8 (2014), research paper no. 32, http://ssrn.com/abstract=2441932.

Mindrup, Matthew. *The Architectural Model: Histories of the Miniature and the Prototype, the Exemplar and the Museum*. Cambridge, MA: MIT Press, 2019.

Minnaert, Jean-Baptiste. *Henri Sauvage*. Gollion, Switzerland: Infolio Éditions du patrimoine, 2011.

Minnaert, Jean-Baptiste. "Henri Sauvage, les brevets et la construction rapide." *Revue de l'Art* 118, no. 4 (1997): 41–55.

Minnesota Mining and Manufacturing Company. *Our Story So Far: Notes on the First 75 Years of 3M Company*. St. Paul: Minnesota Mining and Manufacturing Company, 1977.

Mizuta, Susumu. "Patent Slipways of Bakumatsu and Meiji Japan: 1961–1900s." *Construction History: International Journal of the Construction History Society* 30, no. 1 (2015): 71–91.

Monteil, Annemarie. "Bernhard Luginbühl: Hamlet in tonnenschwerer Rüstung." In *Künstler-Kritisches Lexikon der Gegenwartkunst*, edited by Detlef Bluemler and Lothar Romain, 2–19. Berlin: Weltkunst, 1991.

Moore, Dhoruba. "Strategies of Repression against the Black Movement." *Black Scholar* 12, no. 3 (1981): 10–16.

Morgan, Morris Hickey. *Vitruvius: The Ten Books of Architecture*. New York: Dover, 1960 (orig. 1914).

Moser, Petra. "Do Patents Weaken the Localization of Innovations? Evidence from World's Fairs." *Journal of Economic History* 71, no. 2 (2011): 363–382.

Moser, Petra, Alessandra Voena, and Fabian Waldinger, eds. "German Jewish Émigrés and US Invention." *American Economic Review* 104, no. 10 (2014): 3222–3255.

Mott, Kelsey Martin. "The Standard of Ornamentality in the United States Design Patent Law." *American Bar Association Journal* 48, no. 6 (1962): 548–552.

Mozans, H. J. *Woman in Science: An Introductory Chapter on Woman's Long Struggle for Things of the Mind*. New York: D. Appleton, 1913.

Musciano, Walter A. *Messerschmidt Aces*. New York: Arco, 1984.

Ncube, Caroline B. *Intellectual Property Policy, Law and Administration in Africa: Exploring Continental and Sub-regional Co-operation*. London: Routledge, 2016.

Nettles, Scott H. *Thomas A. Edison's Advancements in Portland Cement Production and Concrete Construction*. Chicago: School of the Art Institute of Chicago, 1996.

Neumann, Dietrich. "'The Century's Triumph in Lighting': The Luxfer Prism Companies and Their Contribution to Early Modern Architecture." *Journal of the Society of Architectural Historians* 54, no. 1 (1995): 24–53.

"Ninety-Year-Old Precast System: 1882 Croydon Houses in Patent Slab Construction." *Concrete* (April 1972): 28–29.

"Norman Foster: Smithsonian Institution Patent Office Building Courtyard Enclosure, Washington, D.C., U.S.A." *GA Document* 85 (May 2005): 64–67.

Nozick, Robert. *Anarchy, State, and Utopia*. New York: Basic Books, 1974.

Nyilas, Agnes. "On the Formal Characteristics of Kiyonori Kikutake's 'Marine City' Projects Published at the Turn of the 50's and 60's." *Architecture Research* 6, no. 4 (2016): 98–106.

Ochsendorf, John. "Los Guastavino y la bóveda tabicada en Norteamérica." *Informes de la Construcción* 57, no. 496 (2005): 57–65.

Ochsendorf, John, and Michael Freeman. *Guastavino Vaulting: The Art of Structural Tile*. New York: Princeton Architectural Press, 2013.

O'Reagan, Douglas M. *Taking Nazi Technology: Allied Exploitation of German Science after the Second World War*. Baltimore: Johns Hopkins University Press, 2019.

Oshima, Ken Tadashi. *Kiyonori Kikutake: Between Land and Sea*. Cambridge, MA: Harvard University Graduate School of Design / Zurich: Lars Müller, 2016.

Osman, Michael. "The Managerial Aesthetics of Concrete." *Perspecta* 45 (2012): 67–76.

Oswald, Ansgar. *Meister der Miniaturen: Architektur Modellbau*. Berlin: DOM, 2008.

Otto, Martin, and Diethelm Klippel, eds. *Geschichte des deutschen Patentrechts*. Tübingen: Mohr Siebeck, 2015.

Otto Hetzer AG Weimar. *Neue Holzbauweisen. Firmenprospekt*. Weimar: Otto Hetzer AG, 1912.

Pabst, Martin. *Willy Messerschmitt: Zwölf Jahre Flugzeugbau im Führerstaat*. Oberhaching, Germany: Aviatic Verlag, 2018.

Pagliari, Francesco. "Un villaggio per i brevetti / A Village for Patents." *Arca* (September 1991): 66–71.

Paid, Abraham. *Subtle Is the Lord: The Science and the Life of Albert Einstein*. Oxford: Oxford University Press, 2005.

Park, Ji-Hyun. "South Korea." In Goldstein and Kraus, *Intellectual Property in Asia*, 259–280.

Parks, Janet. "Documenting the Work of the R. Guastavino Company: Sources and Suggestions." *APT Bulletin: The Journal of Preservation Technology* 30, no. 4 (1999): 21–25.

"Patent of Pre-Fabricated Iron Architecture around 1848." *Architectural Forum* 135, no. 2 (1971): 61.

Pelkonen, Eeva-Liisa. *Alvar Aalto: Architecture, Modernity, and Geopolitics*. New Haven: Yale University Press, 2009.

Penrose, Edith. *The Economics of the International Patent System*. Baltimore: Johns Hopkins University Press, 1951.

Perkins, Nancy. "Women Designers: Making Differences." In *Design and Feminism: Re-Visioning Spaces, Places, and Everyday Things*, edited by Joan Rothschild, 120–124. New Brunswick: Rutgers University Press, 1999.

Pernice, Raffaele. "Japanese Urban Artificial Islands: An Overview of Projects and Schemes for Marine Cities During 1960–1990s," *Journal of Architecture and Planning* 74, no. 642 (2009): 1847–1855.

Petroski, Henry. *Invention by Design: How Engineers Get from Thought to Thing*. Cambridge, MA: Harvard University Press, 1996.

Petroski, Henry. "Shopping by Design." *American Scientist* 93 (2005): 491–495.

Pevsner, Nikolaus. *An Enquiry into Industrial Art in England*. Cambridge: Cambridge University Press, 1937.

Phillips, Maureen. "Mechanic Geniuses and Duckies Redux: Nail Makers and Their Machines." *APT Bulletin: The Journal of Preservation Technology* 27, no. 1–2 (1996): 47–56.

Physick, John. *The Victoria and Albert Museum: The History of Its Building*. Oxford: Phaidon, 1982.

Piano, Renzo. "Modulazione e coordinamento modulare." Master's thesis, Politecnico di Milano, 1964.

Picchi, Francesca, and Renzo Piano. "Prouvé inventore: 32 brevetti," *Domus* 807 (1998): 55–66.

Pila, Justine, and Paul Torremans. *European Intellectual Property Law*. 2nd ed. Oxford: Oxford University Press, 2019.

Pohlmann, Hansjoerg. *Neue Materialien zur Frühentwicklung des deutschen Erfinderschutzes im 16. Jahrhundert*. Weinheim: GRUR, 1960.

Polk, Grace. "Sire of the Skyscraper." *New York Times*, November 21, 1926.

Post, Robert E., ed. *American Enterprise: Nineteenth-Century Patent Models; An Exhibition Organized by the Cooper-Hewitt Museum*. New York: Cooper Hewitt Museum, 1984.

Pottage, Alain, and Brad Sherman. *Figures of Invention: A History of Modern Patent Law*. New York: Oxford University Press, 2010.

Pressman, David, and Stephen Elias. *Patent It Yourself*. Berkeley, CA: Nolo Press, 1985.

Pretel, David. *Institutionalising Patents in Nineteenth-Century Spain*. London: Palgrave Pivot Cham, 2019.

Prus, Victor. *Essays in Architecture of Condition, or What It's Like to Be an Architect*. Montreal: Victor Prus et associés, 1996.

Rabinow, Paul. *French Modern: Norms and Forms of the Social Environment*. Chicago: University of Chicago Press, 1989.

Rahmatian, Andreas. "Neo-Colonial Aspects of Global Intellectual Property Protection." *Journal of World Intellectual Property* 12, no. 1 (2009): 40–74.

Rankin, Bill. "The 'Person Skilled in the Art' Is Really Quite Conventional: U.S. Patent Drawings and the Persona of the Invention, 1870–2005." In *Making and Unmaking Intellectual Property: Creative Production in Legal and Cultural Perspective*, edited by Mario Biagioli, Peter Jaszi, and Martha Woodmansee, 55–78. Chicago: University of Chicago Press, 2015.

Räuschel, Jürgen. *Die BASF: Zur Anatomie eines multinationalen Konzerns*. Cologne: Pahl-Rugenstein, 1975.

Ray, William, and Marlys Ray. *The Art of Invention: Patent Models and Their Makers*. Princeton: Pyne Press, 1974.

Reckwitz, Andreas. *The Invention of Creativity: Modern Society and the Culture of the New*. Translated by Steven Black. Cambridge: Polity, 2017.

Reed, Peter, and Kenneth Frampton. *Alvar Aalto: Between Humanism and Materialism*. New York: Museum of Modern Art, 1998.

Rehm, Robin, and Christoph Wagner, eds. *Designpatente der Moderne 1840–1970*. Berlin: Gebr. Mann, 2019.

Reinhardt, Carsten, and Anthony S. Travis. *Heinrich Caro and the Creation of Modern Chemical Industry*. Dordrecht: Springer / London: Kluwer, 2000.

Renwick, Edward Sabine. *Patentable Invention*. Rochester, NY: Lawyers' Co-operative Publishing Co., 1893.

Rhude, Andreas Jordahl. "Structural Glued Laminated Timber: History of Its Origins and Early Development in the United States." *APT Bulletin: The Journal of Preservation Technology* 29, no. 1 (1998): 11–17.

Richardson, Megan, and Julian Thomas, eds. *Fashioning Intellectual Property: Exhibition, Advertising and the Press 1789–1918*. Cambridge: Cambridge University Press, 2012.

Richardson, Vicky. "Lost Property [Industrial Espionage]." *RIBA Journal* 105, no. 11 (1998): 12–13.

Robinson, Eric. "James Watt and the Law of Patents." *Technology and Culture* 13, no. 2 (1972): 115–139.

Rogers, Holly. "The Audiovisual Eerie: Transmediating Thresholds in the Work of David Lynch." In *Transmedia Directors: Artistry, Industry, and New Audiovisual Aesthetics*, edited by Carol Vernallis, Holly Rogers, and Lisa Perrott, 241–270. London: Bloomsbury, 2020.

Rothschild, Alan, and Ann Rothschild. *Inventing a Better Mousetrap: 200 Years of American History in the Amazing World of Patent Models*. San Francisco: Maker Media, 2016.

Rubin, Robert, and Olivier Cinqualbre. *Jean Prouvé: La maison tropicale*. Paris: Editions du Centre Pompidou, 2011.

Rug, Wolfgang. "100 Jahre Hetzer-Patent." *Bautechnik* 83, no. 8 (2006): 533–540.

Rusak, M. "Wooden Churches, Managers, and Fulbright Scholars: Glued Laminated Timber in 1950s Norway." In *History of Construction Cultures: Proceedings of the 7th International Congress on Construction History, July 12–16 2021, Lisbon, Portugal*, vol. 2, edited by João Mascarenhas Mateus and Ana Paula Pires, 736–739. Boca Raton, FL: CRC Press, 2021.

Sachs, Avigail. "The Postwar Legacy of Architectural Research." *Journal of Architectural Education* 62, no. 3 (2009): 53–64.

Sagar, Rajesh. "Patent Policy in India under the British Raj: A Bittersweet Story of Empire and Innovation." In *Patent Cultures: Diversity and Harmonization in Historical Perspective*, edited by Graeme Gooday and Steven Wilf, 273–301. Cambridge: Cambridge University Press, 2020, https://www.doi.org/10.1017/9781108654333.014.

Saint, Andrew. "Thoughts about the Architectural Use of Concrete." *AA Files* 21 (1991): 3–12.

Sáiz González, Patricio. *Invención, patentes e innovación en la España contemporánea*. Madrid: OEPM, 1999.

Sample, Hilary. "Ambient Maintenance." In *Architecture as Measure*, edited by Neyran Turan, 16. Barcelona: Actar, 2020.

Schallenberg, Richard H. *Bottled Energy: Electrical Engineering and the Evolution of Chemical Energy Storage*. Philadelphia: American Philosophical Society, 1982.

Scheuchzer, Antoine. *Nouveauté et activité inventive en droit européen des brevets*. Geneva: Librairie Droz, 1981.

Schiff, Eric. *Industrialization without National Patents: The Netherlands, 1869–1912; Switzerland, 1850–1907*. Princeton: Princeton University Press, 1971.

Schilling, Alexander. *Architektur und Modellbau: Konzepte, Methoden, Materialien*. Basel: Birkhäuser, 2018.

Schmidt, Alexander K. *Erfinderprinzip und Erfinderpersönlichkeitsrecht im deutschen Patentrecht von 1877 bis 1936*. Tübingen: Mohr Siebeck, 2009.

Schulze, Katrin. "Der Park als Spiellandschaft: Zum Spielkonzept von 1972 für den Olympiapark München." *Gartenkunst* 28, no. 1 (2016): 127–136.

Science Museum. "A Brief History of the Science Museum." Science Museum website, accessed April 17, 2022, https://www.sciencemuseum.org.uk/sites/default/files/2017-10/science-museum-history.pdf.

Scientific American. *The Scientific American Hand-Book: A Treatise in Relation to Patents, Caveats, Designs, Trade-Marks, Copyrights, Labels, etc.* New York: Munn & Co., 1890.

Seckelmann, Margit. *Industrialisierung, Internationalisierung und Patentrecht im Deutschen Reich, 1871–1914.* Frankfurt am Main: Klostermann, 2006.

"Seven Great Wonders of Science and Industry Perfected in 1907." *Chicago Tribune*, December 15, 1907.

Shapin, Steven. *The Scientific Life: A Moral History of a Late Modern Vocation.* Chicago: University of Chicago Press, 2010.

Shaw, R. Norman, and Maurice B. Adams. *Sketches for Cottages and Other Buildings: Designed to Be Constructed in the Patent Cement Slab System of W. H. Lascelles.* London: Batsford, 1878.

Shepard, Adrian. "Victor Prus and the Architecture of Condition." McGill University website, accessed March 30, 2022, https://www.mcgill.ca/architecture/alumni/memoriam/prus.

Sibley, Jessica. *The Eureka Myth: Creators, Innovators, and Everyday Intellectual Property.* Palo Alto, CA: Stanford University Press, 2014.

Simonnet, Cyrille. "The Origins of Reinforced Concrete." *Rassegna* 14, no. 49 (1992): 6–14.

Simpson, Pamela Hemenway. *Cheap, Quick, and Easy: Imitative Architectural Materials, 1870–1930.* Knoxville: University of Tennessee Press, 1999.

Simpson, Pamela Hemenway. "Cheap, Quick, and Easy: The Early History of Rockfaced Concrete Block Building." *Perspectives in Vernacular Architecture* 3 (1989): 108–118.

Siporex. *Dach + Decke + Wand: Bauen und Konstruieren.* Stuttgart: Karl Krämer, 1966.

Skarsgard, Susan. *Eero Saarinen and the General Motors Technical Center.* New York: Princeton Architectural Press, 2019.

Slaton, Amy E. *Reinforced Concrete and the Modernization of American Building, 1900–1930.* Baltimore: Johns Hopkins University Press, 2003.

Sluby, Patricia Carter. *The Inventive Spirit of African Americans: Patented Ingenuity.* Westport, CT: Praeger, 2004.

Smith, Adam. *An Inquiry into the Nature and Causes of the Wealth of Nations.* London: W. Strahan, T. Cadell, 1776.

Smith, Albert. *Architectural Model as Machine.* Oxford: Elsevier, 2004.

Smith, Ryan E. "Connect Homes: Modular Housing Patent and Architect Led Business Model." *Journal of Architectural Education* 70, no. 1 (2016): 168–171.

Sniatkov, Serguey, Semen Souponitsky, and Serguey Grigoriev. "The History of Building Techniques and Russian Patent Literature in the Nineteenth Century." *Icon* 5 (1999): 128–137.

Stalder, Laurent. "Turning Architecture Inside Out: Revolving Doors and Other Threshold Devices." *Journal of Design History* 22, no. 1 (2009): 69–77.

Standing, Guy. *Plunder of the Commons: A Manifesto for Sharing Public Wealth*. London: Pelican, 2019.

Stanley, Autumn. *Mothers and Daughters of Invention: Notes for a Revised History of Technology*. New Brunswick: Rutgers University Press, 1995.

Stix, Gary. "The Patent Clerk's Legacy." *Scientific American* 291, no. 3 (2004): 44–49.

Subcommittee on Patents, Trademarks, and Copyrights of the Committee on the Judiciary, United States Senate, Eighty-Fifth Congress, Second Session. *The Impact of the Patent System on Research*. Washington, DC: Government Printing Office, 1958.

Sudrow, Anne, ed. *Geheimreport deutsches Design: Deutsche Konsumgüter im Visier des britische Council of Industrial Design (1946)*. Göttingen: Wallstein Verlag, 2012.

Sulzer, Peter. *Jean Prouvé: Œuvre complète*. 4 vols. Basel: Birkhäuser, 1999–2008.

Swanson, Kara W. "The Emergence of the Professional Patent Practitioner." *Technology and Culture* 50, no. 3 (2009): 519–548.

"Syntax: Contribution of the Curtain Wall to the New Vernacular." *Architectural Review* 121 (May 1957): 308.

Taylor, Frederick Winslow. *The Principles of Scientific Management*. New York: Harper & Brothers, 1911.

Thomson, Elihu. *Selections from the Scientific Correspondence of Elihu Thomson*. Edited by Harold J. Abrahams and Marion B. Savin. Cambridge, MA: MIT Press, 2003.

Tolhausen, Alexander. *A Synopsis of the Patent Laws of Various Countries*. London: Taylor & Francis, 1857.

Travis, Anthony S. *The Origins of Synthetic Dyestuffs Industry in Western Europe*. Bethlehem, PA: Lehigh University Press, 1993.

Trout, Edwin. "The Deutscher Ausschuß für Eisenbeton (German Committee for Reinforced Concrete) 1907–1945; Part 1: Before World War I." *Construction History* 29, no. 1 (2014): 51–73.

"Two Small Industrial Buildings." *Architect and Building News* 156, no. 43 (1972): 939–954.

"Two Step Foam Home," *Chemical Week Magazine* (June 27, 1964): 61.

Tzonis, Alexander. *Santiago Calatrava: Complete Works*. Expanded ed. New York: Rizzoli, 2007.

Übler, Rebekka. *Die Schutzwürdigkeit von Erfindungen: Fortschritt und Erfindungshöhe in der Geschichte des Patent- und Gebrauchsmusterrechts*. Tübingen: Mohr Siebeck, 2014.

United States Congress, House Committee on the Judiciary, Subcommittee on Courts, Intellectual Property, and the

Administration of Justice. *Architectural Design Protection: Hearing before the Subcommittee on Courts, Intellectual Property, and the Administration of Justice of the Committee on the Judiciary, House of Representatives, One Hundred First Congress, Second Session on H.R. 3990, Architectural Works Copyright Protection Act of 1990, and H.R. 3991, Unique Architectural Structures Copyright Act of 1990, March 14, 1990.* Vol. 4. Washington, DC: U.S. Government Printing Office, 1990.

United States Holocaust Museum. "Antisemitic Legislation 1933–1939." *Holocaust Encyclopedia*, accessed March 26, 2022, https://encyclopedia.ushmm.org/content/en/article/antisemitic-legislation-1933-1939.

United States Treasury Department Procurement Division. *Treasury Department Art Projects, Painting and Sculpture for Federal Buildings, November 17th to December 13th, 1936, the Corcoran Gallery of Art, Washington D.C.* Washington, DC: United States Treasury Department Procurement Division, 1936.

Usselman, Steven W. "Patents Purloined: Railroads, Inventors, and the Diffusion of Innovation in 19th-Century America." *Technology and Culture* 32, no. 4 (1991): 1047–1075.

Vaidhyanathan, Siva. *Intellectual Property: A Very Short Introduction*. Oxford: Oxford University Press, 2017.

van Dulken, Stephen. *British Patents of Invention, 1617–1977: A Guide for Researchers*. London: British Library Science Reference and Information Service, 1999.

van Empel, Martijn. *The Granting of European Patents: Introduction to the Convention on the Grant of European Patents, Munich, October 5, 1973*. Leiden: A. W. Sijthoff, 1975.

van Hoorebeek, Mark. "Patently Obvious." *Architects' Journal* 217, no. 8 (2003): 34–35.

van Ishoven, Armand. *Willy Messerschmitt: Der Konstrukteur und seine Flugzeuge*. Herrsching, Germany: Pawlak, 1986.

Vare, Ann, and Greg Ptacek. *Patente Frauen: Grosse Erfinderinnen*. Vienna: P. Zsolnay, 1988.

Vincent, Jean Anne. "Henri IV's First Royal Patent Holder and Furniture Maker." *Contract Interiors* 114 (1955): 84–89.

Vinsel, Lee, and Andrew L. Russell. *The Innovation Delusion: How Our Obsession with the New Has Disrupted the Work that Matters Most*. New York: Penguin / Random House, 2020.

Violand, Heidi. "Jean Tinguely's Kinetic Art or A Myth of the Machine Age." PhD diss., New York University, 1990.

Vojacek, Jan. *A Survey of the Principal National Patent Systems*. New York: Prentice-Hall, 1936.

Voltaire. *Dictionnaire philosophique*. Vol. 19. Paris: Garnier, 1878 (orig. 1764).

von Borries, Friedrich, and Jens-Uwe Fischer. *Heimatcontainer: Deutsche Fertighäuser in Israel*. Frankfurt: Suhrkamp, 2009.

von Gerkan, Meinhard. *Von Gerkan, Marg und Partner*. Vol. 2, *Architektur 1978–1983*. Basel: Birkhäuser / London: Springer, 2001.

von Vegesack, Alexander. *Jean Prouvé: The Poetics of the Technical Object*. Weil am Rhein, Germany: Vitra Design Museum, 2006.

Wachhorst, Wyn. *Thomas Alva Edison: An American Myth*. Cambridge, MA: MIT Press, 1981.

Wachsmann, Konrad, Christa Grüning, and Peter Reuss. *Holzhausbau: Technik und Gestaltung*. Basel: Birkhäuser, 1995.

Wadlow, Christopher. "The British Empire Patent 1901–1923: The 'Global' Patent That Never Was." *Intellectual Property Quarterly* 4 (2016): 311–346.

Wagenknecht, Antje. *Kiyonori Kikutake als Wegbereiter: Visionen und Realisationen des Bauens mit dem Element Wasser*. Oberhausen, Germany: Athena, 2012.

Wagner, Wolfgang. *Hugo Junkers, Pionier der Luftfahrt: Seine Flugzeuge*. Bonn: Bernard & Graefe, 1996.

Waite, Diana, and Patricia Gioia. "Patents Held by the Rafael Guastavinos, Father and Son." *APT Bulletin: The Journal of Preservation Technology* 30, no. 4 (1999): 59–156.

Walker, Alissa. "Everything You Need to Read about Frances Gabe, Self-Cleaning House Inventor." *Curbed*, July 21, 2017, https://archive.curbed.com/2017/7/21/16008194/frances-gabe-self-cleaning-house-obituary.

Waller, Angie. "Patents for Workstations 1990–2010." *Perspecta* 46 (2013): 248–257.

Weber, Max. "National Character and the Junkers" (1917). In *Max Weber: Essays in Sociology*, edited and translated by H. H. Gerth and C. Wright Mills, 386–395. Abingdon: Routledge, 1991 (orig. 1948).

Weizman, Ines. "Architecture and Copyright: Rights of Authors and Things in the Age of Digital Reproduction." In Lawrence and Miljački, *Terms of Appropriation*, 141–159.

Weller, Susan Neuberger. "Law: Protecting Your Designs." *Progressive Architecture* 71, no. 8 (1990): 63.

Wells, Matthew, Olivia Horsfall Turner, Simona Valeriani, and Terese Frankhänel. *An Alphabet of Architectural Models*. London: Merrell, 2021.

Welter, Volker M. "The Limits of Community: The Possibilities of Society; On Modern Architecture in Weimar Germany." *Oxford Art Journal* 33, no. 1 (2010): 63–80.

Wermiel, Sara E. "California Concrete, 1876–1906: Jackson, Percy, and the Beginnings of Reinforced Concrete Construction in the United States." In *Proceedings of the Third International Congress on*

Construction History, Cottbus 2009, vol. 3, 1509–1516. Cottbus, Germany: Brandenburg University of Technology, 2009.

Wermiel, Sara E. *The Fireproof Building: Technology and Public Safety in the Nineteenth Century American City*. Baltimore: Johns Hopkins University Press, 2000.

Whitehead, Don. *The Dow Story: The History of the Dow Chemical Company*. New York: McGraw-Hill, 1968.

Wibaut, Romain, Ine Wouters, and Thomas Coomans. "Hidden above Church Vaults: The Design Evolution of Early Iron Roof Trusses in Mid-Nineteenth-Century Belgium." *International Journal of Architectural Heritage* 13, no. 7 (2019): 963–978.

Wierzbicki, James Eugene. *Music, Sound and Filmmakers: Sonic Style in Cinema*. London: Routledge, 2012.

Wießner, Matthias. "Das Patentrecht der DDR." *Zeitschrift für Neuere Rechtsgeschichte* 3–4 (2013): 230–271.

"Wilce and Matched Flooring Patent." *Inland Architect and News Record* 41 (June 1903): 43.

Wilhelm, Katrin, Maria Siegmantel, and Simon Kunz. "Bernhard Luginbühl: 'Blauer Ritter,' 1976; Dokumentation, Abbau-Aufbauanleitung." Technical study published online by Technische Universität München, accessed April 25, 2022, https://mediatum.ub.tum.de/doc/1600846/1600846.pdf.

Witcombe, Christopher. *Copyright in the Renaissance: Prints and the Privilegio in Sixteenth-Century Venice and Rome*. Leiden: Brill, 2004.

Woltron, Uto. "Patent Architektur." *Nextroom*, February 17, 2001, https://www.nextroom.at/article.php?id=5611.

"A Woman Clerk." *Washington Sunday Chronicle*, March 11, 1885.

Wong, Winnie Won Yin. "Ambience as Property: Experience, Design, and the Legal Expansion of 'Trade Dress.'" *Future Anterior* 9, no. 1 (2012): 89–105.

Wong, Winnie Won Yin. *Van Gogh on Demand: China and the Readymade*. Chicago: University of Chicago Press, 2013.

Wood, H. T. "Letter to the Editor." *Nature* (August 24, 1876): 349.

World Intellectual Property Organization. "Paris Convention for the Protection of Industrial Properties." Accessed July 23, 2022, https://www.wipo.int/treaties/en/ip/paris/.

Wright, Frank Lloyd. "To the Young Man in Architecture." *Architectural Forum* 68 (January 1938): 37.

Wright, Thomas. "The National Museum of Science and Industry: An Overview." *Technology and Culture* 37, no. 1 (1996): 147–150.

Yabumoto, Masanori. "History of Design Protection in Japan." In Aso, Rademacher, and Dobinson, *History of Design and Design Law*, 19–40.

Zahedieh, Nuala. "Colonies, Copper, and the Market for Inventive Activity in England and Wales, 1680–1730." *Economic History Review* 66 (2013): 805–825.

Zeckendorf, William, and Edward A. McCreary. *The Autobiography of William Zeckendorf*. New York: Holt, Rinehart and Winston, 1970.

Zimmer, Zac. "Between Abundance and Appropriation: Indeterminate Critiques of Global IP Schemes." In Goldgel-Carballo and Juan Poblete, *Piracy and Intellectual Property in Latin America*, 172–190.

Image Credits

Alvar Aalto Museum / Alamy Stock Photo: page 87
Anna Wagner Keichline Biographical Material, Virginia Polytechnic Institute and State University: page 35
AP Photo: page 15
Arcaid Images / Alamy Stock Photo: page 77
ART Collection / Alamy Stock Photo: page 56
BASF: page 136
© Bernhard Luginbühl: page 250
Biblioteca Nazionale Centrale di Firenze, Settore Manoscritti, Rari e Fondi: page 8 and color plate 5
Bibliothèque nationale de France: page 291
British Library / Alamy Stock Photo: page 230
Brunel Institute: page 32
Bundesarchiv Berlin: pages 312–313
Bundesarchiv Koblenz: pages 245, 249
Canadian Intellectual Property Office: page 203
Chris Nash: page 48
Chuck Hoberman fonds, Canadian Centre for Architecture, gift of Chuck Hoberman: page 83
Deutsches Museum: pages 143, 149, 150, and color plate 1
Digital Image © CNAC/MNAM, Dist. RMN-Grand Palais / Art Resource, NY: page 91 and color plate 8
© Domenico Mortellito Trust: page 161
DuPont: page 159
European Patent Office: pages 204, 248, 252 (bottom), and color plate 12
© F.L.C. / ADAGP, Paris / Artists Rights Society (ARS), New York 2024: page 119 (both)
Fondazione Renzo Piano: page 98
© Fonds Sauvage, SIAF/Cité de l'architecture et du patrimoine/ Archives d'architecture du XXe siècle: page 113
The Frank Lloyd Wright Foundation Archives (The Museum of Modern Art / Avery Architectural & Fine Arts Library, Columbia University, New York): page 111
Hagley Museum and Library, Wilmington, DE: pages 157, 162, 164, 179, 180, and color plate 2
Harvard University, Houghton Library: page 11
Harvard University, Loeb Library Special Collections: page 281 and color plate 6
The Henry Ford: page 206 and color plate 9
Imagechina Limited / Alamy Stock Photo: page 306
photograph by Institute for Advanced Study, Princeton, NJ, Library and Archives: page 153

Kjeld Duits Collection / MeijiShowa: page 282
Lebrecht Music & Arts / Alamy Stock Photo: page 57
Library of Congress / Clifford H. Poland: page 52
Los Angeles Times: page 42
© 3M: page 146
Marianna Heikenheimo: page 88
Michael Freeman / Alamy Stock Photo: page 74
National Archives (UK): page 241 and color plate 10
National Archives (USA): pages 192, 194, 196, 197, 198, 199, and color plates 4, 7
National Portrait Gallery, Smithsonian Institution; acquired with a partial gift from Beverly J. Cox: page 185
Nigel Isaacs: page 301
Oxford Science Archive / Heritage Images / Science Photo Library: page 298
Peter H. Christensen: page 175 and color plate 13
Riksarkivet i Stockholm Marieberg: pages 46, 61
Roland Nagy / Alamy Stock Photo: page 171
Science History Images / Alamy Stock Photo: page 13
Smithsonian Institution: pages 181, 182, 183, and color plate 3
Suddeutsche Zeitung Photo / Alamy Stock Photo: pages 134, 252 (top)
Thomas A. Edison Papers, Rutgers University: page 102
Trey Kirk: pages 200, 246
US Copyright Office: page 276
US National Park Service: page 104
US Patent and Trademark Office: pages 19, 49, 64, 103, 107, 110, 122, 140, 270–271
United Archives GmbH / Alamy Stock Photo: page 28
University of Rochester, River Campus Libraries: pages 62, 96, 177, 221, 240
V&A Research Institute (VARI): page 188 and color plate 11
Wikimedia Commons: pages 116, 227, 300
© www.industriesalon.de: page 310

Index

Page numbers in italic indicate illustrations.

A2K (access to knowledge) movement, 315
Aalto, Alvar, 85–89; attention to construction details, 329n48; furniture patents, 85–89; glue-laminated construction, 58, 86; Paimio chair, 85, *87*; response to Finnish patent office, 86, *88*
Accesio principle, in Roman law, 5
Adams, Nathaniel, patent drawing for Roofs, *199*
Advertisements, patents functioning as, 70, 117
Aesthetic authorship screen, in patentability determination, 225, 226
Aesthetics: and design patents, 269; manufacturing culture and, 160–165; Nazi, 212–213; patents as regime for standardizing, 127
African Americans, 39–40; FBI surveillance of, 326n36; women, first patent holder, 30
African states, patent systems, 305–308
Alberti, Leon Battista, 277
Albert of Saxe-Coburg and Gotha (prince consort of the United Kingdom), 187, 190
Allen, Frederick Innes, 269–272
Ambiance, in architecture: intellectual property regimes and, 290, 340n4; Prus on, 121
America Invents Act (2011, US), 234
American Civil War, Confederate Patent Commission, 334n18
American Institute of Architects (AIA): and intellectual property in architecture, 274, 277; on plagiarism vs. inspiration, 327n2
American Systems-Built Homes, 106
Ammonia, BASF patent for production of, 205
Ammonium nitrate, explosion at BASF factory, 133, *134*
Anderson, R. Dean, 156
Anne (queen of the United Kingdom), 265
Antitrust laws, and corporate innovation, 129
Ant-Wuorinen, Jal, 85, 86, 89
Anxiety, and invention, 9, 21
Architectural patents, 14–16, 69–71; business, 71, 99–117; creativity and, 17, 60, 71; evolutionary nature of, 14, 16–17; experimental, 71, 117–123; history of industrialization and, 1–2; individual genius vs. collaborative practices, 2, 14, 100–101, 289, 317; legal reform efforts, 272–279; Lockean (charity) proviso and, 293; macro, 71, 72–85; micro, 71, 85–99; motivations for, 117, 120, 121, 123; nonarchitects holding, 16, 72, 100–101, 114–115. *See also* Design patents; Utility patents
Architectural work: AIA definition of, 274; broader notion of, 277, 278–279
Architectural Works Copyright Protection Act (1990, US), 3, 272–273, 339n53; congressional hearing for, 275–279
Architecture: as artistic expression, 69–70, 112; copyright in, 16, 69, 272–273, 275–279; "gentleman's understanding," 12, 70, 315; innovation, patents and, 60; intellectual commons in, 278–279, 315, 317; intellectual property law and, 3, 12, 14, 17, 273; invention in, 12, 320; relation to patenting, 1, 12–14, 16, 69–70, 112–114, 261–262, 319; and utility, 14, 273, 340n4; women in, first, 36. *See also* Architectural patents
Architecture firms, patents held by, 115–117
Argentina, patent system, 284, 308
Armat, Thomas, 176
Armstrong, Sir William, 266–267
Artificial building materials, 53–58
Artistic expression: and architecture, 69–70, 112; copyright associated with, 16, 69; and utility, 14, 20, 273, 340n4
Asia: copyright and patent infringement, 263, 285, 304, 305; patent systems, 308. *See also specific countries*
Aspirin, patent for, 209
Astor Tennis House (Rhinebeck, New York), *74*
Aureli, Pier Vittorio, 317
Australia: and German patent technology after World War II, 208; patent model repository, 334n2
Austria, patent model repository, 334n2
Authorship screen, in patentability determination, 225, 226
Automobile roulette, 51
Aviation industry: contributions to building industry, 129, 147–156; patenting tradition in, 148
Avon Gorge (England), designs for bridge across, 30–33, *32*
Avonius, Asko, 86–89

375

"Badalone, Il" (Brunelleschi), 6, *8, plate 5*
Badosa, Paul, 242
Baldwin, James, 27
Ballauf, Daniel, patent model workshop, 176, *177*
Ban, Shigeru, 14
Barton, Clara, 229
BASF: Caro's innovations, 128, 130–131, 331n3; corporate research lab, 129; dye research and patenting, 131; first European patent, 243; and German Patent Act of 1877, 132; Haber-Bosch process patented by, 205; indigo laboratory, *136*; Oppau facilities, ammonium nitrate explosion at, 133, *134*; patent activity, 128, 130–131, 132, 133; patents lost after World War I, 205; research scientists at, 132, 133
Bauhaus, and Junkers AG, 148, 152
Bavaria: patent model repository, 334n2; patent system, 259
Bayer (firm): innovation and patenting, 331n3; post–World War II seizure of assets, 209
Beaux-Arts architecture, 72, 73, *74*
Belgium: patent model repository, 334n2; patent system, 259
Bell, Alexander Graham, 176
Benjamin, Miriam, 30
Benton, Tim, 118, 120
Berg, Max, 142
Bergdoll, Barry, 321
Berliner, Emile, 176
Berne Convention for the Protection of Literary and Artistic Property (1886, 1893), 10, 274
Bessemer, Henry, 9, 130
Betonpfähle Strauss (Strauss concrete pile), 142–145, *143*

Biagioli, Mario, 4, 226
Biomimetic form: and Metabolist movement, 82; and Mortellito's Growing Shelters Synthetically scheme, 163
BIOS. *See* British Intelligence Objectives Subcommittee
Biosphere (Expo '67, Montreal), 77, 78
BIRPI. *See* United International Bureau for the Protection of Intellectual Property
BKF chair design, 280–284, *281, plate 6*
Blades, Herbert, 156
Blauer Ritter sculpture (Luginbühl), 247–251, *250*
Boehlert, Sherwood, 273, 339n53
Bohn, René, 131
Bonet, Antonio, 280
Bouis, John, patent drawing for Tin Copper and Zinc Roofs, *192, 193*
Brekke, Guttorm N., 58
Breslau (Germany), Centennial Hall, 142
Brett's Colonists' Guide and Cyclopaedia (Leys), 299, 300
Brezhnev, Leonid, *15*
Bricks: optimized (K Brick), 37–38; rapid manufacture of, 326n72; for roofs, patent drawing for, *196*
Bridges: I-beam used in, 58; patent models for, 178, *181, 183*; suspension, 31–33
Bridges, William, design for bridge across Avon Gorge, 30–33, *32*
Briquetterie d'Alfortville, Le Corbusier and, 120
Bristol (England): designs for bridge across Avon Gorge, 30–34, *32*; patent holders in, 30
Britain: Building Research Station, 238–242; German patent technology acquired after World War

II, 207–208, 210–215; patent model repository, 334n2; technological progress, historiographic narrative of, 58–59. *See also* British Patent Office; British patent system; England; *specific cities*
British Intelligence Objectives Subcommittee (BIOS), 210–211; Pevsner as industrial spy for, 211, 212–215, 336n93
British Patent Office: Building Research Station, 238–239; establishment of, 3, 265; library, 338n70; museum, 187–191, *188, plate 11*
British patent system: acquisition of foreign technology and, 211; colonial footprint, 272, 296–302; competition with German patent system, 10; Dickens on, 219–222; evolution of, 265; and Indian patent system, 272, 296, 308; model of, 260; and patent abolition movement, 266–267; rate of patent approval, 235; reforms (1852), 265, 297; seventeenth-century, 191
Bronze powder, invention of, 130
Brunel, Isambard Kingdom: design for bridge across Avon Gorge, 33, *34*; friendship with Guppy, 31; position on patents, 9, 31, 266
Brunelleschi, Filippo, 6, 12: "Il Badaloné," 6, *8, plate 5*
Bryant, William, patent drawing for Supplying Houses with Water, *198*
Bryce, Hugh G., 138
Buffington, Leroy, patent for Iron Building Construction, 202, *203*, 335n58
Building Arrangement Maximizing Views, patent for, 341n4

Building materials: British certification for, 238–239; imitative/artificial, 53–58; innovation of, 53–60; rapid manufacture of, patents for, 326n72. *See also* Prefabricated construction systems

Building Research Station (UK), 238–242

Burbank, Elizabeth, 202

Burbank, Luther, 202, *206*, *plate 9*

Business patenting, in architecture, 71, 99–117

Busselmeier, Rudolf, 333n76

Butterfly chair, 280–284, *281*

Calatrava, Santiago: "ornamental" used in patent applications of, 108, 115; patents held by, 14, 115; Turning Torso building, 115, *116*

Cambodia, patent system, 308

Canada: Expo '67, *77*, 78; German patent technology after World War II and, 208; patent model repository, 334n2

Cannetis, 239–242, *241*, *plate 10*

Capitalism, patent system and, 4, 316; satirical critique of, 319

Capsule Tower (Tokyo, Japan), 82

Carney, Richard, 275, 278

Caro, Heinrich, 127–128, 331n3; patent system exploited by, 130–131

Carpenter, Paul, 145

Castells, Manuel, 316

Catalan timbrel vaulting techniques, 72, 73, *74*

Caveat system, inauguration of, 265

Cavity Wall System, invention of, 30

Çelik Alexander, Zeynep, 169

Centennial Hall (Breslau, Germany), 142

Centre Pompidou (Paris, France), 97

Chair designs: Aalto and, 85–86; Cole and, 187; Ferrari Hardoy and, 280–284, *281*, *plate 6*; women and, 30

Chareau, Pierre, 89

Charity proviso, 292–293

Charter of the Forest (1217, England), 290

Chemical arts: German patents used during World War I, 205; inventions, patentability requirements for, 337n8; propensity for patenting, 265

Chemical Foundation (US company), 209

Childs, David, 340n85

Chile, patent system, 308

China: comprador system, 10; copyright and patent infringement, 263, 285, 304, 305; indanthrone dyes, popularity of, 131; Parisian-style architecture in, *306*; patent applications from, and US patent system, 234; patent system, 304–305; Song dynasty, copyright system under, 5; TRIPS and, 304

Christoph & Unmack (company), 92

CIAM. *See* Congrès Internationaux d'Architecture Moderne

Clear-span timber arch construction, origins of, 55

Clifton Suspension Bridge (Bristol, England), 33, 34

Coignet, François, 109

Cold War: and corporate patent research, 135; and German émigrés, 207; Nixon-Khrushchev kitchen debate, 12, *15*; and patent systems, 12

Cole, Isaac, patent for One-Piece Plywood Chair, 187

Collaborative invention practices: in architecture/construction sector, 2,

14, 100–101; vs. genius inventors, 100, 289

Collins, Kevin Emerson, 14, 16, 224–225, 226, 264, 324n51, 341n4

Colman, Charles, 22, 260, 261, 280

Colonialism: and international intellectual and creative activity, 289; and patent regimes, 2, 272, 296–302, 305. *See also* Neocolonialism

Color patents, 128, 131–132

Comerma, John, 73

Commons: of information vs. knowledge, 315, 316–317; as legal concept, origins of, 290. *See also* Intellectual commons

Comprador system, in China, 10

Comptroller General of Patents (UK), 265

Concrete: color penetrant for, 133; in Edison's Single-Pour Concrete House, 101–105, *102–104*, 109, 330n96; furniture made out of, Edison's efforts to patent, 105; in Heifetz's Domecrete building system, 60, *62*; in Lascelles's Improved Method of the Construction of Buildings, 47, *48*; in Messerschmitt housing system, *153*, 154; in Nelson's building structure, 108, *110*; precast, German factories for, 142; reinforced, patent system and evolution of, 16–17; Siporex block construction system, 239, *240*

Concrete block, genealogy of, 325n26

Confederate Patent Commission, during American Civil War, 334n18

Confucianism, 304, 309

Congrès Internationaux d'Architecture Moderne (CIAM), 82

Index 377

Conservatoire Nationale des Arts et Métiers (Paris, France), 97
Constitutions, national, patent systems enshrined in, 259–260
Construction sector: 3M auxiliary patents for, 145–147; aviation industry contributions to, 129, 147–156; innovations coming from, 141–145; patents, ownership of workplace knowledge and, 263–264
Cooper-Hewitt Museum, exhibition of patent models at, 335n43
Copper House System, 151
Copyism: in Asian countries, 263, 285, 304, 305; intellectual commons and concern about, 314. *See also* Infringement
Copyright: for architectural drawings vs. for buildings, 273; in architectural work, 16, 69; in architecture, US laws regarding, 3, 272–273, 275–279; arts and, 16, 69; of educational material, 292; of ideas, 273–274; vs. patents, 69, 225, 226, 264, 273
Copyright Act of 1909 (US), 273
Copyright Act of 1976 (US), 273
Corian (product), 135
Corporate patents, 127–165; in construction sector, 263–264; patent scientists, 130–131
Corporations: as beneficiaries of patent system, 238, 294, 296; research laboratories created by, 129, 138–139
Cotton Gin, invention of, 184
Council of Industrial Design (COID, Britain), 211
Courts: gender stigmatization of design patents, 260–262; role in patent system, 258–259. *See also* Patent lawyers; Patent litigation

Craft patronage, English system of, 5, 6–7
Crafts, architects with background in, 85
Craft secrecy, preindustrial tradition of, 130
Creativity: architectural patents and, 17, 60, 71; corporate-sponsored, 128–129; vs. inventiveness, 17; Kant on, 20–21; novelty imperative and, 21; organic idea of, 20–21; origins of concept, 20; patents and insights into, 319; patents as roadblocks to, 267; patent system and, 21, 65; problem solving and, 18, 21; as product of modernity, 18, 20, 319; psychology of, 21; Reckwitz on ("creativity dispositive"), 17–20, 21. *See also* Invention
Croydon (England), Sydenham Row, *48*
Crypto-patent system, during World War II, 210
Culinary recipes, copyright for, 274
Curtain wall, innovations in, 115
Cuthbertson, Jim, 340n85

Dahlström, Gustav, 54–55
Daileda, David, 275, 277
Databases, for patents, 169–170. *See also* Repositories
Davey, Peter, 251–253
Davis, Henry S., 334n18
DBA. *See* Deutsche Bauakademie
De architectura (Vitruvius), 5, 145
Declaration on the Rights of Indigenous People (United Nations, 2007), 303
Defensive patenting, 60; America Invents Act (2011) and, 234; corporate patent strategy and, 135; early fireproof patents and, 73; Edison and, 99–100
Demetriades vs. Kaufman, 340n85

Demountable House, Prouvé drawing for, *91, plate 8*
Design patents: architectural patents filed as, 106–108, 115; biases in patent law regarding, 260–264; Calatrava and, 108, 115; gender stigmatization of, 260–262; infringement cases related to, 260–261, 268–269, 280; Italian patent system and, 237–238; notion of novelty, 22; obscurity until late nineteenth century, 267–268; US legal reforms regarding, 269–272; vs. utility patents, 22, 71, 268–269, 280; Wright and, 106–108, 330n108
Design Rules, Le Corbusier patent for, 118
Dessau (Germany): Bauhaus relocation to, 148; Junkers company, 147, 148, 152
Deutsche Bauakademie (DBA, East Germany): patent activity of, 309; and *Raumzellenbauweise* (cellular room construction), 311, *312–313*
Deutscher Patentschutzverein (German Patent Association), 267
Deutsches Museum, and EPO, 247, 251
Dickens, Charles, 219–222, *221*
Digital design, rise of, 3, 272
Digitalization, and patent offices, 3, 251
"Disposition of space," patentability of. *See* Spatial configurations
Dixon, John, 80
Domecrete building system, 60, *62*
Domestic inventorship: men and, 43–53; women and, 27–29, 34, 40–44
Dom-ino project (Le Corbusier), 118–120, *119*
Doors: fireproof, 30; revolving, 139–141, *140*

Drawings. *See* Patent drawings
Drescher, Hans, 151
Drexler, Arthur, 187, 335n43
Du Bois, Max, 118
Duell, Charles, 295
DuPont Company: contributions to construction sector, 135, 156; corporate research lab, 129, *162*; decentralization of corporate research at, 138; Fuller and, 79; and German émigrés, 336n68; Ludox colloidal silica foam insulation, *162*; patent regime at, 135–138; scheme for "growing" houses with foam, 158, 160–165; Tyvek, 156–158
Dyckerhoff, Eugen, 142
Dyckerhoff, Wilhelm Gustav, 142
Dyckerhoff & Widmann AG (DYWIDAG), 129; Betonpfähle Strauss (Strauss concrete pile) system used by, 142–145, *143*; innovations, 142, 144
Dyes, patents for, 128, 131
Dymaxion family of patents, 75, 76, 78

East Germany: patent activity, 309–314, *310*, 342n76; *Raumzellenbauweise* (cellular room construction), 311, *312–313*
École des Beaux-Arts (France), position on patents, 112, 114, 118
Edison, Thomas, 99–105; approach to patenting, 99–100, 130; caveats used by, 265; inventions of, 99; and Schneider, 176; Single-Pour Concrete House, 101–105, *102–104*, 109, 330n96; team assisting, 100–101; and US Plant Patent Act of 1930, 202
Edward II (king of England), 211

Eiermann, Egon, 244
Eiffel Tower (Paris, France): Chinese copy of, *306*; patent model preceding, 178, *179*
Einstein, Albert: as émigré to US, 336n79; as patent examiner, 229–231; and Wachsmann, 329n13
Ekman, P. J., 45
Elevators, patent models for, 178, *179*, *182*, *plate 2*, *plate 3*
Ellsworth, Henry, 334n19
Elton, Julia, 31, 33–34
Emmons, Paul, 170–172
Employees, as owners of workplace knowledge: Japan's Patent Act of 1921 on, 272; in nineteenth vs. twentieth century, 263–264
Engineers, architectural patents held by, 16, 72
England: artificial building materials, development of, 55; Avon Gorge, designs for bridge across, 30–33, *32*; Charter of the Forest (1217), 290; concrete prefabricated housing, 47, *48*; craft patronage (kingly privileges), system of, 5, 6–7; patent models, 187–191, *188*; profession of patent examiner, 231; Statute of Monopolies (1623), 7. *See also* Britain; *specific cities*
Ennis house (Los Angeles, California), Wright design for, 108
EPO. *See* European Patent Office
Equivalence, doctrine of, 341n14
Ericsson, E., prefabricated house design by, 44–45, *46*
Escher, Otto, 213–214
Europe: architectural firms patenting in, 117; artisanal traditions and philosophy of design, 85; measurement of nonobviousness in patent offices, 226; patent systems enshrined in national constitutions, 259–260; patent systems, antecedents of, 5–7, 259; property rights principle, 5; unification of national patent systems, 9–12, 242. *See also specific countries*
European Coal and Steel Community, 242
European Patent Office (European Patent Organization; EPO), 243–244; advent of computing and, 251; Deutsches Museum and, 247, 251; headquarters (Munich), 237, 242, 244–253, *245*, *246*; interpretation booths, 247, *248*, *plate 12*; Luginbühl *Blauer Ritter* sculpture, 247–251, *250*; origins of, 12, 242–243; patent litigation meeting, *248*, *plate 12*; patent records, *252*; Telelift pneumatic tube system proposed for, 247, *249*
Exclusive rights, as patent antecedents, 1, 2–3, 5–6
Experimental patenting, in architecture, 71, 117–123
Expo '67. *See* Montreal (Canada), Expo '67
Expo '70. *See* Osaka (Japan), Expo '70

Fairbairn, William, 58, 59
Fashion houses, design patents in collaboration with, 71
Fees, patent: annual renewal, 234–235; filing, 219–220, 226, 232
Ferrari Hardoy, Jorge, 280–284; BKF chair, 280–284, *281*, *plate 6*
Fiddler, Robert, 257–258
Finley, James, 33
Fire escapes: patent model for, 178, *180*; women inventors and, 30
Fire Ladder, patent drawing for, *197*, *plate 4*

Index 379

Fire safety: Guastavinos' patents for fireproof buildings, 72–75; used as veil for form or style, 73, 79, 94; women's inventions for, 30
Fisk, Catherine, 264
Fleischmann, C. L., patent drawing for Follet's Self-Balancing Sashes, 193–195, *194, plate 7*
Floating aquatic city, patent for, 81–82
Floor plans, patents for, 70–71
Florence cathedral, construction of dome of, 6, *8*
Foam insulation, DuPont research on, *162*
Follet, Francis, patent for Self-Balancing Sashes, 193–195, *194, plate 7*
Ford, Henry, patent model collection of, 187, 335n41
Ford, The (museum, Dearborn, Michigan), patent models at, 187
Fordist model of manufacturing, 120
Forestek Plating and Manufacturing Company vs. Knapp-Monarch Company, 268–269
Förster, Friedrich, 151
Foster, Josiah R., patent model for Wooden Truss Bridge, 178, *183*
Foster, Norman, 71
Foullon, Abel, holometer invention, 7, *11*
Fowke, Francis, 190–191
France: Centre Pompidou, 97; Conservatoire Nationale des Arts et Métiers, 97; École des Beaux-Arts, position on patents held by architects, 112, 114; and German industrial patents after World War II, 208, 336n78; mass housing after World War I, 109; patent models, 170, *171*, 334n2; patents encouraged by, in post-World War I period, 109, 112. *See also* French patent system; *specific cities*
Franklin Knitting Mills vs. Gropper Knitting Mills, 261
Free use, patent law and delay in, 315
French patent system, 208; characteristics, 260, 308; Latin American patent systems modeled on, 308; after World War I, 109
Frugal innovation, 29, 325n3
Fuller, R. Buckminster, 75–81; 4D House, 75, 76; Dymaxion inventions, 75, 76, 78; Geodesic Dome, 75, 76, *77*, 78, 79, 85, 235, 236; Hoberman's kinship with, 84; *Inventions*, 78–79, 328n16; Octet Truss, 75, 76, 335n43; patent applications, 235–236; patent lawyer of, 236; patents held by, 14, 75; Prus's work compared to, 121; relations with academy, 80–81; relations with industry, 79–80; self-fashioning as inventor, 76; student following, 328n28
Functionality screen, in patentability determination, 224–225, 228. *See also* Utility
Furniture design: Aalto and, 85–89; Cole and, 187; Edison and, 105; Escher and, 213–214; Ferrari Hardoy and, 280–284, *281, plate 6*; German patent system and, 213–214; Knoll and, 214; women and, 30

Gabe, Frances, 40–44, *42*
Galileo Galilei, 7
Ganea, Peter, 305
Garvey, Glenwood, 43
Gehler, Willy, 144
Gehry, Frank, 14
Gender. *See* Men; Women
Genius: vs. collaborative practices, in architectural patenting, 2, 14, 100–101, 289; and development of intellectual property concept, 6
Geodesic Dome (Fuller), 76, *77*, 78, 79, 85; patent application for, 235, 236
Gerkan, Marg & Partners, 244, *245*, 253
German émigrés: architects, 85, 92–95, 207; companies discouraging hiring of, 336n68; in patent modelmaking business, 176; World War II and, 39, 92–95, 207, 209, 336n79
German Patent Association (Deutscher Patentschutzverein), 267
German Patent Office: in late nineteenth century, 231; in Munich, 237; records repository at, 252; after World War II, 208, 210, 237
German patent system: competition with British patent system, 10; critique of, 238; European inventors and, 228; on furniture and interior architecture (*Innenarchitektur*), 213–214; nonobviousness threshold, 263; reforms, 267; seventeenth-century, 191; strength of, 10, 228; after World War II, 208, 210, 213
German states, premodern: patent model repositories, 334n2; patent systems, 259; unification of patent laws, 10
Germany: aviation industry, contributions to building industry, 147–156; collaboration between private inventors and big business, 129; color and dye patents, 128, 131–132; concrete construction, innovations in, 142–145; corporate patent activity, 127–135; economic miracle after World War II, 237;

industrial patents after World Wars I and II, 205–215, 237; Patent Act of 1877, 132; post–World War I mass housing, 147–148; post–World War II industrial landscape, Pevsner's survey of, 212–215; post–World War II mass housing, 152–155, *153*; post–World War II patent activity, 237; profession of patent examiner, 231; research scientists, complaints from, 132–133; tar paint companies, 131–132. *See also* East Germany; German Patent Office; German patent system; National Socialism; *specific cities*
Gibbs, Joseph, 325n26
Giedion, Sigfried, 16, 58–59
Gilbreth, Frank, 142
Ginsburg, Jane, 275–277, 278
Globalization: and patent infringement, 284–285; and intellectual commons, 296; and patents, 9, 12, 304–308
Global South: intellectual property monopolies and, 292–293; patent systems, 272, 305–308
Glue-laminated construction: Aalto's use of, 58, 86; patent for, 55–58
Goberna, Christina, 317
Goulthorpe, Mark, 123
Graham, Aaron, 296–297
Grant, Thomas, 325n26
Grau, Urtzi, 317
Graves, Michael, 71: congressional testimony by, *275, 276,* 278–279, 315
Gray, Eileen, 89
Great Britain. *See* Britain
Great Exhibition of 1851 (London, England), 187, 190
Gridiron, patent for, 38–39
Griset, Pascal, 244
Grocery store, self-service, *52*: Saunders's patent for, 47–51, *49*
Gropius, Ise, 92

Gropius, Walter: design patent for teapot, 95; and Wachsmann, 92, 93–95
Grossman, Marcel, 229
Grothe, Hermann, 338n6
Grove, William Robert, 266
Growing Shelters Synthetically scheme (Mortellito), 160–165, *164*
Grunwald, Henry Anatole, 27
Guadet, Julien, 112, 118
Guastavino, Rafael, 72, 73–75: and Catalan timbrel vaulting techniques, 72, 73, *74*; fire safety as veil for form/style by, 73, 79, 94; herringbone pattern of tile layout, 72; and patent system, 95, 327n14
Guastavino, Rafael, Jr., 72
Guastavino Fireproof Construction Company, 72–75
Guilds, trade secrets of, as antecedent to patent system, 1, 2–3, 5
Günschel, Günter, 328n28
Guppy, Sarah, 30–34
Gutch, John Matthew, 31–33

Haber-Bosch process, BASF patent for, 205
Habsburg patent system, 260
Haertel, Kurt, 243
Hagley Museum (Wilmington, Delaware), 41, *175, 179, 180, 187, plate 13*
Hague, The (Netherlands): Dutch patent office, 251; International Patent Institute, 12
Handelskammer (German trade union), 238
Harrison, Wallace, 14
Hart, Franz, 237
Hegel, Georg Wilhelm, 23
Heifetz, Haim, Domecrete building system, 60, 62
Heikenheimo, Marianna, 86, 89
Helix plan, patent for, 114–115
Hennebique, François, 16
Hennebique pile system, 340n85

Henry II (king of France), 7
Henry VI (king of England), 6
Herbert, Gilbert, 92–93, 94
Hetzer, Otto Karl Friedrich, patent for glue-laminated construction, 55–58
Hilten, Heinz, 207
Hirsch Kupfer- und Messingwerke, Copper House System, 151
Hirschman, Sarah M., 277, 278–279
Hoberman, Chuck, 82–84: Folding Structure, *83*; kinship with Fuller, 84; Retractable Iris Dome, 335n43
Hoberman sphere, 84
Hollow wall construction, K Brick for, 37–38
Holometer, invention of, 7, *11*
"Home Delivery: Fabricating the Modern Dwelling" (exhibition), 321
Homes: men's inventions related to, 43–53; mobile, patents for, 51–53; patents manifested in, 12; women's inventions related to, 27–29, 34, 40–44. *See also* Houses
Hoover, J. Edgar, 326n36
Hounshell, David A., 139
Houses: catalog and mail-order, 71; DuPont scheme for "growing" with foam, 158, 160–165; Edison's Single-Pour Concrete, 101–105, *102–104,* 109, 330n96; Fuller's 4D House, 75, *76*; Junkers AG housing system, 147–152, *149*; Lazy Susan Home, 53; Le Corbusier designs for, 118–120, *119, 121*; made from railway gauges, patent for, 59–60, *61*; Messerschmitt AG housing system, 148, 152–156, *153*; metal, 147–151; mobile, 51–53; Prouvé designs for, *91, 92, plate 8*; self-cleaning, 40–44, *42,* 326n42;

Houses (cont.): with unfinished bonus space, 63, *64*; Wachsmann's Packaged House system, 93–95, *96*; Wright designs for, 106, 108. *See also* Mass housing; Prefabricated construction systems
Hull, Orson, 147
Hume, David, 22
Huntsville (Alabama), Marshall Space Flight Center, 207
Huxtable, Ada Louise, 54

I-beam, used in steel construction, 58–60, *61*
Ideas: copyright of, 273–274; patents as way of transmitting, 120
Idiota (Nicholas of Cusa), 20
Imitative building materials, 53–58
Immigration, and patents, 209. *See also* German émigrés
Indanthrone dyes, 131
India: German patent technology after World War II and, 208; patent model repository, 334n2; patent system, British colonialism and, 272, 296, 308
Indigenous cultural expression, and intellectual property rights, 293–294, 302–303, 307
Indigo dye, patent for, 128
Indonesia, patent system, 308
Industrialization, history of, and history of patents, 1–2, 319
Industrial Revolution: and concentration of creative activities in large companies, 128; and invention, 1, 7, 319; and profession of patent examiner, 231; third, 295
Information: architectural, ability to access, 315; commons constituted of, 316

Informationalism, 316–317
Infringement, patent, 279–285; in architecture and construction, 280, 285, 340n85; in Asian countries, 263, 285, 304, 305; claims against Wright, 108–109; doctrine of equivalence and, 341n14; globalization and, 284–285; trespassing compared to, 279; TRIPS as bulwark against, 305. *See also* Patent litigation
Innenarchitekt (interior architect), in post-World War II Germany, 213
Innovation. *See* Creativity
Intellectual commons, 290, 292–296, 314; in architecture, 278–279, 315, 317; denigration of, 289; imperial framework and, 302; informationalism and, 317; international character of, 296
Intellectual property (IP): A2K movement on, 315; architecture and, 17, 273; Dickens's interest in, 220–222; global standardization of, 304; indigenous rights and, 293–294, 302–303; Kant on, 22; Merges on, 292–294; national constitutions and, 259; origins of formal concept, 4, 6; real property compared to, 279; as right vs. privilege, 21; short-run inefficiencies created by, 316; socioeconomic conundrum of, 292–293; three pillars of, 9; TRIPS Agreement and, 304; vital vs. cultural, 292–293. *See also* Copyright; Patents
International Convention for the Protection of Industrial Property (1883), 265
Internationalization, and patents, 9–12
International Patent Congress (1874), 129

International Patent Institute (The Hague, Netherlands), 12
International patent offices, 242–253. *See also* European Patent Office
International Style, 211
Invention: in architecture, 12, 320; collaborative/nonhierarchical, in construction sector, 2, 14, 100–101; experimentation and, 117; Industrial Revolution and, 1, 7, 319; as knowledge, 4; legal vs. architectural understanding of, 14; patent system and rhetoric of, 9, 23; premodern practices encouraging, 6–7; premodern privileges vs. modern patents and, 4; problem solving and, 9, 141; psychology of, 9, 21. *See also* Creativity; Inventors
Inventions (Fuller), 78–79, 328n16
Inventors: and general public, 315; genius vs. collaborative forms of invention, 100, 289; nineteenth-century, 219–222; patent applications by, 222, 223; and patent examiners, 231–232; twentieth-century advantages of, 222
IP. *See* Intellectual property
Iron Building Construction, Buffington patent for, 202, *203*, 335n58
Israel, Domecrete building system and, 60, *62*
Israel, Paul, 100
Italy: Florence cathedral, construction of dome, 6, *8*; patent model repositories, 334n2; patent system, 237–238; Renaissance city-states, and origins of patents, 4, 6, 7

Jackson, Andrew, 173, 174
Jagdmann, Astrid, 30

Jamaica, patent system, 296
Japan: copyright and patent violations, history of, 263, 284–285; Expo '70, 97; furniture store, *282*; intellectual property law, 304, 308; Kansai Airport, 340n85; Metabolist movement, 82; Nakagin Capsule Tower, 82. *See also* Japanese patent system
Japanese patent system, 342n64; and Kikutake, 81; nonobviousness threshold, 81, 263; reforms, 272
Jarvis, Philip, patent model for Bridge, 178, *181*
Jeanneret, Chares-Édouard, 118. *See also* Le Corbusier
Jeanneret, Pierre, 89
Jefferson, Thomas: and American patent system, 7, 224, 229; as first patent examiner of US Patent Office, 229; and "functionality mandate" of nonobviousness, 224
Jin, Haijun, 305
John of Utynam, 6
Johns, Adrian, 9, 18, 267, 289, 297
Johnson, James, patent drawing for Fire Ladder, *197, plate 4*
Jones, Louise, 156
Junkers, Hugo, 148
Junkers Flugzeug- und Motorenwerke AG, 148: Bauhaus and, 148, 152; Metal House Construction system, 147–152, *149, plate 1*; Nazi regime and, 152; Office of Alien Property (OAP, US) and, 210; patenting strategy, 148–152, *150,* 154
"Junk" patents, 234

Kahn, Louis, 97
Kansai Airport (Japan), Piano design for, 340n85
Kant, Immanuel: on creativity, 20–21; on property, 22

Kapczynski, Amy, 316
Kastenmeier, Robert W., 273, 275, 339n53
K Brick, patent for, 37–38
Keichline, Anna Wagner, *35*, 36–37; optimized brick invented by, 37–38, 325n26
Kempe, John, 5
Keramik (product), 133
Kessler, Jacob, patent model for Wooden Truss Bridge, 178, *183*
Kew Gardens (London, England), herbarium at, 169
Khrushchev, Nikita, 12, *15*
Kiesler, Frederick, 14
Kikutake, Kiyonori, 81–82
Kinetic structures, patents for, 84
Kingly privileges *(privilegio)*, as antecedent to modern patents, 1, 2–3, 5–7, 259, 265
Kivenson, Gilbert, 21
Kletzin GmbH, 151
Knapp, Andrew S., 268; Combined Sandwich Toaster and Tray, 268, *270, 271*
Knapp-Monarch griddle case, 268–269
Knoll, Wilhelm, 214
Knowledge: ability to access, 315–316; architectural, recombined, 317; commons constituted of, 290, 316–317 (*see also* Intellectual commons); and economic growth, 315–316; freedom of, vs. freedom of information, 315; implicit, creativity dispositive and, 17, 18, 191; informationalism and, 316–317; invention as, 4; patents as form of, 176, 191, 222, 316; related to prior art, in patent applications, 226; repositories of, 169; technical, feedback loop of, 317; traditional/indigenous, 302–303, 307; workplace, ownership of, 263–264, 272

Koolhaas, Rem, 14
Korea, intellectual property law, 308
Kornerup-Koch, H. J., 58
Krafft, Robert, 151
Krupp AG, Nazi regime and, 210
Kurchan, Juan, 280
Kurokawa, Kisho, 82; Nakagin Capsule Tower (Tokyo, Japan), 82

Lan & Co., 142
Lascelles, William Henry, 47; concrete prefabricated houses built by, 47, *48*
Latin America, patent systems, 308
Laws. *See* Patent laws; *and specific laws*
Lawyers. *See* Patent lawyers
"Lazy Susan," invention of, 29
Lazy Susan Home, 53
Le Corbusier, 118–121; approach to patenting, 120–121; and Briquetterie d'Alfortville, 120; Design Rules patent, 118; Maison Dom-ino, 118–120, *119*; Maison Monolith (Eternit), 121; Modulor Man, 118, *119,* 120; patents held by, 14, 118; and Prouvé, 89–90; and UAM, 89–90; Unité d'Habitation, 121
Legal systems, and patent system, 4. *See also* Patent laws; Patent lawyers
Leischner, Margaret, 336n93
Lewis, J. C., 299–302
Leys, Thomson W., *Brett's Colonists' Guide and Cyclopaedia,* 299, *300*
Libraries: British Patent Office and, 338n70; German Patent Office and, *252. See also* Repositories
Lincoln Logs toy set, 109
Linoleum, patent for, 55, *56, 57*
Locke, John: charity proviso, 292–293; and concept of property, 22–23

London (England): Great Exhibition of 1851, 187, 190; Kew Gardens herbarium, 169; South Kensington Museum, 187–191, *188, plate 11*
London Agreement (1946), 205
Long, Pamela, 20
Longchambon, Henri, 242, 243
Los Angeles (California), Ennis house, 108
Ludox colloidal silica foam insulation, *162*
Luehrsen, Hannes, 207
Luginbühl, Bernhard, *Blauer Ritter* sculpture, 247–251, *250*
Luxfer company, and Wright prism block designs, 105–106, *107*
Lynch, David, 27

MacFie, Robert Andrew, 20, 266
Machlup, Fritz, 315, 316
MacLeod, Christine, 4, 9, 324n37
Macro patenting, in architecture, 71, 72–85; philosophical underpinnings, 78–79, 84
Magna Carta (England), 290
Maison Dom-ino, Le Corbusier and, 118–120, *119*
Maison Monolith (Eternit) project, Le Corbusier and, 121
Maisons à Meudon, Prouvé design for, 92
Maison Tropicale, Prouvé design for, 92
Maki, Fumihiko, 82
Makowski, Zygmunt Stanislaw, 97
Mallet-Stevens, Robert, 89
Malmö (Sweden), Turning Torso building, 115, *116*
Malone, G. B., 135–137
Marble, artificial, patent for, 55
Marcantonio, Vito, 40

Marine City project (Kikutake), 81–82
Marshall Space Flight Center (Huntsville, Alabama), 207
Massachusetts Institute of Technology (MIT), and experimental patenting, 123
Mass housing: Le Corbusier interest in, 118; in post–World War I France, 109; in post–World War I Germany, 147–148; in post–World War II Germany, 152–155, *153*. *See also* Prefabricated construction systems
Matrix patenting, 148–152, *150*
McHarg, Joe Albert, Jr., "Building" for Pizza Ring Enterprises, *19*
McQuaid, Matilda, 335n43
Melman, Seymour, 257, 264
Men: and domestic inventorship, 43–53; ties for, design patent infringement case, 261
Merges, Robert, 21, 22, 23, 84, 128–129; interview with, 292–294
Merlien, Anna, 30
Merritt, Deborah, 29
Messerschmitt, Willy, 152
Messerschmitt AG, 129; housing system of, 148, 152–156, *153*; patent activities of, 154, 333n76
Metabolist movement (Japan), 82
Metal House Construction system, Junkers AG and, 147–152, *149, plate 1*
Mgbeoji, Ikechi, 307
Micro patenting, in architecture, 71, 85–99
Mies van der Rohe, Ludwig, I-beam used by, 59
Military-industrial complex, and patent laws, 10
Miljački, Ana, 12
Mills, Robert, 174, 334n18
Mining, propensity for patenting, 265
Minnaert, Jean-Baptiste, 109

Minnesota Mining and Manufacturing (3M): auxiliary patents for construction sector, 145; corporate research lab, 129, 138–139; Fuller and, 79–80; sandpaper, 145–147, *146*
Mitchell, Charles E., 186
Mobile homes, patents for, 51–53
Models. *See* Patent models
Modern Color Inc., 280–284
Modernism, architectural, 72; patents forming technological backbone of, 14, 16–17
Modernity, creativity as product of, 18, 20
Modulor Man (Le Corbusier), 118, *119*, 120
Monopolies: culture of, and craft secrecy, 5, 6–7; deployment to encourage innovation, 6; on innovative design solutions, 114–115; patent abolitionists' arguments regarding, 266, 267
Montreal (Canada), Expo '67, Biosphere, *77, 78*, 121
Moorhead, Carlos, 275
Morgenthaler, Ottmar, 176
Mortellito, Domenico, 158–165; experimental art techniques of, 333n95; Growing Shelters Synthetically housing system of, 160–165, *164*; *Man with Horse*, *161*
Moser, Petra, 209
Mott, Catherine, 30
Mott, Kelsey Martin, 268, 269
Mount, Christopher, 335n43
Multinational patents, 89, 148–152
Multistory Building Structure, patent for, 114–115
Munich (Germany): EPO headquarters, 237, 242, 244–253, *245, 246*; German patent office, 237
Munich Convention (1973), 12, 243

Musée des Arts et Métiers (Paris), patent models at, 170, *171*
Museum of Modern Art (MoMA): "Home Delivery: Fabricating the Modern Dwelling" exhibition, 321; paper clip in design collection, 337n22; patent models at, 187, 335n43
Museums, patent records at, 169; in Britain, 187–191, *188, plate 11*; in France, 97, 170, *171*; in Germany, 247, 251, 252; in US, 174–184, *175, 179, 180,* 187, 335n43. *See also specific museums*
Muthesius, Hermann, 211

Nails: introduction in New Zealand, 297–299, *301*; rapid manufacture of, patents for, 326n72
Nakagin Capsule Tower (Tokyo, Japan), 82
National constitutions, and patent systems, 259–260
National patent offices, 236–242; critiques of, 238, 244; European, unification of, 9–12, 242–243; variations in interpretation of patentability, 263. *See also under specific countries*
National Socialism (Nazism): aesthetics of, and German design and manufacturing practices, 212–213; industrial companies' relations with, 152, 210; intellectual property laws under, 10, 38, 324n45
Natural law: and patent system, 4; and property rights, Locke on, 22
Ncube, Caroline, 307
Negroponte, Nicholas, 29, 325n3
Nelson, William E., Concrete Building Structure of, 108, *110*
Neocolonialism, TRIPS and, 307

Netherlands: globally significant inventions, 308; patent abolition movement, 7; patent office, digitalization of, 251; patent system, origins of, 259. *See also* Hague, The
Neuererwesen/Neuererbewegung (innovator movement), in East Germany, 309–314, *310*
Newness. *See* Novelty
Newton, Ernest, house design by, *48*
New York City: Seagram Building, 59; St. Patrick's Cathedral, 262; World Trade Center reconstruction, 340n85
New Zealand: British patent system and, 297–302; frontier settlement, 297–299, *300*; independence, 302; nail patents, 297–299, *301*; patent system, 308
Nicholas of Cusa, 20
Nixon, Richard, 12, *15*
Nobel, Alfred, 130
Nondisclosure agreements, 264
Nonobviousness, threshold of, 223–228: different standards applied to, 294; in European patent offices, 226, 263; experimentation and, 117; functionality mandate of, 224, 225–226; identifying, patent office's role in, 222, 223; in Japanese patent system, 81, 263; novelty necessitated by, 225; patent lawyers and, 222; and unified European patent system, 243
Norway: invention of paper clip, 228; patent model repository, 334n2
Novelty (newness): architectural patents and, 17; demand for, 21–22, 230, 319; examination in British patent system, 265; examination in Italian

patent system, 237; patents and insights into, 319; premodern privileges vs. modern patents and, 4; Reckwitz on, 21; state as arbiter of, 6; as threshold of patentability, 21–22, 23, 224, 225, *230*; unified European patent system and new notions of, 243
Nozick, Robert, 22–23

Ochsendorf, Jonathan, 72, 327n14
Octet Truss, Fuller patent for, 75, 76, 335n43
Offensive patenting, 60; corporate patent strategy and, 135; Edison and, 99
Office of Alien Property (OAP, US), 209–210
Oliver Patent, cartoon of, *13*
Oman, Ralph, 277, 279
"One Laptop Per Child" (OLPC) project, 325n3
One-Piece Plywood Chair, Cole patent for, 187
Open access movement, 315
Operation Paperclip, 207, 336n68
O'Reagan, Douglas, 208, 211
"Ornamental" (term): Calatrava's use in patent applications, 108, 115; stigmatization of, in design patent infringement cases, 260, 261; use in US law on design patents, 269–272; Wright's use in patent applications, 108
Osaka (Japan), Expo '70, Pavilion of Italian Industry, 97
Osman, Michael, 16
Otis, Charles and Norton, patent model for Elevator, 178, *182, plate 3*
Owen, L. D., 325n26

Packaged House, Wachsmann system of, 93–95, *96*
Paimio chair, Aalto design for, 85, *87*

Index 385

Palatine orange dye, patent for, 131
Palmer, Harmon S., 325n26
Papal States, patent system, 259
Paper clip: in design collection at MoMA, 337n22; invention of, *227*, 228; symbolism in World War II, 228
Paris (France): Chinese copy of architecture, *306*; Eiffel Tower, patent model preceding, 178, *179*; Sauvage terraced residential building, *113*
Paris Convention for the Protection of Industrial Property (1883), 10
Parker, James, patent drawing for Bricks for Roofs, *196*
Patentability: criteria for, 21–23, 224–225, 228; and national patent offices, 263
Patent abolition movements, 7–9, 266–267, 289; arguments in support of, 308–309; cartoon representative of, *13*; idea of "organic" creativity and, 20
Patent Act of 1902 (UK), 265
Patent Act of 1921 (Japan), 272
"Patent, a Poem, The" (Woty), 319
Patent applications: backlog of, 232–233; fees associated with, 219–220, 226, 232; initial, 222; length of, 233; prior art required in, 43, 84, 226, 316; rate of approval of, 234–235; revisions of, 222, 235
Patent disclosures, strategic opacity in, 226
Patent dossier, changes over time, 170
Patent drawings, 191–202; in black and white ink, 195, *203–204*, 335n60; color, 191–195, *192*, *194*, *196–199*, 202; dialectic of proximity and ambiguity, 195–201; earliest, 191–193, *192*; heterogeneity prior to standardization, 191; importance in patent examination, 233; by inventors, 191, *192*, 193; by professional draftspeople, 191, *194*; US Patent Office requirements for, 195–201, 200, 335n54
Patent examination: importance of drawings in, 233; process, 232
Patent examiners, 223, 228–229; in Britain, 265; critique of, 294; on futility of patent system, 257–258; in Germany after World War II, 237; interviews of, 232–234; low public profiles of, 231; profession of, and Industrial Revolution, 231; questions asked by, 235–236; relationship with inventors, 231–232; relationship with patent lawyers, 233; in US, famous figures acting as, 229–231
Patent infringement. *See* Infringement
Patent laws: biases in, 260–263; military-industrial complex and, 10
Patent lawyers, 233; corporate patents and, 132, 139; defense of patent system by, 258; Edison and, 100; Fuller and, 235, 236; interviews with, 294–296; role of, 43, 84, 222, 233, 236; strategies of, 201
Patent litigation, 258–259; cost of, 296; design patent cases, 260–261, 268–269; design vs. utility patent cases, 280; at EPO, *248, plate 12*; Wright and, 108–109
Patent models, 170–191; cost of, 176; didactic value of, 176; in England, 187–191, *188*; at Hagley Museum, *175, 179, 180, 187, plate 13*; loss during fire of 1877, 184; Luginbühl *Blauer Ritter* sculpture compared to, 251; at Musée des Arts et Métiers, 170, *171*; at Museum of Modern Art, 187, 335n43; premodern repositories for, 334n2; private collections of, 187; professional makers of, 174–176, *177*; at Smithsonian Museum, 186, 187; in US, 172–187, *175, 179–183*, 334n19; US Patent Office requirements regarding, 173–174, 176, 178; War of 1812 and, 172–173; worldwide repositories of, 334n2
Patent Museum (US), 174, 176–178, 334n19, 335n39; fire of 1877 and, 178–184, *185*
Patent offices: digitalization and, 3, 251; European, unification of, 9–12, 242–243; international, 242–253, 247; interpretation of patentability, 263; mystery shrouding, 222; national, 236–242, 244; nonjuridical character of, 223; as repository of knowledge, 169; role of, 222–223. *See also under specific countries*
Patents: as advertisements, 70, 117; aesthetic qualities, standardization of, 127; annual renewal fees for, 234–235; architects' complex relationship with, 1, 12–14, 16, 69–70, 112–114, 319; and architectural innovation, 60; in architecture, 14, 69–71 (*see also* Architectural patents); arguments in favor of, 266–267; as bargain between inventor and general public, 315; beneficiaries of, 294, 295, 296; celebrating technology for its own sake, 178; colonial system and, 2,

296; vs. copyright, 69, 225, 226, 264, 273; corporations and, 127–165; cultural and aesthetic novelty and, 18; databases for, 169–170; filing fees for, 219–220, 226, 232; globalization and, 9, 12, 304–308; and idea transmission, 120; immigration and, 209; incentives for, in premodern era, 324n37; industrialization and, 1–2, 319; initial, strategy for prolonging, 141; and intellectual commons, 289; intellectual products represented by, 3–4; and knowledge, 176, 191, 222, 316; manifestation in built environment, 12; motivations for seeking, 117, 120, 121, 123; multinational, strategy for, 89; offensive vs. defensive, 60; origins of, 1, 2–3, 4–7; and property rights, 1, 22; scholarship on, 4–5, 9, 17; women and, 2, 29–44. *See also* Design patents; Utility patents

Patents, Designs, and Trade Marks Act (1883, UK), 265

Patent systems: antecedents of, 1, 2–3, 4–7, 259; in Asia, 308; blind spots, 84; capitalism and, 4, 316, 319; colonialism and, 2, 272, 296–302; continuities in, 9; corporations as primary beneficiaries, 238, 294, 296; courtroom's role in, 258–259; creativity and, 21, 65; dialectics of progress and conservatism, 7; in Europe, 5–7, 259–260; futility of, patent examiner's perspective on, 257–258; in Global South, 272, 305–308; internationalization and, 9–12; in Latin America, 308; major models of, 259–260; modern, characteristics of, 319; modern, origins of, 7, 259–260; natural law and, 4; notion of novelty, 21–22; patent lawyers' defense of, 258; power dynamics, 280; as quasi-democratic institution, 7, 92; reforms advocated, 257–258, 264–279; and rhetoric of invention and progress, 9; under socialism, 309–314, 342n76, 342n79; and wars, 10, 308–309; women's participation, 2, 29–44, 202, 260. *See also under specific countries*

Paxton, Joseph, 190

Pei, I. M., patent for Multistory Building Structure, 114–115

Pennycuick, John, 106

Perret, Auguste, 14, 330n114

Perriand, Charlotte, 89

Peru, patent system, 308

Peterson, Russell W., 158

Petroski, Henry, 9

Pevsner, Nikolaus, 211; survey of postwar German industrial landscape, 212–215, 336n93

Pharmaceutical patents, 292, 293

Phillips, William C., patent model for Improvement in Fire-Escapes, 178, *180*

Phoebus, Gaston, *Le livre de la chasse*, 291

PHOSITA (person holding ordinary skill in the art), 225, 226

Piano, Renzo, 85, 95–99; design for Kansai Airport (Japan), 340n85; study for shelter made with inflatable polyethylene, *98*

Piggly-Wiggly self-service grocery store, *52*

Pizza Ring Enterprises, McHarg "Building" for, *19*

Plage, Wilhelm, 284

Plagiarism vs. inspiration, AIA on, 327n2. *See also* Copyism

Plant Patent Act (1930, US), 202

Plant species, patents for new varieties of, 202, *206*, *plate 9*

Platonic form, macro patenting and, 75, 79

Poelzig, Hans, 92

Polk, Grace, 335n58

Polyethylene shelter, Piano study for, *98*

Pomet, Pierre, "Slave labour on a sugar plantation in the West Indies," *298*

"Poor Man's Tale of a Patent" (Dickens), 219–222, *221*

Portugal, patent system, 259

Prefabricated construction systems: concrete, English patents for, 47, *48*; concrete, German system for, 152–156, *153*; metal, German aviation industry and, 147–152; Museum of Modern Art exhibition, 321; multinational patent protection for, 148; *Raumzellenbauweise* (cellular room construction), East Germany, 311, *312–313*; timber, Scandinavian patents for, 44–45; Viviani, 202, *204*; Wachsmann, 93, 94–95, *96*, 321

Prior art: in British patent applications, 235; citations compared to, 223; digitalization and, 3, 251; EPO access to, 247, 251; multinational patenting and, 151; patent models and, 186; patent repositories and, 169, 205; patent system requirements for, 43, 84, 226, 316; Wright's use of, 106

Privilegio (kingly privileges), system of, as antecedent to modern patents, 1, 2–3, 5–7, 259, 265

Problem solving: and creativity, disentangling, 18, 21; and invention, 9, 141

Property: Hegel on, 23; Kant on, 22; as key feature of intellectual property, 17; Locke on, 22–23; patents and insights into, 319. *See also* Intellectual property

Property rights: patents and, 1, 22; premodern practices regarding, 5; women and, common law on, 29; women's use of patents and, 2

Prouvé, Jean, 85, 89–92; Demountable House, *91, plate 8*; design systems, 92, 329n61; and Piano, 97

Prus, Victor, Building Structure patent, 121–123, *122*

Prussia, patent system, 259

Publicity, patenting and, 70, 117, 120, 121, 123

Pulte, William J., patent for House with Unfinished Bonus Space, 63, *64*

Putilov, Nikolay Ivanovich, patent for House Made from Railway Gauges, 59–60, *61*

Rahmatian, Andreas, 307

Railway gauges, reuse as structural members in construction, 58, 59–60, *61*

Raj patent system, 272

Ranger, William, 325n26

Rankin, Bill, 195, 201–202

Ransome, Ernest, 142

Ransome, Frederick, 325n26

Real estate developers, architectural patents held by, 16, 114–115

Reckwitz, Andreas, 17–20, 21

Recognition: need for, patents as fulfilling, 117, 120, 121, 123; research scientists' desire for, 138, 139

Reinforced concrete: patent system and evolution of, 16–17; Perret's work with, 330n114

Renaissance city-state, Italian, and origins of patents, 4, 6, 7

Renwick, Edward Sabine, 262

Renwick, James, 262

Repetto, Grazia, 340n85

Repositories, patent, 169–170; in Germany, 252; need for, 169, 205; premodern, 334n2. *See also* Museums

Research and development (R&D): at 3M, 129, 138–139; at BASF, 129, 131, 132, 133; corporations and, 127, 128–129, 138–139; at DuPont, 129, 138, *162*

Research scientists: corporate, 130–131; desire for recognition, 138, 139; and experimental patents, 123; rhetoric of autonomy regarding, 138–139

Retractable Iris Dome (Hoberman), 335n43

Revolving doors: environmental benefits, 141; invention of, 139–141, *140*

Rhinebeck (New York), Astor Tennis House, *74*

Rhodes, Rufus B., 334n18

Rhodes, T. B., 325n26

Robb, Alfred, roofing nail patent, 299, *301*

Robertson, Donald W., 236

Robeson, Paul, 39–40

Robie, Wendy, 27, *28*

Rogers, Richard, 97

Roman law, principle of *accesio*, 5

Roman pile foundation, Vitruvius on, 145

Roofs: bricks for, *196*; flat, modernism and, 14; nails for, 299, *301*; patent drawings for, *192, 193, 196, 199*; stone system for, 30; tin, copper, and zinc, *192,* 193

Roussel, Raymond, 51–53

Royal Society (Britain), as patent examiner, 265

Rubber, substitute for, 55

Ruf, Sep, 155

Ruggles, John, 176

Russia, patent system, 259. *See also* Soviet Union

Sackur, Lenore, 30

Sandpaper: 3M packaging for, *146*; patent for, 145–147

Sandstone, artificial, patent for, 54–55

Saunders, Clarence, patent for Self Serving Grocery Store, 47–51, *49*

Sauvage, Henri, 109–114, *118*; building construction system using telescopic jacks, 112–114; influences on, 330n114; Stepped Building scheme of, 109–112, *113*

Savignon, François, 243

Sawyer, Lemuel, patent model for Improvement in Elevator-Towers, 178, *179, plate 2*

Sazarin, Charles, 109; terraced residential building, *113*

Scandinavia, timber construction in: Hetzer glue-laminated construction and, 58; patents related to, 44–45

Scharoun, Hans, 309

Schaumarin Faneeritehdas Company, 86

Schindele, Dorothy, 283–284

Schneider, Rudolf, 176

Scientific American (journal), on patent models, 184

Scientific management theory, 36–37

Seagram Building (New York City), I-beam used in, 59

Secretary of State (US), patent examination duties of, 229, 337n23

Seidl, Gabriel von, 251

Self-Balancing Sashes, Follet patent for, 193–195, *194, plate 7*

Self-cleaning home (Gabe), 40–44, *42*, 326n42

Self Serving Grocery Store, patent for, 47–51, *49*

Sewing machine, history of, 304

Shaw, R. Norman, 47; house designs by, 47, *48*

Shepard, Alexander, 184
Sherman Antitrust Act (1890, US), 129
Shine, Thomas, 340n85
Shine vs. Childs, 340n85
Shufelt, Helen, 30
Simpson, Pamela, 53, 325n26
Single-Pour Concrete House, Edison design for, 101–105, *102–104,* 109, 330n96
Siporex block construction system, 239, *240*
Skidmore, Owings & Merrill (SOM): design for reconstruction of World Trade Center, 340n85; patents held by, 115–117
Skyscrapers: Buffington patent application for, 202, *203,* 335n58; Calatrava Turning Torso, 115, *116;* revolving doors deployed in, 141
Smith, John Kenly, 139
Smithsonian Institution Building (Washington, DC), 262
Smithsonian Museum, patent models at, 186, 187
Socialism, patent systems under, 309–314, 342n76, 342n79
Société Centrale des Architectes (France), 112
Société des Architectes Diplômés par le Gouvernement (SADG, France), 112
Société des Constructions Rapides (SCR), 112–114
SOM. *See* Skidmore, Owings & Merrill
Song dynasty (China), rudimentary copyright system, 5
South Africa: British patent system and, 302; German patent technology after World War II and, 208
Southampton (England), buckling of piles for dock, 340n85
South Kensington Museum (London, England), patent collection, 187–191, *188, plate 11*
Soviet Union: fabrication systems from, and East German patents, 311; and German industrial patents after World War II, 207, 208; patent system, 342n79
Spain: patent model repository, 334n2; patent system, 259
Spatial configurations, patents relating to, 16, 60–63, 225; ambiguities in, 264; floor plans, 70–71; in Guastavinos' fireproof buildings, 75; House with Unfinished Bonus Space, 63, *64;* programmatic affordance of, 341n4; Self Serving Grocery Store, 47–51, *49;* shortcomings of, 226
Stalder, Laurent, 139, 141
Stam, Mart, 120
Standing, Guy, 290
Statute of Monopolies (1623, England), 7
Steel construction, use of I-beam in, 58–60, *61*
Steelmaking process, inventor of, 130
Stepped Building, Sauvage design for, 109–112, *113*
Stewart, Potter, 337n3
Stock, Richard, 43
Stoke, William, roofing nail patent of, 299, *301*
Storm-Door Structure (Van Kennel), *140*
St. Patrick's Cathedral (New York City), 262
Strasbourg Patent Convention (1963), 243, 244
Strauss, Anton, 142
Strauss concrete pile (Betonpfähle Strauss), 142–145, *143*
Strowich, Maria, 30
Structural steel, invention leading to, 130
Sullivan, Mark J., patent model for Wooden Truss Bridge, 178, *183*
Supplying Houses with Water, Bryant patent drawing for, *198*
Suspension bridge: first patent for, 33, 34; Guppy's design, 31–33, 34; origin of, 34–36
Sweden: artificial building materials, development of, 54–55; Turning Torso building, 115, *116;* patent model repository, 334n2; patent system, 259; prefabricated house patents, 44–45; Siporex block construction system, 239, *240;* steel production, 60
Switzerland: Einstein as patent examiner, 229–231; patent abolition movement, 7; patent for glue-laminated construction, 55; patent system, creation of, 132
Sydenham Row (Croydon, England), *48*
Szegö, Ilse, 38–40, 326n29, 326n31
Szegö, Stefan, 38–40, 208, 326n29

Taylor, Frederick Winslow, 36–37
Telelift pneumatic tube system, 247, *249*
Teller, Myron S., 53
Terraced residential building, Sauvage design for, 112–113, *113*
Tesla, Nikola, 130
Textiles: design patent infringement case, 261; and patent activity, 7; Wright designs for, 108, *111*
Thomson, Elihu, 130
Thomson-Houston Electric Company, 130
Thornton, William, 172–173
Threshold devices, innovation and patenting of, 139–140
Tiles: glass prismatic, Wright designs for, 105–106, *107;* Guastavinos and, 72

Index   389

Timber construction:
 clear-span, origins of, 55;
 Scandinavian patents for,
 44–45
Timbrel vault, 72, 73, *74*
Tin Copper and Zinc Roofs,
 patent drawing for, *192, 193*
Tinguely, Jean, 251
Tokyo (Japan), Nakagin
 Capsule Tower, 82
Tolhausen, Alexander, 334n2
Torroja, Eduardo, 14
Trade dress, legal mechanism
 of, 264, 340n4
Trade secrets, of guilds and
 artisans, as antecedent to
 patent system, 1, 2–3, 5, 264
Trading With the Enemy Act
 (TWEA, US), 205
Traditional cultural
 expression, and intellectual
 property rights, 293–294,
 302–303, 307
Treasury Relief Art Project
 (TRAP, US), 158
TRIPS (Agreement on
 Trade Related Aspects
 of Intellectual Property
 Rights): and African
 countries, 305–307; as
 bulwark against patent
 piracy, 305; and China, 304;
 critique of, 307
Tubular Construction System,
 invention of, 30
Turning Torso building
 (Malmö, Sweden), 115, *116*
Twin Peaks (TV series),
 domestic inventorship in,
 27–29, *28*
Tyvek Home Wrap: in con-
 temporary construction,
 159; development of, 156–
 158; photomicrographic
 analysis of structure, *157*

UNESCO, on intangible
 heritage, 303
Union des Artistes Modernes
 (UAM), 89–90
Unique Architectural
 Structures Copyright Act
 (1990, US), 272, 275–279

Unité d'Habitation, 121
United International Bureau
 for the Protection of
 Intellectual Property
 (BIRPI), 10, 12
United Kingdom. *See* Britain
United Nations: and
 African patent law, 307;
 Declaration on the Rights
 of Indigenous People
 (2007), 303
United States: architectural
 firms patenting, 117; concrete construction, 142;
 copyright laws regarding
 architectural works, 3, 272–
 273, 275–279; corporate
 patent activity, 127, 135–139;
 Fordist model of manu-
 facturing, 120; German
 émigrés, World War II and,
 39, 92–95, 207, 209, 336n68;
 and German industrial patents after World Wars I and
 II, 205–210, 237; and modern patent system, 259–260;
 Office of Alien Property
 (OAP), 209–210; Operation
 Paperclip, 207, 336n68;
 patent drawings, 191–193;
 patent models, 172–187;
 Sherman Antitrust Act
 (1890), 129; Trading With
 the Enemy Act (TWEA),
 205; Treasury Relief Art
 Project (TRAP), 158. *See also*
 US Patent Office; US patent
 system; *specific cities*
US Constitution, intellectual
 property clause, 259
US Patent Office: American
 Civil War and, 334n18;
 creation of, 3, 337n23; and
 European Patent Office,
 243; fire of 1836, 193; fire
 of 1877, 178–184, *185*; first
 patent examiner, 229; first
 superintendent, 172–173;
 housing of, 172, 173, 174;
 interview with patent
 examiner at, 232–235; patent drawing requirements,
 195–201, *200*, 335n54;

patent model require-
 ments, 173–174, 176, 178;
 patent models displayed at,
 174, 176–178, *179–183*, 186,
 334n19, 335n39; War of 1812
 and, 172–173
US patent system: critique of,
 257–258, 294; early, 172; and
 European national patent
 systems, 259–260; foreign
 patent applications,
 increase of, 234; German
 perspective on, 338n6;
 Jefferson and, 7, 224, 229;
 and Latin American patent
 systems, 308; Melman
 report on, 257, 264; reforms,
 234, 264, 269–273
US Senate Subcommittee on
 Patents, Trademarks, and
 Copyrights, Melman report
 to, 257, 264
US Supreme Court, design
 patent cases and, 280
Utility: of appearance vs.
 function, 268; and artistic
 expression, 14, 20, 273,
 340n4; divergent notions
 of, 268, 269; premodern
 privileges vs. modern
 patents on, 4; problem
 solving embedded in, 18; as
 threshold of patentability,
 18, 224–225, 228, 337n8
Utility patents: architectural
 patents filed as, 112–114;
 vs. design patents, 22, 71,
 268–269, 280; fire safety
 used as veil for form in, 73;
 infringement cases related
 to, 280; notion of novelty
 in, 22

Vaaler, Johan, paper clip
 invention, *227*, 228
Vaidhyanathan, Siva, 14, 304
Van Depoele, Charles, 176
Van Kennel, Theophilius, 139:
 revolving door invention,
 139–141, *140*
Venice (Renaissance city-
 state): patent statute issued
 (1474), 6; *privilegio* system, 7

Vievering, Bill, 147
Vitruvius Pollio, Marcus, 5, 145
Viviani, Giuliano, prefabricated construction system of, 202, *204*
Voena, Alessandra, 209
Voltaire, on women inventors, 30

Wachsmann, Konrad, 85, 92–95: approach to patenting, 94, 99; and Einstein, 329n13; and Gropius, 92, 93–95; Packaged House system, 93–95, *96*, 321
Waldinger, Fabian, 209
Walls: Cavity Wall System, 30; curtain wall, SOM and innovations in, 115; hollow, K Brick for, 37–38
Walton, Frederick Edward, 55
War of 1812, 172–173
Wars, patent systems and, 10, 308–309. *See also specific wars*
Washington (DC): Smithsonian Institution Building, architect of, 262; Smithsonian Museum, patent models at, 186, 187
Water, large-scale construction on, 81–82
Water, supplying houses with, patent drawing for, *198*
Watt, John Francis, 325n26
Wealth, monopoly-driven, patent system and, 267
West Indies: British patent system and, 296; patent model repository, 334n2; sugar plantation, *298*
"WetDry" sandpaper, 145
White, J. R., 156
White, Stanford, *74*
Whitla, Dean, 43
Whitney, Eli, 184
Widmann, Gottlieb, 142
Wildhagen, K., 45
Wilhelm Schaumarin Faneeritehdas Company, 86
William the Conqueror (William I of England), 290
Wilson, Woodrow, 205, 209

Window glass, Wright patents for, 105–106, *107*
Windows, Follet's Self-Balancing Sashes, 193–195, *194*, plate 7
Winkler, Helmuth, 237
Women: in architecture, first, 36; design patents associated with, 260–262; and domestic inventorship, 27–29, 34, 40–44; innovations in architecture, design, and construction, 30–44; participation in patent system, 2, 29–44, 202, 260; patent examiners, first, 229; scientists, discrimination against, 156
Wong, Winnie Won Yin, 340n4
Wood, H. T., 190
Woodcroft, Bennet, 187, 189, 191
Wood Joinery System, invention of, 30
World Intellectual Property Organization: and African patent law, 308; and Berne Convention for the Protection of Literary and Artistic Property, 274; and European Patent Office, 243; origins, 10, 12; on Paris Convention for the Protection of Industrial Property (1883), 10; on traditional and indigenous knowledge, 303
World Trade Center (New York City), design for reconstruction of, 340n85
World War I: French mass housing after, 109; German industrial patents after, 205, 209; German mass housing after, 147–148
World War II: crypto-patent system during, 210; German economic miracle after, 237; and German émigrés in US, 39, 92–95, 207, 209, 336n68, 336n79;

German industrial patents after, 205–215; German mass housing after, 152–155, *153*; Operation Paperclip after, 207, 336n68; paper clip symbolism, 228; patent systems after, 12, 237; and Wachsmann's Packaged House system, 93–95, *96*
Woty, William, 319
Wright, Frank Lloyd, 105–109: on buildings as assets, 275; drawings by, licensing of images of, 278; Ennis house, 108; infringement claims against, 108–109; patents held by, 14, 105, 106–108, 330n108; standardized house designs by, 106; textile block design by, 108, *111*; window glass block design patents of, 105–106, *107*
Wright, John Lloyd, 109

Zeckendorf, William, patent for Multistory Building Structure, 114–115
Zeisel, Eva, 30

© 2024 Massachusetts Institute of Technology

This work is subject to a Creative Commons CC-BY-NC-ND license. Subject to such license, all rights are reserved.

The MIT Press would like to thank the anonymous peer reviewers who provided comments on drafts of this book. The generous work of academic experts is essential for establishing the authority and quality of our publications. We acknowledge with gratitude the contributions of these otherwise uncredited readers.

This book was set in Haultin by Jen Jackowitz. Printed and bound in the United States of America.

Library of Congress Cataloging-in-Publication Data

Names: Christensen, Peter H., author.
Title: Prior art : patents and the nature of invention in architecture / Peter H. Christensen.
Description: Cambridge, Massachusetts : The MIT Press, [2023] | Includes bibliographical references and index.
Identifiers: LCCN 2023018021 (print) | LCCN 2023018022 (ebook) | ISBN 9780262048958 (hardcover) | ISBN 9780262378352 (epub) | ISBN 9780262378345 (pdf)
Subjects: LCSH: Architecture—Patents. | Intellectual property and creative ability.
Classification: LCC NA2755 .C49 2023 (print) | LCC NA2755 (ebook) | DDC 720.2/7—dc23/eng/20231115
LC record available at https://lccn.loc.gov/2023018021
LC ebook record available at https://lccn.loc.gov/2023018022

10 9 8 7 6 5 4 3 2 1

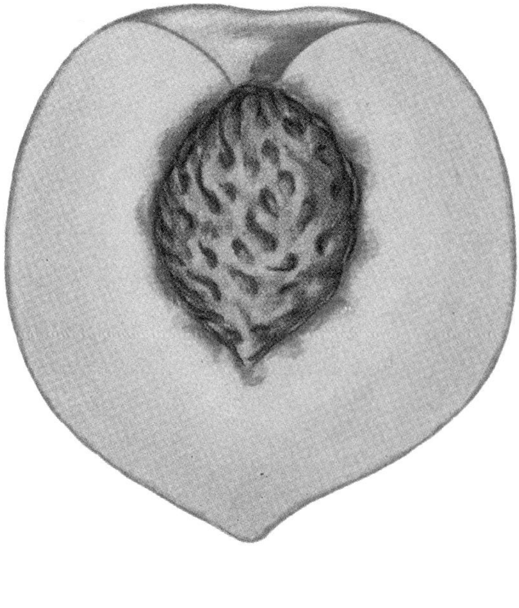